CLINICIAN AND CHEMIST

The Relationship of the Laboratory to the Physician

CLINICIAN AND CHEMIST

The Relationship of the Laboratory to the Physician

Proceedings of the First Arnold O. Beckman Conference in Clinical Chemistry

EDITORS
Donald S. Young
David Uddin
Henry Nipper
Jocelyn Hicks

EXECUTIVE EDITOR
J. Stanton King

The American Association for Clinical Chemistry
1725 K St., N.W.
Washington, D.C. 20006
301 Miller St., Box 5218
Winston-Salem, N.C. 27103

Other books recently published by the American Association for Clinical Chemistry

Clinical Immunochemistry: Chemical and Cellular Bases and Applications in Disease

Second International Symposium on Clinical Enzymology

A National Understanding for the Development of Reference Materials and Methods for Clinical Chemistry: Proceedings of a Conference

Methods for the Centrifugal Analyzer

Library of Congress Catalog No. 78–72880
ISBN 0–915274–08–6

Preface

Donald S. Young, *Conference Chairman*

The first Arnold O. Beckman Conference, sponsored by the American Association for Clinical Chemistry, was held in Annapolis, Maryland, September 28–30, 1977. These conferences are intended to address important applications of clinical chemistry in medicine. Clinical chemistry has developed as an analytical discipline in this country, whereas it has had a different role elsewhere in the world, where the same field's description as clinical biochemistry or chemical pathology implies greater emphasis on application than analysis. In this first conference, the organizers decided to address the interaction between physicians and clinical chemists and the proper use of clinical laboratory services, especially those of the clinical chemistry or biochemistry laboratory.

Proper use of the clinical laboratory has become a matter of great concern to health planners and economists as well as to clinicians and clinical laboratory scientists. The number of tests performed in clinical laboratories continues to increase from year to year. It is estimated that about 4.5 billion tests are now performed annually and there appears to be little likelihood of a reduction in the usual 10–15% annual increase. The increase affects both tests that have been used for many years and tests that have been recently introduced in research laboratories and subsequently into service laboratories. While the quality of laboratory work was a matter of great concern several years ago, national surveys show a distinct improvement—in those laboratories covered by such surveys—in the quality of performance of the common tests, although there is still considerable variability within a laboratory over a long time and between laboratories at the same time for many of the more complicated and technically demanding tests.

With an increase in number of tests done in most hospitals,

without a proportional increase in the number of patients, it must be assumed that more tests are now required for the management of a patient than previously. This affects traditional tests as well as recently added new tests. The newer tests generally have greater specificity for particular diseases than have many of the conventional tests, yet they also tend to be more expensive and less amenable to mechanization. The introduction of new tests without adequate education of the potential users, both physicians and medical students, has in effect invited their overuse. Invariably, publications concerning laboratory tests speak of their usefulness and not of the appropriate application of the tests in given situations. The overall sensitivity and specificity of the tests tend to be overlooked.

Examples of such overglamorized tests are the tumor markers. Several of these tests were initially hailed as being specific for malignant disease, and even disease of specific organs. They have subsequently been demonstrated to be much less specific. Yet their association with cancer has prompted their use by physicians to assist them in making the diagnosis of cancer, even though few laboratory tests, let alone tumor markers, have great utility in diagnosis.

Appropriate use of the clinical laboratory still is taught inadequately in most medical schools, perhaps because laboratory medicine is not often recognized as a separate discipline, and test utilization and interpretation is not taught by individuals who are experts in this area. With the continued introduction of new tests there is also a need to ensure that physicians no longer in training are educated to the proper use of these tests. This can come about through balanced publications written from an educational standpoint, but also from close interaction between clinicians and laboratory scientists, so each becomes aware of the other's problems and needs. In this way the provider of laboratory services may be able to ensure more logical use of the clinical laboratory and the laboratory component of the patient's bill may be reduced to a level that still is consistent with quick and efficient diagnosis and monitoring of treatment without unnecessary wasted effort and cost.

During the first Arnold O. Beckman Conference many of the issues involved in proper use of the laboratory were discussed. Background concepts were first discussed and these were followed by a session on screening for disease, as applied to a nominally well adult population, newborn children and a population exposed to environmental pollutants. Diagnostic strategies were presented and the organizers deliberately invited different viewpoints. The current typical response to abnormal data was reviewed and the meeting concluded with a session on proper use of clinical laboratories and how this might be improved by education.

In preparing these proceedings of the meeting, the editors have attempted to maintain the spontaneity of the meeting by including discussions that followed each session. On all occasions the discussion period had to be kept short and the questions and answers have been carefully edited to further reduce their length. Nonetheless it was apparent to all the participants in the meeting that a very valuable exchange of ideas took place between the clinicians and chemists who attended the meeting and that further meetings of this type must help in providing the background necessary for reducing the cost of laboratory services, while still providing all the tests required for diagnosing and treating patients. We hope that these ideas will be received by the reader much as if he were present at the conference itself.

If you believe [that medicine has become too scientific], you are entitled to think that perhaps it is time now to call a halt to inquiry, stop the science, and settle down to apply today's store of knowledge with more intelligence, so that more people can be benefited, and with equity. Clearly, if the science of medicine has moved as far as it is likely to go, there must be something appallingly wrong with the way we are delivering its benefits, for there is still a formidable roster of incapacitating and fatal diseases, and people seem to be dying from these at about the same rate as twenty-five years ago. Why is this so? Is it because we have not learned how to apply today's information? Or is it because there is something wrong with the information?

—Lewis Thomas, "Future Directions in Biomedical Research," in *Beyond Tomorrow; Trends and Prospects in Medical Science* (Rockefeller Univ. Press, 1977), p. 111

The future function of the hospital is already foreshadowed in most of our university teaching hospitals—basically it is the application of biological science and particularly of biochemistry to the prevention of death and the recovery of health. The hospital may well become much like the "intensive care" wards of today in which the whole panoply of electronic instrumentation and biochemical control can be set in action at a moment's notice. I suspect that the hospital will be a highly effective, rather inhuman place where very highly trained specialists will work in teams guided to an increasing extent by EDP machines.

We may well find that the men who staff the hospitals of next century will include many who are much more mathematicians and biochemists than physicians as we know them today, but there will still be wide range of surgical and other specialists. And despite all the instrumentation, I fancy that those men will still need to be able to apply common sense, courage and compassion in dealing with all the human difficulties that escape the machines.

—Sir Macfarlane Burnet, in *Changing Patterns, An Atypical Autobiography* (American Elsevier Pub. Co., Inc., New York, 1969), pp. 248, 252

Contents

Preface v

I. Basis for Laboratory Testing

Why There Is a Laboratory 3
Donald S. Young, M.B., Ph.D.

The Physician's Logic in Making a Diagnosis 23
David L. Sackett, M.D.

Problem-Oriented Laboratory Testing 31
Lawrence I. Weed, M.D.

The Role of the Laboratory in Monitoring Therapy and Disease 56
Joseph E. Rall, M.D., Ph.D.

Selection of Appropriate Laboratory Tests 69
Robert S. Galen, M.D., M.P.H.

Discussion 106

II. Screening for Disease

Cost Effectiveness of Multiphase Health Testing 121
Morris F. Collen, M.D.

Cost Effectiveness of Laboratory Testing in Genetic Disease 131
Harvey L. Levy, M.D.
Ann J. E. Bennett
Elizabeth D. K. McDevitt

Discussion 142

III. Strategies for Laboratory Testing

Organ Panels and the Relationship of the Laboratory to the Physician 157
John Bernard Henry, M.D.
Peter J. Howanitz, M.D.

Use of the Laboratory: Dollars and Sense 175
Edwin L. Overholt, M.D.

The Application of the Laboratory to Hypertensive Patients 191
Harold D. Itskovitz, M.D.

Unraveling the Laboratory Model of a Syndrome: The Osmolality Model 200
Harry F. Weisberg, M.D.

Discussion 244

IV. Response to Abnormal Test Data

The Clinical Usefulness of Multiple Result Ranges for a Laboratory Test, as Exemplified by Serum Urea Nitrogen 265
G. William Cole, M.D.

The Physician's Response to Unsolicited Tests 273
Lawrence J. Schneiderman, M.D.

Usefulness of Laboratory Information in Medical Monitoring 284
C. H. Altshuler, M.D.

Discussion 297

V. Education in Laboratory Utilization

Current Practices in Laboratory Utilization in the Teaching Hospital: Possible Impact of Educational Programs 307
Paul F. Griner, M.D.

Use of a Computer to Issue Interpretive Reports: Do Clinicians Find This Form of Reporting Useful? 322
Richard L. Reece, M.D.
Russell K. Hobbie, Ph.D.

Teaching Proper Laboratory Use to Medical Students and Physicians 334
Patrick C. J. Ward, M.D.
Charles A. Horwitz, M.D.
M. Desmond Burke, M.D.

Teaching Laboratory Utilization by Use of Computer-Based Clinical Simulations 348
Richard B. Friedman, M.D.
Donald R. Korst, M.D.

Discussion 360

Optimizing the Usefulness of Simple Laboratory Tests 366
Thomas P. Whitehead, Ph.D., M.R.C. Path.

Postscript: *The Chemyst's Tale* 376
J. Stanton King, Ph.D.

I

BASIS FOR LABORATORY TESTING

Why There Is a Laboratory

Donald S. Young, M.B., Ph.D.

Department of Laboratory Medicine, Mayo Clinic and Foundation, Rochester, Minn. 55901

Nobody knows for certain how many clinical laboratory tests are done each year in this country, but some estimate it to be 4.0 billion *(1)*. We do know that the number is increasing.

More than half are clinical chemical tests. In spite of the great number of tests and the associated costs and manpower involved, the usefulness of laboratory tests is seldom assessed. One reason for this is the difficulty perceived in differentiating those benefits to patients that originate from the use of laboratory tests and those due to such things as other advances in diagnosis and treatment. A second is the understandably greater concern with potential benefits to patients than with costs of laboratory tests. Nevertheless, there is good evidence that certain procedures are efficacious—e.g., monitoring digoxin concentrations in the serum does lead to fewer instances of drug toxicity *(2)*.

The greater number of conventional tests performed on patients with the same disease *(3)* and the availability of additional tests both contribute to the steadily increasing workload in the typical hospital laboratory. Unfortunately, this has not shortened the patient's stay in the hospital. The cost of health care continues to increase, with the charge for laboratory tests not uncommonly comprising as much as a quarter of a patient's hospital bill, and as much as a third if he is an outpatient.

Why Are Laboratory Tests Requested?

Whatever the reason for ordering tests, an increased workload for the clinical laboratory is the ultimate result. Table 1 lists the

Table 1. **Reasons for Ordering Laboratory Tests**

Diagnosis of disease
Screening for disease
Determination of severity of disease
Determination of appropriate management of patient
Monitoring progress of disease and monitoring therapy
Monitoring drug toxicity
Predicting response to treatment

main reasons for ordering laboratory tests. To these should be added "gamesmanship" *(4):* an intern or resident requests certain tests, mainly to rule out some obscure and improbable disease, only to have the attending physician order additional tests to eliminate further possible causes of a patient's illness. Laboratory tests may also be used to determine a baseline or reference state before a diagnostic procedure is performed or course of treatment is instituted, or to determine the fitness of a patient to undergo another medical procedure.

Unfortunately, physicians now often feel compelled to order laboratory tests that they do not really think are clinically justified, to protect themselves against legal repercussions if traditional practice is not followed. As recently demonstrated, some physicians order laboratory tests because of arrangements whereby they gain financially.

Another factor contributing to an increased workload is that physicians often fail to take note of laboratory results already obtained and repeat the ordering of tests when the values do not justify this. Or physicians may ignore test results *(5),* even when efforts are made to encourage their interest in the results. Such lack of attention to reported values evidently reflects the importance a physician attaches to laboratory data relative to his clinical impression or findings. Reliance on a faulty memory may prompt a physician to order a test the results of which he vaguely recollects may be abnormal in association with a particular disease. Yet he should consider that if he had some uncertainty about the value of the test it probably has no great diagnostic usefulness. The practice of allowing standing orders for laboratory testing to be continued without question until countermanded by the patient's physician frequently leads to more tests being ordered than are required for the patient's management.

Physicians may be unaware of the physiological mechanisms underlying a test and thereby select the wrong test to evaluate a specific problem. When a physician uses laboratory tests to confirm a clinical impression, and the results do not conform to his original hypothesis or preconceived ideas, he often assumes that the results are in error and repeats his requests. In candor, we must agree that not many years ago such an assumption was more often justified than not. The performance of most clinical laboratories is now sufficiently good that, with certain exceptions of which both physician and laboratory should be aware, analytical results should be assumed to be correct in the first instance. An unfortunate residuum of the past is that a physician often is still likely to prefer his clinical impression to laboratory results that are not in harmony, unless such results are consistent when the tests are repeated.

Not all physicians appreciate the interrelationships between test results and normal physiological processes. Some values that appear to be abnormal may indeed reflect a normal response to a physiological change. For instance, the concentration of thyroxine-binding globulin in serum normally increases during pregnancy and automatically leads to an increased concentration of total thyroxine. Indeed, failure of the concentration of either compound to increase would reflect an underlying clinical problem. Because the binding protein is metabolized in the liver, certain hepatic diseases are likely to be associated with an increased serum thyroxine concentration that actually does not indicate abnormal thyroid metabolism *(6)*. In both of these situations, data on the free thyroxine concentration in serum would indicate the true state of thyroid metabolism.

Different tests that measure the same physiological function contribute to the overall increase in workload without necessarily providing additional clinical information. Such tests include serum urea nitrogen and creatinine, both of which reflect renal function; often both are requested, even when there is little reason to expect one test to yield additional information over the other. In many uncomplicated cases of cholestasis, measurement of alkaline phosphatase (EC 3.1.3.1) or γ-glutamyltransferase (EC 2.3.2.2) activity in serum yields as much information as does measurement of leucine amino-peptidase (EC 3.4.11.1) or 5′-nucleotidase (EC 3.1.3.5), for which automated procedures are not usually available in a laboratory. Certainly in the follow-up of individuals in whom a simple cause of an abnormality has been identified and the disease is progressing as anticipated it is only necessary to measure the analyte that best indicates the status of the disease.

Physicians can hardly be expected to remember how each drug they prescribe may affect each test value. Thus side effects of a drug or its artifactual in vitro effect on a test result may be the cause of misinterpretation. For some drugs (e.g., oral contraceptives) as many as 100 different in vivo effects on clinical laboratory tests have been reported *(7)*. Usually the therapeutic effects of a drug exceed its side effects but the therapeutic response, most easily followed in the clinical laboratory by monitoring the alteration of the concentration of a test constituent in serum, may vary considerably from one person to another. Interindividual differences in response should always be considered before extensive workups are undertaken. Many seemingly inexplicable test results which lead to repeat testing and directly contribute to a laboratory's increased workload could be avoided by the use of more specific analytical procedures in the clinical laboratory.

Usually, when a physician orders clinical laboratory tests he has in mind one or more of the specific objectives listed in Table 1. These are discussed below.

Diagnosis

Bases for a diagnosis. Although a physician may request many laboratory tests when he first sees a patient—and in some centers they may even be ordered *before* he sees his patient—he will use few of the resulting data in making his diagnosis. He uses the other data mainly to confirm his clinical suspicions or, if he is more confident of his diagnosis, to substantiate that diagnosis.

About three-quarters of all the diagnoses that a physician makes on clinically ill patients are made from the patient's history, perhaps another 10 to 15% after the physical examination. The proportion of primary diagnoses that can be made exclusively from the results of laboratory tests ordinarily is very small, but in one study from a university hospital laboratory the data suggested clues to unsuspected disease in about 12% of the hospital patients studied *(8)*.

In individuals who ostensibly are well but are being examined (e.g.) prior to employment or are receiving a periodic health evaluation, the proportion of diagnoses from clinical laboratory tests may be substantially greater. Thus, in a population of this type, Franco *(9)* found that the history was primarily responsible for 34% of such diagnoses, the physical examination for another 22%, and laboratory tests for 14%. The lower yield of the history probably reflects the absence of an overt problem or complaint to cause an individual to seek medical help. In a more recent study, Bates et al. *(10)* confirmed that the importance of clinical evaluation is less in pre-employment examinations, although they emphasized the complementary nature of history and laboratory tests in the making of a diagnosis. The yield of abnormal information from a history is probably directly related to the time spent by the physician with a patient—usually brief in the absence of a recognized problem. The skill of a physician in eliciting and elucidating a good history also influences the yield of diagnoses, and if the history is obtained by questionnaire its design may greatly affect the yield of diagnostic information.

Test result as related to diagnosis. Few clinical laboratory tests have such specificity that they, by themselves, may be used to establish a diagnosis. Even the measurement of carcinoembryonic antigen, initially hailed as a test that could establish the diagnosis of carcinoma of the colon or rectum, not only has been shown to be nonspecific for tumors of the large bowel, but also is not even specific for cancer. Abnormal values may occur in diseases such as ulcerative colitis or if multiple polyps are present in the large bowel. Abnormal values may occur also in pregnant women and in some healthy adults, especially those who are heavy smokers. Specificity in diagnosis of tumors of the gastrointestinal tract may be enhanced by measurement of CEA-S, a fraction of carcinoembryonic agent isolated by Edgington et al. *(11)*.

High concentrations of calcitonin are frequently found in the serum of patients with medullary carcinoma of the thyroid gland, but this hormone also may be secreted by other tumors, and other tumors may secrete different hormones. Abnormal values do not necessarily indicate disease of an endocrine gland. Other tests, such as measurement of alpha-fetoprotein, are not by themselves diagnostic of a single disease, although in the pregnant woman a persistently high value strongly suggests a neural crest defect in the fetus and in the non-pregnant adult a high value strongly suggests cancer—although not necessarily the primary hepatocarcinoma of which it was initially presumed to be characteristic. High values for choriogonadotropin in a non-pregnant woman strongly suggest the presence of choriocarcinoma, but this hormone may also be secreted by other tumors. The presence of the hormone at a detectable concentration signifies severe disease, even though it may not be clear which organ is actually the one affected. Even serum lysozyme activity (EC 3.2.1.16), an abnormally high value for which was thought to be highly suggestive of acute myelomonocytic leukemia, is now recognized as being only a measure of reactive leukocytosis.

Results for urinary cystine or electrolytes in sweat can specifically indicate cystinuria and cystic fibrosis, respectively. Other tests, if done at certain times in the course of a disorder, may have great value in establishing a diagnosis. These include the measurement of phenylalanine and thyroxine in the serum of newborn infants; abnormal values can establish the diagnosis of phenylketonuria or congenital hypothyroidism, respectively. At other times in an individual's life, measurement of these compounds does not have the same potential diagnostic impact. The specificity of a test may greatly improve the diagnostic specificity of a group of tests, as has been demonstrated for serum folate in the diagnosis of alcoholism *(12)*. γ-Glutamyltransferase activity measurements in serum may also be of value in assisting in the diagnosis of alcoholism. The measurement of the triiodothyronine concentration in serum is required before the conclusive diagnosis of triiodothyronine thyrotoxicosis can be established *(13)*.

None of the tests in the above examples, except those for inborn errors of metabolism in the newborn, can be considered "primary" or are likely to be requested when a physician first sees a patient. None of the typical tests that a physician requests when he first sees a patient is likely to establish a conclusive diagnosis, although abnormalities in urinalyses can provide strong evidence for the presence of diabetes or renal disease, and abnormal serum glucose and uric acid values may provide clues to diabetes or gout. Measurement of hemoglobin is useful in detecting anemia, and other common tests may also provide clues to a possible diagnosis, even though, singly or collectively, they cannot be considered capable of establishing diagnoses.

From the preceding, one should conclude that, in general, laboratory information should not be used to make a diagnosis without supporting information from the history or physical examination, and that weaknesses in any laboratory test should be recognized. The less specific the test, the less likely is it to be of diagnostic value.

Screening

Purposes. The term "screening" has been widely applied to the search for disease in individuals in whom its presence has not been previously established. Wilson *(14)* has described screening as a medical investigation that does not arise from a patient's request for advice concerning a specific complaint, that endeavors to identify persons with early disease and thus improve patient care, and that by early intervention seeks to prevent chronic disability requiring long-term support. The definition of screening proposed by the United States Commission on Chronic Illness, and accepted by the World Health Organization, defines screening as the presumptive identification of unrecognized disease or defect by the application of tests, examinations, or other procedures that can be rapidly applied. Neither of the above definitions differentiates between non-laboratory and laboratory forms of screening; each is applicable to both, and to screening for a specific disease and for disease in general. Another purpose of screening, not encompassed by the above definitions, is to identify individuals who are at high risk of a particular disease.

Screening strategies. Henderson and Sherwin *(15)* classified screening strategies according to whether a single test or several tests are done. Either approach may be used to search for a single disease or disease in general, and either may be used to investigate a population as a whole (indiscriminate approach) or a group of individuals selected according to (e.g.) sex, age, or race, or some combination of these (discriminate approach). The same approach may be used to screen for the risk of disease. Table 2 illustrates some examples of the screening strategies.

For a well-defined population, screening tests with a high

Table 2. **Outline of Screening Strategies**
(modified from reference *15*)

	Discrete	
	Specific	*Nonspecific*
Discriminate	Phenylalanine in serum of new-born infants for detection of phenylketonuria	ESR measurements in all patients with fever
Indiscriminate	Mass screening of urine for glycosuria in a population of non-hospitalized individuals	Serum uric acid as part of a health-screening program
	Multiple	
	Specific	*Nonspecific*
Discriminate	Amino acid screening, by chromatography, of all newborn infants for inborn errors of metabolism	Urinalyses in all pregnant women on routine hospital visits
Indiscriminate	Complete blood count for detection of anemia in a population on all visits to a physician	Multiple laboratory testing, e.g., with SMA 12/60 on a non-selected population of individuals

predictive value also may serve a diagnostic function. Indeed, screening is of little value unless there is a high potential yield of abnormal results, except when other considerations, such as economic impact, counterbalance the low yield of positive findings. However, screening not only may be useful in indicating the presence of a disease, it may be equally useful in ruling out the likelihood of a disease. Thus, Reynoso et al. *(16)* have demonstrated that a low serum lactate dehydrogenase (EC 1.1.1.27) activity militates against a diagnosis of malignant disease.

Screening may be used periodically to evaluate an individual's health, or only when an individual is admitted to the hospital. As already mentioned, in some centers a group of tests that have some diagnostic specificity are performed even before a physician interviews his patient. Several of the tests frequently included in such a group are mandated by agencies that accredit clinical laboratories. Typically, urinalysis and measurement of hemoglobin are included.

Interpretation. There is the same need for critical evaluation of screening-test results as for other test results. When the mean

± 2 standard deviations is used to define a normal or reference range, as is done for most tests with a gaussian distribution of results in health individuals, some abnormal results will necessarily be observed, even in healthy individuals, if enough are tested. Moreover the probability that an individual will give an abnormal test result increases with the number of tests performed. This unquieting fact has led to some unwarranted loss of confidence in the clinical usefulness of abnormal results obtained from screening profiles. Confidence is undoubtedly greater when the range of abnormal values differs greatly from that for healthy individuals, or when results are abnormal for two or more tests that in effect measure the same physiological or pathophysiological process. Nevertheless, in spite of the shortcomings of screening tests there is evidence that early identification of disease with prompt follow-up decreases the mortality of patients over several years *(17)*.

Value of screening. Laboratory screening is now often thought of as the performance of several tests, even though only one test might be indicated clinically, because 12, 17, 20, or more tests happen to be packaged together in one analytical system and it is as easy to perform many tests as only one. There is still considerable debate concerning the cost/benefit aspect of laboratory screening. Unexpectedly abnormal tests will surely be observed in some patients and the occurrence of such abnormal values may lead to considerable work if the responsibility of tracking down the cause of abnormal data is accepted. Paulus et al. *(18)* found that 13.2% of all male patients had a high serum uric acid concentration on laboratory screening. Azotemia caused this in one fifth of the patients, acidosis was the cause of another fifth of the abnormalities, and as many were attributable to the use of diuretics. Only 20% of all the patients with a high serum uric acid concentration had either a history of gout or a value sufficiently high to warrant therapy. Sisson et al. *(19)* have also drawn attention to the deterioration in management of patients that may occur if test data are too plentiful.

In a five year follow-up of abnormal but unexplained screening test values Bradwell et al. *(20)* determined that only 1.5% of such values could be interpreted as suggestive of presymptomatic disease. A third of all repeat values could provide an explanation for the earlier abnormal values, and a third of all the initial abnormal values were normal at the subsequent examination. Obviously, the number of abnormal results is a function of what range of values is considered "normal." Thus selection of cutoff points between normal and abnormal is perhaps more critical in screening than in hospital practice, because patients with values in a grey-zone are readily re-checked, but screened individuals are often more difficult to follow up. That conventional assessments of normal may be inappropriate for screening is evidenced by the yield of abnormal results in individuals undergoing either pre-employment physical examinations or annual check-ups. In one study *(21)*, 74.4% of individuals excreted in their urine what was considered to be an abnormally large number of erythrocytes, 57.9% excreted an abnormal number of leukocytes, and 38.3% excreted abnormal numbers of epithelial cells. In this study 90% of an apparently healthy population had some physical aberration or clinical disorder.

Much of the screening for inborn errors of metabolism in newborn infants is governed by tradition, and in some cases by law. The diseases for which screening is commonly performed are not necessarily those with the greatest individual or community impact. Thus, while screening for phenylketonuria is mandatory in many countries, other diseases that are potentially as treatable are more prevalent but are not specifically investigated in the newborn.

One report *(22)* of the prevalence of various abnormalities in newborns for which biochemical screening is feasible indicates

1 per 6000–20000 for phenylketonuria
1 per 7 to 20 for heterozygotes for α_1-antitrypsin deficiency
1 per 2000 for cystic fibrosis
1 per 8000 for congenital hypothyroidism
1 per 40000 or less for galactosemia

The prevalence of neural crest defects in human fetuses is uncertain but low α-fetoprotein measurements may be used to screen either amniotic fluid or maternal serum for this disorder. The incidence of glucose-6-phosphate dehydrogenase (EC 1.1.1.49) deficiency among Negroes is 1 in 8.5 *(23)*, and its presence should be determined early in life to reduce later problems. Although the prevalence of aminoacidopathies other than phenylketonuria is low, it is comparatively simple to screen for some of these disorders such as maple-syrup urine disease, homocystinuria, histidinemia, or Hartnup disease.

Screening is obviously indicated for specific problems in certain high risk situations. Down's syndrome (mongolism) in infants is more common for children of mothers older than age 35 years. Tay-Sachs disease is more common in the children of Ashkenazic Jews than in any other population group. Certain other diseases in children are related to high-risk environmental situations; e.g., lead poisoning is common in urban areas and dietary deficiency of a trace element may be endemic in a certain area.

On a restricted scale, screening for the presence of a toxic compound and for the damage caused by it is appropriate in chemical workers and others exposed to toxic chemicals. For instance, individuals who work in battery factories are exposed to large amounts of lead and should be examined both for the anemia that may be caused by lead poisoning and for evidence of a high concentration of lead in their blood. Workers exposed to high concentrations of pesticides or organic chemicals should also be monitored at regular intervals. Screening of workers in some other industries should also be considered.

Because certain risk factors have been identified with particular diseases, these should be monitored in individuals at greatest risk of developing these diseases. For example, plasma lipid concentrations should be determined frequently in the hypertensive smoker with the hope that dietary modification might diminish his risk of a heart attack. Obviously, it is desirable that other risk factors are identified for the common diseases and suitable screening programs introduced to monitor them.

Estimation of Severity of Disease

In general, the higher the concentration or activity in serum of a component that has been associated with damage of a tissue, the more severe the disease. But, at best, measurement of compounds in serum only indirectly reflects the damage to an organ, and measurement of a single component rarely indicates the type of damage, although by judicious selection of an appropriate panel of tests it is sometimes possible, as in the case of liver disease, to get an indication of both the type and severity of the disease. However, direct histological or histochemical examination of biopsies of an affected organ, which would provide the best measure of type and extent of damage of a tissue, is now feasible but not yet routinely done, except for renal and liver biopsies in large medical centers. Even these examinations do not necessarily help the physician in his prognosis, the most-desired information. Few current laboratory tests assess the function of an organ; most only show what the extent of the damage may be, and few indicate the metabolic capabilities or reserve of the tissue. The glucose tolerance test is relatively unusual in that it does estimate the ability of an organ to respond to a specific challenge and thus serves to indicate overall pancreatic function, even though the body's response does involve other endocrine glands.

The continuum between normal and abnormal in the concentration of most serum constituents renders it difficult to differentiate mild disease in some individuals, whose normal homeostatic set-point for an analyte might be low, from healthy individuals with a higher normal set-point. This ignores normal biological variability, which affects even the day-to-day value in all individuals. Prior knowledge of typical values for an individual is required if values for him that are still within the group normal range, or are only slightly abnormal, are to be interpreted properly. Such baseline values, obtained when an individual is healthy, are of much help in evaluating the significance of later abnormal values found when he is ill—if it can be guaranteed that differences of analytical origin are small enough to be ignored.

Usually, severe disease is easily recognized, both clinically and by the laboratory. Certain test values can be expected to be considerably different from normal. Few tests that are indicators of disease give quite normal results until severe disease is present, then become quite abnormal. An exception to this is the semiquantitative test for ketones in urine, which ordinarily is negative until diabetes becomes severe or some other disease is present that causes a shift from predominantly carbohydrate metabolism to fat metabolism. Likewise the concentration of creatinine in serum only becomes abnormal when about half of the nephrons have stopped functioning.

With knowledge of the biological half-lives of various enzymes in serum one can, by extrapolation, guess the probable size of a myocardial infarction from serial determinations of the activity of one or more cardiac enzymes in serum. The most reliable information is obtained when the time of the infarction is also known. Measurement of the total activity of an enzyme suffices in most clinical situations, but measurement of the specific cardiac isoenzyme creatine kinase-MB (EC 2.7.3.2), for example, is more specific, and indeed can indicate the presence of tissue damage even when the total activity of the enzyme is within normal limits. Measurement of total creatine kinase activity—but better, measurement of the cardiac isoenzyme, because factors such as intramuscular injections of drugs may cause release of creatine kinase-MM from skeletal muscle *(24)*—allows detection of a new infarct in a person who recently has had a previous infarct. In theory, the same approach may be used to measure the extent of damage to other organs, the degree of accuracy with which this can be done depending on the availability of a readily measured organ-specific isoenzyme with known degradation characteristics in serum. The type of treatment of an organ is rarely modified as a result of information on the extent of damage, but the duration and/or intensity of such treatment may be altered. Serial determinations of an analyte do allow the change from chronic to acute state, or vice versa, to be assessed, and may be useful in determining a different therapeutic approach.

Laboratory data should always be considered in conjunction with a physician's clinical impression, and interpreted accordingly. Thus, if clinical impressions strongly suggest a myocardial infarct and laboratory data do not, the patient should still be treated as if he had had an infarct. On the other hand, when the laboratory results are abnormal, but not substantiated by clinical findings, a physician should thoroughly evaluate the data before dismissing the results as an artifact.

Determination of Management of a Patient

Clinical laboratory tests may assist in determining a differential diagnosis, and thus indirectly influence the management of many patients. For example, measurement of hormone-receptor activity in excised tumors of the breast may affect management by providing information concerning the hormone dependency of the tumors and thus the need to remove the adrenal or pituitary glands; measurements of serum prolactin concentration may aid in the management of cases of secondary amenorrhea *(25)*; the serum bilirubin concentration is used to determine if a newborn infant should be subjected to an exchange transfusion; and information on the serum urea nitrogen concentration is used to determine the need for dialysis in patients with renal failure.

Monitoring Progress of a Disease

In many cases laboratory tests, even though they may lack organ or disease specificity, more sensitively indicate a slight change in a patient's status than do clinical impressions. They may provide a quantitative index for measuring change, and so they are widely used in monitoring response to treatment or the rate of deterioration of the patient. While measurement of a very specific component, such as a tissue-specific isoenzyme, would provide the best yardstick for following the progress of a disease, measurement of the total enzyme activity usually suffices if it is assumed that additional causes of a change in activity are unlikely to supervene on an already-established disease. If additional causes of an alteration in the activity of an enzyme occur, these usually can be detected clinically and by observing changes in the concentration or activity of other components of serum. In monitoring the change in status of a patient's disease one should take into account the normal turnover of the compound that is being followed, so that useless measurements are avoided.

In assessing results of surgery, components should be measured that are uninfluenced by the stress of surgery or anesthesia and are unaffected by the concomitant tissue damage. If a compound is secreted by or present in a tumor, as for example carcinoembryonic antigen, its concentration in serum may sensitively indicate the success of surgery. Yet frequently the concentration of the antigen increases immediately after surgery and the best laboratory indications of the response to surgery are not obtained until several weeks after the operation.

Most laboratory tests performed in a clinical chemistry laboratory are made to enable change in the disease state of a patient to be followed, although the same test frequently is used in a dual role of monitoring both response to treatment and the presence of toxic changes induced as a complication of treatment.

Drug Administration and Laboratory Testing

Because of possible toxicity that may occur with certain drugs, monitoring serum components that have been associated with tissue damage is often desirable. When there is a narrow margin between therapeutic and toxic concentrations in the blood, measurement of the drug in serum may help prevent possible toxic reactions, especially when the toxic symptoms of the drug are similar to those of the condition for which the drug was administered. Monitoring is also important when a drug is first introduced and there is little clinical experience of its toxic actions. It should be remembered, however, that biological variability between individuals may be so great that a dose (or concentration) that is barely therapeutic in one person may be toxic in another.

Monitoring is useful for drugs such as lithium, antiepileptics, and digoxin, for which the therapeutic index is small and response to treatment depends on an adequate concentration being maintained in the serum. Even with these drugs, changes in a patient's general state should always be considered in conjunction with the drug concentration in serum. For example, changes in the serum concentration of potassium or calcium, or in the patient's acid–base status, may markedly affect the penetration of digitalis and its derivatives into the cells of the myocardium.

For the patient who responds atypically to drug administration, measurement of the drug concentration in serum is indicated to determine whether the response is caused by abnormal metabolism altering the pharmacokinetics of the drug. These atypical responses may be a result of differences in absorption, excretion, or even idiosyncratic or genetic differences in metabolism. Some of these responses may be related to the disease for which the drug was administered. The bioavailability of different preparations of a drug may also affect a patient's response.

In many cases, measurement of a drug concentration is most useful for determining whether the patient is complying with his prescribed regimen. This need is greater in outpatients than when an individual is hospitalized, but some patients attempt to avoid their prescribed medications, even in hospital.

To ensure uniform monitoring of drug concentrations, the time of day and conditions under which specimens are collected should be standardized, and, ideally, measurements should be made when the drug concentration is presumed to have become stabilized.

Measurement of drug concentrations of those drugs normally excreted through or metabolized in the liver is of particular value in the presence of liver disease; similarly in renal disease, measurement of the urinary excretion of those drugs and their metabolites that normally are eliminated via the kidneys is useful. In the presence of renal failure the same dose of drug that is nontoxic in a healthy individual or in a person without renal disease may cause a sudden and sustained increase in serum concentration that may be toxic. Chronic administration of many drugs may lead to cumulative toxicity. The pharmacokinetics of many drugs are altered by the presence of renal disease or liver disease, or both. Uremia may also alter the metabolism of certain drugs in addition to altering their excretion. The administered drug is not always itself (or solely) pharmacologically active, so a disease that affects the transformation of a drug into an active metabolite may markedly alter the therapeutic response. In such a situation the concentration of the parent compound and the active metabolite(s) or the metabolite rather than the parent compound should be measured, e.g., phenobarbital rather than mephobarbital.

Precise control of drug concentrations in serum is also required in pregnant women, because the fetus may be more sensitive to a drug than an adult is. The elderly may respond differently to a drug than do younger adults, and measurement of the serum concentration may be useful in eliminating differences in gastrointestinal absorption as a cause of this. The concentration in serum of potentially dangerous drugs such as gentamicin, which does not produce symptoms until toxic damage has actually occurred, should be monitored regularly *(26)*.

In some situations it is better to monitor the action of a drug rather than its concentration. This is especially true when a drug is co-administered with others that might prolong or shorten its action. Typically such effects arise through stimulation or inhibition of enzymes in the liver that metabolize the drug, or through displacement from transport proteins in the plasma. Drugs for which this is especially true are anticoagulants, the metabolism of which is frequently affected by drugs that affect liver metabolism. Fortunately it is quite simple to measure the prothrombin time, which provides a very convenient measurement of the action of the drug. For the antiepileptics, which also are affected by the induction of liver enzymes, there is no simple laboratory test for monitoring effectiveness. It should be noted that the current practice of measuring only the total concentration of a drug in serum is not ideal, as only a drug that is not bound to protein will penetrate into cells and exert a pharmacological effect. Most of a drug in serum is protein bound, so that it is often impossible to recognize changes in the concentration of the pharmacologically active component through measurement of the total drug concentration. Measurement of the concentration of a drug in saliva may help indicate its concentration in serum in unaltered form.

For many drugs it is important to monitor for toxic effects, for example methotrexate and other potent cancer chemotherapeutic agents that may cause considerable tissue destruction. One should examine both effects that might arise from destruction of tumor—e.g., the release of nucleic acids and hence uric acid with its potential for renal damage—and effects due to damage of the liver and other organs. A regular toxicity monitoring schedule should be included with all such therapeutic regimens. By the time a drug is introduced into clinical practice the most common types of toxicity associated with the drug have usually been established. The monitoring schedule should be oriented to identification of damage to the organs most commonly affected. The tests that should be used are those that are best for the anticipated type of toxicity, e.g., the aminotransferases when hepatotoxicity is anticipated or bilirubin if cholestatis is most likely. Although frequently a single test is all that is required, a group of tests is most often used to provide an indication of both the organ affected and extent of damage.

Predicting Response to Treatment

The use of laboratory tests to predict possible response to treatment is relatively recent and has not been fully explored. A high carcinoembryonic antigen concentration in serum suggests a poorer prognosis after surgical removal of a tumor than does a low concentration. If the antigen concentration is low or undetectable sometime after surgery, this generally indicates that removal of the tumor has been complete. Measurement of estrogen, progesterone, and prolactin receptor activity in excised tumors of the breast provides a good index to the possible response of the patient to ablative surgery of the pituitary or adrenal glands, or to replacement hormone therapy.

Measurement of thyroid-stimulating immunoglobulins has proved useful in some instances in predicting the likelihood of response of thyrotoxicosis to treatment. Sullivan et al. *(27)* have recently demonstrated that measurement of platelet monoamine oxidase activity may be used to predict the response of patients with manic–depressive disease to therapy with lithium.

Future Applications of Laboratory Tests

One further current application of laboratory tests, which is likely to have greater future application to an increasingly well-informed general public, is to reassure the "worried well" that they have no serious disease. This application will probably develop most rapidly with the introduction of disease-specific tests.

The precision and accuracy now achievable for most clinical chemical tests exceeds the physician's ability to use the data. Multivariate analysis may result in more use of the data generated in a typical laboratory, computers being used to detect and interpret small changes in laboratory data that might not be considered significant by a physician. Still, this approach is likely to be successful only if analytical techniques are used that are highly accurate, precise, and specific, and if patients are studied under standardized conditions. This is both a great challenge and a great opportunity for clinical laboratory physicians and scientists.

References

1. Kosowsky, D. I., The clinical laboratory industry: Future outlook. *Med. Mktg. Media,* 46 (1976).
2. Duhme, D. W., Greenblatt, D. J., and Koch-Weser, J. Reduction of

digoxin toxicity associated with measurement of serum levels. *Ann. Intern. Med.* **80,** 516 (1974).
3. Griner, P. F., and Liptzin, B., Use of the laboratory in a teaching hospital. Implications for patient care, education and hospital costs. *Ann. Intern. Med.* **75,** 157 (1971).
4. Russe, H. P., The use and abuse of laboratory tests. *Med. Clin. N. Am.* **53,** 223 (1969).
5. Williamson, J. W., Alexander, M., and Miller, G. E., Continuing education and patient care research. Physician response to screening test results. *J. Am. Med. Assoc.* **201,** 118 (1967).
6. Zieve, L., Misinterpretation and abuse of laboratory tests by clinicians. *Ann. N.Y. Acad. Sci.* **134,** 563 (1966).
7. Young, D. S., Pestaner, L. C., and Gibberman, V., Effects of drugs on clinical laboratory tests. *Clin. Chem.* **21,** 1D (1975). (Special issue)
8. *Anon.*, Deluge of laboratory data becomes doctor's dilemma. *Med. World News* **8,** 82 (1967).
9. Franco, S. C., The early detection of disease by periodic examination. *Ind. Med. Surg.* **25,** 251 (1956).
10. Bates, B., Parker, R. C., Jr., and Reifler, C. B., Clinical evaluation and multiphasic screening: A comparison of yields. *Ann. Intern. Med.* **75,** 929 (1971).
11. Edgington, T. S., Astarita, R. W., and Plow, E. F., Association of an isomeric species of carcinoembryonic antigen with neoplasia of the gastrointestinal tract. *N. Engl. J. Med.* **293,** 103 (1975).
12. Drum, H. E., Cited by McNeil, B. J., and Adelstein, S. J., in Determining the value of diagnostic and screening tests. *J. Nucl. Med.* **17,** 439 (1976).
13. Shalet, S. M., Beardwell, C. G., Lamb, A. M., and Gowland, E., Value of routine serum triiodothyronine estimation in diagnosis of thyrotoxicosis. *Lancet* **ii,** 1008 (1975).
14. Wilson, J. M. G., in *Mass Health Examinations.* Public Health Papers, No. 45. World Health Organization, Geneva, 1971.
15. Henderson, M., and Sherwin, R., Screening—a privilege, not a right. *Prev. Med.* **3,** 160 (1974).
16. Reynoso, G., Chu, T. M., Aungst, W. C., and Priore, R. L., Automated multiphasic screening in cancer population. *N.Y. State J. Med.* **2,** 1251 (1972).
17. Collen, M. F., Dales, L. G., Friedman, G. D., et al., Multiphasic check-up evaluation study. 4. Preliminary cost benefit analysis for middle-aged men. *Prev. Med.* **2,** 236 (1973).
18. Paulus, H. E., Coutts, A., Calabro, J. J., and Klinenberg, J. R., Clinical significance of hyperuricemia in routinely screened hospitalized men. *J. Am. Med. Assoc.* **211,** 277 (1970).

19. Sisson, J. C., Schoomaker, E. B., and Ross, J. C., Clinical decision analysis: the hazard of using additional data. *J. Am. Med. Assoc.* **236,** 1259 (1976).
20. Bradwell, A. R., Carmall, M. H. B., and Whitehead, T. P., Explaining the unexpected abnormal results of biochemical profile investigations. *Lancet* **ii,** 1071 (1974).
21. Zola, I. K., Culture and symptoms—an analysis of patients' presenting complaints. *Am. Soc. Rev.* **31,** 615 (1966).
22. Carter, R. A., The value of screening in paediatrics. *J. R. Coll. Physicians London* **10,** 153 (1976).
23. Uddin, D. E., Dickson, L. G., and Brodine, C. E., Glucose-6-phosphate dehydrogenase deficiency in military recruits. *J. Am. Med. Assoc.* **227,** 1408 (1974).
24. Batsakis, J. G., Preston, J. R., Briere, R. O., and Giesen, P. C., Iatrogenic aberrations of serum enzyme activity. *Clin. Biochem.* **2,** 125 (1968).
25. Shearman, R. P., and Fraser, I. S., Impact of new diagnostic methods on the differential diagnosis and treatment of secondary amenorrhoea. *Lancet* **i,** 1195 (1977).
26. Werner, M., Sutherland, E. W., III, and Abramson, F. P., Concepts for the rational selection of assays to be used in monitoring therapeutic drugs. *Clin. Chem.* **21,** 1368 (1975).
27. Sullivan, J. L., Cavenar, J. O., Jr., Maltbie, A., and Stanfield, C. Platelet monoamine-oxidase activity predicts response to lithium in manic–depressive illness. *Lancet* **ii,** 1325 (1977).

The Physician's Logic in Making a Decision

David L. Sackett, M.D.

Department of Clinical Epidemiology and Biostatistics, and Department of Medicine, McMaster University Medical Centre, Hamilton, Ontario, Canada

Introduction

This presentation will summarize the four modal strategies that clinicians use in making diagnoses, show how these strategies use the clinical laboratory, and point out some of the ways in which the clinician and the clinical laboratory may be working at cross purposes. The focus is on how we *do* it, *not* on how we *ought* to do it, or *will* do it at some future date or technologic state.

In this presentation, *making a diagnosis* will mean *the process of recognizing the class or group to which an illness belongs so that, based on our prior experience with the class, the subsequent clinical acts we can afford to carry out will maximize the patient's health (1).*

Diagnostic Strategies

A mix of anectodal and more recent systematic observations suggests that clinicians use one or more of the following strategies in making a diagnosis:

(I) Pattern Recognition (Gestalt)

If one pictures a thin, wide-eyed, tremulous woman with an obvious mass in her anterior neck; a barrel-chested, red-faced old man with prolonged expiration through pursed lips and cigarette-stained fingers; a hotel guest with a linear pattern of erythematous wheals (each with a central haemorrhagic punctum) on his ankles, wrists, and buttocks, despite the inadequacy of these brief work-pictures, the diagnoses thyrotoxicosis, emphysema, and bedbugs leap to mind.

This is *pattern recognition, the realization (usually instantaneous) that the patient conforms to a previously learned picture (or pattern) of disease.*

This diagnostic strategy usually is triggered by visual stimuli (the patient's face, skin, or gait) but auditory (cleft-palate speech) or olfactory (the breath of diabetic ketoacidosis) stimuli may also initiate it. The use of pattern recognition increases with clinical experience and, because it is executed in an instantaneous non-verbal process, it is difficult to teach the novice *(2).*

Clinical laboratory tests, when they are used at all in the pattern-recognition strategy, are ordered for confirmation only.

(II) Multiple Branching (Arborization)

A middle-aged woman, with sustained diastolic blood pressure in excess of 100 mmHg, is being assessed for secondary forms of hypertension. She is asked whether she is taking oral contraceptives or other hormones. If she answers "yes," further diagnostic studies stop until her blood-pressure response to hormone withdrawal is determined. A "no" answer prompts the initiation of an inquiry into the next possible cause for her hypertension, the results of which will, in turn, determine the selection of subsequent diagnostic procedures.

This is the *multiple branching method: the progression of the diagnostic process down but one of a large number of potential, preset paths by a method in which the response to each test automatically determines the next test to be performed.*

This strategy is logical and precise, and tends to maximize the clinician's efficiency while minimizing the ordering of unnecessary laboratory tests. Often executed with the use of a written protocol or "algorithm," the multiple-branching method has found application in the evaluation of rare disorders or when responsibility for the diagnostic process is delegated to nonphysicians *(2, 3).*

(III) Exhaustion

In my first clerkship as a third-year medical student, I followed the "Outline for the Complete History and Physical" provided by the faculty. The resulting 35-page entries into the patients' clinical records were both praised ("The best work-ups ever performed on this service!") and condemned ("Who the hell would ever wade through them!")—significantly, for the purposes of this presentation, both verdicts were passed by the same staff man.

I was practicing the *method of exhaustion: the painstaking, invariant search for (but paying no immediate attention to) all medical factors about the patient, followed by sifting through the data for the diagnosis.*

As suggested by the example, this strategy is inefficient, a property which partly explains why its use tends to be limited to the clinical novice. Seasoned clinicians have often called for its abandonment *(4, 5)*, and recent quantitative research into the diagnostic process has documented that it is seldom used *(6)*.

These studies of the diagnostic process have, however, noted the insertion of abruptly launched and terminated segments of the exhaustion strategy during the taping of patients' encounters with seasoned clinicians. Replays of these tapes for the involved clinicians have revealed that the exhaustion strategy was introduced to keep the patient occupied while the clinician was thinking about possible diagnoses, to reassure the patient and gain rapport or, rarely, to fish for additional diagnostic possibilities when the original candidates have been ruled out *(6)*.

A prominent property of the method of exhaustion is that it is considered the "correct" strategy for diagnosis by many clinical students and by at least some clinical faculty. As a result, the latter are frequently embarrassed to admit that they routinely utilize other diagnostic strategies and excuse their failure to use the "proper" method on the basis of time pressure. The apparent conclusion by some clinical laboratory directors that the exhaustion strategy is the correct diagnostic method will be discussed later.

(IV) Hypothetico-deductive

Imagine that you have been telephoned by the emergency room and requested to see a 54-year-old man who has presented there with chest pain and shortness of breath. What goes through your mind as you hear this news?

It is most likely that you are formulating some possible diagnoses (does he have myocardial infarction, pulmonary embolus, or something else?) or possible courses of action (shall I put him in the coronary care unit, on the ward, keep him in the emergency room for observation, or send him home?). In addition, and on the basis of these possible diagnoses and courses of action, you are deciding which items of the history, physical examination, and laboratory evaluation should be done first (indeed, you probably ordered some of them before you hung up the 'phone). This behavior typifies what is currently thought to be the most common strategy of diagnosis *(6)*, the *hypothetico-deductive method: the formulation (from earliest clues about the patient) of a "short list" of potential diagnoses or actions, followed by the execution of those clinical (history and physical) and paraclinical (laboratory, x-ray, etc.) manoeuvres which will best reduce the length of the list.*

The hypothetico-deductive strategy has been proposed as the explanation for the development and testing of "short lists" in "pure" science *(7)* and, indeed, in the interpretation of art *(8)* as well as in clinical medicine, although the fashion in which the hypotheses are formulated in the latter application remains largely unexplored. Suggestions that the "short list" can arise through pattern recognition or even through exhaustion appear reasonable, but will not be discussed further here.

Investigations of medical students and both general and specialist physicians indicate that the hypothetico-deductive strategy is used by clinicians of all degrees of experience and that seasoned clinicians differ from novices in the specificity and accuracy of their lists of hypotheses and the efficiency of their selection of the "best" manoeuvres to reduce the length of this list *(6)*. The definition of the "best" manoeuvre includes considerations of validity, accessibility, and cost, and has been the subject of a great deal of recent investigation (for example, *9–11*) and debate *(12)*, as seen in several of the other contributions to this monograph.

Before turning to the implications of the physician's logic in making a diagnosis upon the operation of the clinical laboratory, one comment should be made about the training of clinicians. Although the mastery of the "complete history and physical" is of little value as a model for diagnosis *(13)*, the ability to carry out individual subroutines is essential to the successful execution of the hypothetico-deductive process. The clinician-in-training must master the subroutines, remembering all the while that the whole is less than the sum of its parts.

The Diagnostician and the Diagnostic Laboratory

The increasing scope and power of diagnostic laboratory procedures have enormously expanded the capabilities of the diagnostician, and this debt is insufficiently acknowledged. At the same time, insufficient communication between those who provide and those who use these sophisticated services, coupled with increasing economic pressures which may drive these diagnostic partners in opposite directions *(14)*, sometimes lead to situations in which the diagnostician and the diagnostic laboratory are working at cross purposes.

Four examples of working at cross purposes are:

(1) The performance of unsolicited laboratory tests.

(2) Confusion about the range of normal *(15, 16)*.

(3) The consequences of repeating abnormal tests *(17)*.

(4) The inappropriate introduction of new tests *(18)*.

The last three examples are discussed in detail elsewhere *(15–18)*, but the first is a logical extension of the previous section of this presentation and will be discussed here.

The automation of several clinical laboratory procedures, and their combination into groups of tests such that the ordering of any causes of the performance of all, has led to apparent improvements in the efficiency of the clinical laboratory *(19)*. In the hope that the results of the unsolicited portion of these laboratory test groupings would also benefit the diagnostician and patient, and without regard to the clinician's diagnostic process, these additional test results have become part of the routine laboratory report.

Does this scheme of performing and reporting unsolicited laboratory tests aid the physician's logic in making a diagnosis? A review of Table 1 reveals that it does not. When the physician is using the pattern recognition method, a specific laboratory test is ordered for purposes of confirmation only, and unsolicited laboratory test results are merely irrelevant. However, when the diagnostician is using the multiple branching or hypothetico-deductive strategies, the appropriate choice of laboratory tests is selective and sequential, and unsolicited laboratory test results

Table 1. **Relation between the Diagnostic Process and the Clinical Laboratory**

Diagnostic strategy	*Choice of lab. tests*	*Diagnostic value of unsolicited lab. results*
Pattern recognition	selective, confirmatory	irrelevant
Multiple branching	selective, sequential	interferes
Exhaustion	all available and affordable	always helpful
Hypothetico-deductive	selective, sequential	interferes

may interfere with the diagnostic process by presenting extraneous information out of sequence and context. It is only in the exhaustive strategy that unsolicited laboratory results are always welcome.

The preceding objections to the performance and reporting of unsolicited laboratory tests may be faulted as theoretical in nature. However, they are bolstered by empirical observations. For example, Daniels and Schroeder *(20)* found that the correlation coefficient relating the quality of care received by hypertensive patients to the number of laboratory tests performed on them was both negative and sizeable *(20)*. Furthermore, a major randomized trial of the value of extensive unsolicited laboratory testing found no benefit of this manoeuvre upon the speed, process, cost or clinical outcome of the diagnostic process *(21)*. Thus, there are both theoretical and empirical grounds for abandoning the performance and reporting of unsolicited laboratory tests.

In summary: In making a diagnosis, the physician may use one of four strategies: pattern recognition, arborization, exhaustion, or the hypothetico-deductive method. Although the physician's diagnostic capabilities have benefited enormously from advances in the clinical laboratory, the performance and reporting of unsolicited laboratory tests has not been shown to help—and frequently hinders—the diagnostic process.

Several of the ideas presented here emerged from meetings of the "Diagnosis Study Group" held over the last two years at McMaster University. The special contributions of Drs. E. J. M. Campbell and V. R. Neufeld are acknowledged.

References

1. Adapted from: Cohen, H., The nature, methods and purpose of diagnosis. *Lancet* **i,** 23–25 (1943).
2. Murphy, E. A., *The Logic of Medicine,* The Johns Hopkins University Press, Baltimore, Md., 1976, p 153.
3. Grim, C. E., Weinberger, M. H., Higgins, J. T., and Kramer, N. J., Diagnosis of secondary forms of hypertension. *J. Am. Med. Assoc.* **237,** 1331–1335 (1977).
4. Michaels, L., A plea for the abandonment of the complete history and physical examination. *Can. Med. Assoc. J.* **108,** 229–303 (1973).
5. Dudley, H. A., Clinical method is all wrong. *Med. Hypotheses* **2,** 164–167 (1976).
6. Feightner, J. W., Barrows, H. S., Neufeld, V. R., and Norman, G. R., Solving problems: How does the family physician do it? *Can. Fam. Physician* **23,** 457–461 (1977).
7. Popper, K. R., *Logic of Scientific Discovery,* Hutchinson, London, 1968.
8. Gombrich, E. H., *Art and Illusion: A Study in the Psychology of Pictorial Presentation,* 2nd ed., Princeton University Press, Princeton, N.J., 1961.
9. Lusted, L., *Introduction to Medical Decision Making,* Charles C Thomas, Springfield, Ill., 1968.
10. McNeil, B. J., Keeler, E., and Adelstein, S. J., Primer on certain elements of medical decision making. *N. Engl. J. Med.* **293,** 211–215 (1975).
11. Galen, R. S., and Gambino, S. R., *Beyond Normality: The Predictive Value and Efficiency of Medical Diagnoses,* John Wiley and Sons, New York, N.Y., 1975.
12. Ingelfinger, F. J., Decision in medicine. *N. Engl. J. Med.* **293,** 254–255 (1975). Editorial
13. Leaper, D. J., Gill, P. W., Staniland, J. R., et al., Clinical diagnostic process: An analysis. *Br. Med. J.* **iii,** 569–574 (1973).
14. Bailey, R. M., From professional monopoly to corporate oligopoly: The clinical laboratory industry in transition. *Med. Care* **15,** 129–146 (1977).
15. Murphy, E. A., *op. cit.,* pp 123–133.
16. Sackett, D. L., Clinical diagnosis and the clinical laboratory. *Clin. Invest. Med.* **1,** 37–43 (1978).
17. Sackett, D. L., The usefulness of laboratory tests in health-screening programs. *Clin. Chem.* **19,** 366–372 (1973).
18. Sackett, D. L., and Holland, W. W., Controversy in the detection of disease. *Lancet* **i,** 357–359 (1975).
19. Whitehead, T. P., and Wootton, I. D. P., Biochemical profiles for

hospital patients. In *Screening for Disease,* The Lancet, London, 1974.

20. Daniels, M., and Schroeder, S. A., Variation among physicians in use of laboratory tests: II. Relation to clinical productivity and outcomes of care. *Med. Care* **15,** 482–487 (1977).

21. Durbridge, T. C., Edwards, F., Edwards, R. G., and Atkinson, M., An evaluation of multiphasic screening on admission to hospital. *Med. J. Aust.* **i,** 703–705 (1976).

Problem-Oriented Laboratory Testing

Lawrence L. Weed, M.D.

University of Vermont, Burlington, Vt. 05401

In presenting a logical approach to medical problem solving with attendant benefits of speed, efficiency, and reduction in cost, we must recognize that new approaches to laboratory utilization are needed. My recent article, "A New Paradigm for Medical Education" *(1)*, dealt with many of these same issues from the standpoint of medical education, and I would like to relate that information to problem-oriented laboratory testing.

Problem-oriented laboratory testing must begin with a precise statement of the problems themselves. The solutions to the four problems stated below are intended to create a framework in which laboratory scientists, pathologists, and physicians can interrelate in a productive manner.

The four problems are:

1. The lack of coordination of activities and information among providers;
2. Failure of the medical record unambiguously and consistently to preserve the logic of actions, as well as evidence of the actions taken;
3. Unrealistic use of and dependence upon human memory; and
4. Lack of meaningful feedback loops on everyday medical actions, i.e., the system is blind.

First, a framework is needed in which to think about these problems. In medical care, as in any other organized endeavor, there should be a goal (defining comprehensive or episodic care) defined as:

1. A system or a structure with rules;

2. Audit of performance within the rules and correction of the deficiencies; and
3. Audit of the system itself, by examining outcomes—and a change in the system when necessary.

What is our present medical system and how did it develop? Let us go back to the days of a patient who was cared for by one physician in a small town, for example. In this system the doctor often recorded little, but because he lived near the patient, he gleaned through frequent contact all that he needed to know about his or her social, psychiatric, and medical history. Furthermore, there was much less precise scientific knowledge and fewer techniques for him to master and few if any specialists interacted with the patient of the family doctor.

Then, as the body of knowledge grew and doctors continued to believe they should learn it, memorize it, and pass examinations, they started to specialize. They also legitimately specialized to develop new surgical skills and new techniques in pathology and radiology. So now a single patient relates to several physicians.

Some of these physicians, such as the pathologist, dealt with the patient mainly in a technical sense and did not presume to provide total care. Then the various fields of medicine developed ancillary medical personnel, and a single patient was surrounded by a large number of people. It was inevitable that a problem of coordination of medical information would develop. The right hand did not know what the left hand was doing. The specialist prescribing a drug was unaware that the drug was invalidating a test being done in the clinical laboratory. Indeed, the laboratory specialist did not even know the patient was being seen by the specialist who ordered the test. However, even if the laboratory scientist was aware that the drug had been administered, it is unlikely that either specialist would have known that the drug invalidated the test because of the mass of other details that need to be learned, but cannot be learned, with modern medicine, because there is too much information for the human memory to maintain.

To make matters worse, one patient may not see the same doctor, nurse, or medical worker over a long period of time, even within a given area of specialization, because of the increased mobility of both medical personnel and patients and the frequent "rotating" of personnel in educational institutions. The resulting provincialism, in time as well as in space, has led, in le Corbusier's phrase, "to spectacle of fragments of intention." In such a situation, cost/benefit analysis applied to one of the fragments can be very misleading. One hasty, episodic bit of care in a clinic to save a few minutes and dollars may generate costs of hundreds of dollars for care in an expensive hospital months later. The cost/benefit analysis of the clinic alone would not present a true picture. Emergency rooms, laboratories, and x-ray departments in hospitals are often not thought of as part of the feedback loop and a source of valuable end-stage information that should be used to modify behavior at other points in the loop. Providers at one point in a cycle rarely try to control the inflow to that point by dealing with basic causes. They just expand and seek more resources for that particular service. Medical faculties and specialty groups position themselves at various disconnected points in these potential cycles and elaborate on their expertise to handle the problem at hand.

In an effort to solve the problem of coordination, medical personnel started to keep records, both to relate to one another and to support their own memories. Furthermore, in the old multiple records, the data were kept in a source-oriented way, so that the logic of what was done was never preserved, even though there was an effort to record statements as to "what" was done. The average laboratory, for example, may have statistics on what procedures were done and how many, but not for what purpose and with what effect. Because the logic was not preserved and because multiple copies were not fully coordinated, there was little feedback, and bad practices in medicine could go undetected and uncorrected. Any system without a feedback loop runs wild, and conditions were such that medicine could indeed run wild. We cannot sort the good from the bad; we run the risk of trying to do all and going bankrupt and hurting people unreasonably. Or we can stop it all irrationally and throw the baby out with the bathwater.

I shall not review in detail the available data on other examples of actual inadequacies in the performance in modern medicine; I shall go directly to the proposed solutions with new tools and premises.

The Tools

I intend to show here how the problem-oriented record and the computerized problem-oriented record are useful tools in dealing with the problems of coordination, logic, memory, and feedback loops described above.

The Manual Problem-Oriented Medical Record

The coordination problem. For coordination to work, it is essential that the record be in a standard form that preserves the logic of all the providers, so all the interconnecting links are intelligible to the patient and to the providers. In this way, the patient's problems and capabilities become the basis for integrating all actions in the interest of the total care of an individual.

The problem list immediately puts each contributor into context. Everyone must be taught to pause and look at the list of problems and ask how they could possibly relate to one another in meaningful ways. No plans for solving one problem should be concluded without some thought about its implications for the management of the others.

A patient has health problems in the first place because of contradictory forces working within his mind and body; the medical profession, through specialization, may add further contradictory forces and make him worse unless they have tools to prevent this from happening.

The logic problem. Plans and progress notes should include the appropriate problem and thereby preserve the logic of each provider so others can see it. The notes must be randomly audited for reliability in representing the patient's true situation, as well as for their intrinsic logic and efficiency of thought and action in the pursuit of the single problem. When the problem-oriented system is first used, it becomes apparent that goals have not been set; problems lists are inadequate to the point of being irritating and misleading; progress notes do not relate to original plans; random thoughts are left dangling; excessive laboratory work is performed; unnecessary risks are taken; and patients are confused about the process in which their cooperation is necessary for the best results. At this point, one should not become angry with the problem-oriented medical record. Do not destroy the messenger because you do not like the message;

thank the messenger and proceed to deal with the poor thinking that has always been there but only now explicitly revealed.

In the initial plan, the medical personnel and patient are to be the architects; in the titled progress notes, they become the contractors. The teacher is the auditor of the process. All should learn from the process as human and biological situations unfold.

The memory problem. If each provider sees the same problem-oriented medical record (POMR) (and the clinical laboratory should think of itself as an independent, thinking provider and not the unquestioning servant of other providers) and if the patient has a copy of the record, there is far less chance that a given problem will be dealt with out of context or overlooked altogether. A provider should not take pride in how much he remembers about a patient; he should be continually aware that his mind alone cannot recall all he should know at each encounter in order to give the best possible care. Because one banker may remember figures better than another does not justify giving up the use of bookkeeping systems in banking. Furthermore, the POMR facilitates the development of the physician's capacity to formulate and solve problems, using books, journals, and consultants as necessary. It allows us to audit the logical application of knowledge to patient's needs without wasting time on determining whether the physician actually "knew" the knowledge or simply had developed the habits and skills to use references and consultants effectively.

Feedback loops. I stated at the outset that we should have: 1) a system with rules; 2) audit of performance within the system, with corrective action; and 3) audit of the system itself with outcomes, making corrections when necessary. The system is made up of loads, resources, personnel, messages, and communication tools. The POMR represents the structure and the rules for communicating over time; it is the basis for the feedback loop. Outcomes should not be interpreted until, in terms of the rules, performance is good. Otherwise the system may be prematurely criticized and altered. Continuously operating feedback loops with a well-defined system form the basis for reaching goals. The POMR is the principal tool within the system. Although the tool itself guarantees nothing, without it, consequences cannot be related to means, and the loop is broken.

The Computerized Problem-Oriented Medical Record

The role of the computerized POMR is major and its implications are far-reaching. (To have perfect coordination and reliable feedback control, factual recall by human minds must be removed from the loop, and that can only be done by establishing logical connections among all elements in a computerized system. New elements can be entered, but not at random; their place in the structure is determined by the logic by which the structure is constructed.)

How Does the System Work?

A cathode ray tube (CRT) terminal provides the user with the computerized displays by which information is transmitted and decisions are recorded. All the resources of medical knowledge may be utilized to organize the displays in the computer from which choices can be made in logical sequence. If patients' data are accumulated in a highly organized manner and are properly analyzed, they will become one of the principal sources of feedback with which we correct the system. This can be done when the medical-care system has been brought to a predefined level of thoroughness and reliability in the context of one whole individual and not just when one of his organs has been dealt with by a specialty service.

The feedback loop is complete. The role of human memory has been changed—at no point is it directly in the loop, and no longer does the system depend on human memory. The isolation and provincialism of separate documents in the medical-care system have been broken. The boundaries among coordination, memory, logic, and feedback also fall away: they merge into one another.

The overall integrity of the logical structure is forced by a single characteristic of the system: you can proceed only one step at a time by choosing one display at a time. The vast amount of medical knowledge that has been fed into the computer is available in no other way. One step can represent the synthesis and application of a great deal of medical knowledge, just as one mathematical formula can represent an extensive understanding of mathematics. For this reason, the users' pathways can be efficient, rather than annoying and cumbersome. At the same time, the precise and full documentation for each action can be ascertained. And no computer display from which you begin, and by which you must proceed, can be independent of the structure of the POMR.

The logical sequencing of the displays has its origin in the basic structure of the POMR—each level of content in a given structure becomes the structure of the next level of content. For example, the physical examination is content for the data-base structure, the heart examination is content for the physical-examination structure, and the auscultation is content for the heart-examination structure.

The computer becomes essential if intelligent application of expanding knowledge is ever to be a reality. The relatively incomplete decision tables we now have, because of poor medical records, can be updated and improved in the future from accumulated computer data, with use of the appropriate mathematical techniques to allow the maximum yield from a minimum amount of data economically acquired. It is pointless to bemoan our ignorance and say we do not yet know enough to program the computer. We never will know enough if we do not start, make first approximations using present knowledge and structured input, analyze our results, update our pathways, and bootstrap our way up. We should have started long ago—but never mind; let us get on with it. The studies that have been made simply do not bear out the notion that the human mind, when unaided, is up to the task. The mind is willing to proceed without adequate data and feedback loops, and, having done so, is much more apt to conclude it was correct without the evidence; that is, clinical judgement is the willingness to make decisions on inadequate data.

We have only barely begun to exploit the enormous capabilities for manipulating complex data structures in a highly interactive computer terminal environment. It is to this area that the efforts of the PROMIS Laboratory are being directed within the framework of the problem-oriented philosophy for comprehensive care. The question should never be minds or computers, but rather what is the most effective combination for getting the job done.

The clinical chemist may well ask if he and all the technical people involved in doing a day's laboratory work can afford the time to think about each test and its context and indications. Shouldn't the computer just grind out work lists for each worker independent of any knowledge of the problem or the actual interpretation of the result? The computer can, of course, be programmed to generate such work lists, and laboratory results can go directly from the automated laboratory equipment into the patient's record. But these latter considerations are not the important ones. The important questions are: who will provide random audits of the use of the guidance system in the computer, who will write the guidance in the first place, and who will study the outcome and review the literature so that the guidance can be up-dated appropriately? It is in these areas that the clinical chemist can assume a leading role and thereby facilitate consistently meaningful use of the laboratory in the solution of clinical problems.

The Role of the Clinical Laboratory in the Computerized Problem-Oriented Medical Record System

Having discussed four main problems and their proposed solutions, I will now proceed to discuss two major roles the laboratory can play in the system. The laboratorian must bring his expertise to the task of building computer displays and analyzing displays built by others for the purpose of solving specific medical problems. He must also show the leadership in the solving of the coordination problem by seeing that medical computer systems are developed with a total information architecture in mind, so that completed feedback loops are present within the system. If this latter step is not taken, proliferation of separate computer speciality systems occurs, efficient comprehensive care to an individual becomes all but impossible, and costs skyrocket, thereby endangering the viability of the whole medical enterprise.

Let me now deal with each of these functions more specifically.

A. Building displays:

Background: There is already a detailed sequence of computer displays in the computerized problem-oriented medical record system for guidance through the four phases of medical actions, i.e., the data base, the problem list, the plans, and the progress notes. These computer displays contain an approach to problem solving in general and proceed directly to displays containing content for the pursuit of specific problems. The basic display is shown in Figure 1. In a brief discussion such as this, it is impossible to illustrate all the thousands of frames now available in the system. Laboratory scientists can first examine the basic approach to problem solving made explicit on the frames to see if they agree with it, and then proceed to examine the content that is already there for updating purposes, and then finally see that new, additional content is added to the appropriate places. They then should get involved with the whole concept of building mechanisms for keeping the material up to date.

There is no substitute for a person new to the system (whether he is a builder or user) sitting down at the terminal and actively working in a dynamic way with the displays. One visit is worth a thousand articles.

Taking each of the four phases of medical action in turn, we can now proceed to outline the sorts of things

Doe, June .... F 59 000–000–0

——— 4 Phases of Medical Action ———

——— Retrieve: ———		——— Add to: ———	
—Data Base		—Data Base	
—Problem List		—Problem List	—Exp—
—Initial Plans ←		—Initial Plans	—Exp—
—Progress Notes		—Progress Notes	
—Other Retrievals		—Other Actions	
—Flowsheet Retrievals	—Exp—	—Emergency Management	—Exp—
—Graph Retrievals	—Exp—	—Consult Reply	
—To Printer		—Audit	

—Choose other ward/other functions —Choose other patient on this ward

Fig. 1. The above is the basic display from which one can enter into any of the four phases of medical action. At the top is the number of the display, followed by the patient's name. In the center of the top line is the main title of the display. Below the top line are choices which, when touched, lead you immediately to the next display that is a logical progression from that particular choice. For the purpose of showing a series of choices, a given choice will be marked with an *arrow*. The arrow is, of course, not present in the actual display.

pathologists can work on. I have previously discussed the relationship of the autopsy to the POMR *(2)*.

I. THE DATA BASE: In the comprehensive POMR there is a defined data base, which is obtained from each patient for the purpose of comprehensive care. There are many "mini" data bases for episodic care. The emphasis should be on comprehensive care, managing episodes within that context. Persistent episodic care of an individual should not be allowed to go on indefinitely before one gets a comprehensive data base. The pathologist should review the laboratory section of the comprehensive data base at regular intervals to see that all providers using the system are using the laboratory in the most effective and efficient way possible for the discovering of significant problems. The clinical history, the physical examination, and the laboratory should not be thought of as competing with one another for scarce resources. The nature of the problem being screened for should determine the choice of data to be assembled in the data base for the purpose of discovering that particular problem or a crucial pointer to that problem. Certain problems are best detected by historical facts, such as a history of a transient visual disturbance pointing to a neurologic problem; others by physical examinations, such as a breast examination for cancer of the breast; and others by laboratory determinations, such as serology for finding syphilis. Furthermore, one should not even seek problems if there is no imaginable benefit from finding the problem early or before the patient has symptoms attributable to it. General statements about the value of screening are meaningless. One must have specific problems in mind and very specific ideas as to the value of detecting them.

No procedure should be used in a data base if it isn't of specific value in pointing to a problem or group of problems. Furthermore, one should not fill up the data base with a series of procedures that are useful for diagnosing and delineating problems in detail; that is, well beyond just discovering their presence. For example, one blood pressure measurement points to a group of 50 or 60 possible problems. If a blood pressure is high, it is in the specific plan section of the problem-oriented record, not in the data base, that we should consider pursuing which of the 50 causes is present in a particular individual. It is not always appropriate to pursue all problems in depth, but it is always valuable to know about their existence, each in the context of the others. Without this overall understanding of an individual and all his problems, one cannot organize goals and treatments for any single problem that are appropriate to his needs and way of life.

The next question that arises is, should the data base be large enough to have pointers to all possible problems in a given population? In answering this question, one should take into consideration the load or number of patients, the resources available, including personnel, and the value to the patients or the community of uncovering the problem in the first place. It is better to have a very small defined data base that uncovers or points to the five or six major causes of sickness and death and for which there are specific treatments than it is to have a larger defined data base that cannot be applied to the whole population because of the simple limits of resources in the presence of a very large load. In the presence of limited resources, one should make a maximum effort to utilize the patient himself in getting crucial information.

Having established a working data base in the computer, it is then the job of the laboratorian to point out how the performance of the tests in the laboratory section of the data base should be monitored, and then at regular intervals he should look at the outcomes to see if the problems that the data base was designed to reveal or point to are actually being discovered, and if this leads to a worthwhile outcome. If the only outcome of finding an abnormal value is for the provider to order another determination and then ignore it, then either this behavior should be stopped or the procedure removed. If all one harvests is anxiety on the part of the patient without corrective action, then one should cease in that situation also. There is no such thing as the perfect defined data base for a given community, available from outside sources. You can make your own first approximation, use it, and slowly improve it. In a given area for a given population of people, there should not be multiple data bases, each promulgated by a different specialized provider. The providers should get together and develop a single data base, have a single problem-oriented record for the patient, which is available for any visit to any provider. Our efforts should be coordinated and cumulative, and patient at all times should have a copy of the record.

II. THE PROBLEM LIST: The pathologist should review the basic problem-formulation scheme in the computerized problem-oriented record. He should pay particular attention to those problems that are identified simply as abnormal results in a given laboratory determination. It will be remembered that in the problem-oriented system, the provider should review the data base and then identify problems at the level for which he has evidence and not go beyond that. Different providers may achieve different levels of synthesis, one listing four separate findings from the data base as four separate problems, whereas another would combine them logically into a single problem. This is to be expected since we all have different levels of sophistication. But no problem list should have guesses and impressions in it.

In the system just described, even the most sophisticated providers are frequently confronted with findings in the laboratory data base that are abnormal but not easily attributed to a problem already on the problem list or to a well-understood medical diagnosis. To record these problems one merely makes the choice "Laboratory Abnormality" in the problem formulation sequence and then from the choices available one makes a specific selection to identify a particular problem. For some of these problems there are accompanying definitions of what is abnormal and how interpretations should be made and how problems should be stated—for example, in the hyperlipoproteinemias. The pathologist or clinical chemist should review all these displays to make sure that they are accurate and appropriate for his area. There should be an effort made on the part of all of us to have the basic data base be similar over the largest possible area and the widest possible population, so that a common language and series of problems can be studied in terms of outcomes having a maximally significant result.

For each problem in the computerized problem-oriented medical record system there is a basic table. In each table there are categories of information such as manifestations, risk factors, treatments, and so forth. It is from these tables that information is electronically and dynamically sought as various actions are taken in other parts of the computerized system, such as writing a plan or a progress note. The pathologist should review the contents of these tables at regular intervals so that any user is automatically up to date as he goes about his daily work. Each of us is responsible for coupling his expertise to the needs of a maximum number of people at a minimum cost. This coupling is not accomplished by the hit-or-miss reading and writing of articles in books, or the occasional meeting or death conference on occasional patients. Such coupling must be built into the tools that every provider uses on every patient, and one tool that has not proved to be up to the coupling task is the limited, fallible human memory, which may have been programmed in a limited and provincial manner in the first place.

III. THE PLAN: Plans written for the typical problem for the typical patient in this country fall short of organization and precision that good care and meaningful feedback loops require. Indeed, in many cases they are only a scribbled order for a drug or a test without a precise statement as to whether the object is to rule out something, to follow the course of something, to measure the disability from something, to detect a complication, to determine a baseline for therapy, and so forth. Rules for the quantitative aspects of plans—such as the frequency of determinations—are either extremely variable from provider to provider or not made explicit at all. An overall structure and philosophy for detailed planning and careful follow-up is not available in many places. Indeed, one might call the Plans section the disaster area of American medicine. It is so easy to order things off the top of the head and from a specialized point of view with an inconsistency and parochialism that makes growth of medical science practically impossible.

In the computerized problem-oriented record there is a detailed superstructure for planning. Although anyone can use the detailed sequences easily while sitting in front of a terminal, it becomes impractical to describe all the branching logic. A small segment of such a logic sequence is illustrated in Figures 2 to 11. However, one can be assured that there are many displays in that superstructure and many points of departure whose original construction requires the expertise of the clinical pathologist. He, like others, will occasionally be frustrated as he goes to the literature to get the specific information to complete the displays for specific problems, since our past record-keeping habits have not provided us with the detailed, rigorous experience that is necessary for making precise interpretations of tests and providing specific guidance as to how often determinations should be obtained. In those instances where the present medical literature does not provide specific, unambiguous guidance, it is very useful for the busy provider to know this from experts. Through the medium of the computerized problem-oriented record system he can instantly find it out at the time he is confronted with a specific problem. Furthermore, experts can provide definite first approximations of what to do under circumstances of ignorance and ambiguity, and from that well-defined starting point and subsequent audit of outcomes we can gradually work our way up to better understanding through organized cumulative efforts on the part of all of us.

/26.142–A Doe, June ....

/Doe, June .... F 59 000–000–0 ...../Initial Plans

—16 ———— Active Problems for Initial Plans: ————

— * Hospitalization 8/1

— 1. → Histiocytic lymphoma. 8/1

Cond: unknown

— 3. Enlarged thyroid, onset approximately (15 years) euthyroid chem . . . 8/1

— 4. Fever, onset approximately (3 weeks ago.) 8/1

— 5. Low hematocrit, onset date/age not known, etiology unknown. 8/1

Cond: unknown

— 6. Histiocytic lymphoma. 8/1

Guarded & getting worse.

— 7. Blood in urine (hematuria). 8/2

— 8. Coronary atherosclerosis. 8/2

— 9. Anemia: chronic, hypochromic, microcytic. ← 8/2

—10. (Hematuria) 8/3

Cond: unknown

Fig. 2. It will be noted that having touched "Plan," as the arrow shows in Fig. 1, the patient's list of problems immediately appears. In this way the logic of the provider is made explicit by forcing all actions to be problem-oriented.

/4.171 Doe, June ..... 11. Anemia: Chronic, Hypochromic, Microc . F 59 000–000–0 ..../Initial Plan 11. Anemia: Chronic, Hypochromic, Microcytic

———— Initial Plan for DX Entity or Pathophys. Abn. ————

—1. State aims for problem management —Exp—

—2. Check how problem may be contributing to patient's "sickness" —Exp—

—3. Check for effects/disabilities produced by problem —Exp—

—4. Check function/status of systems that may be involved with problem —Exp—

—5. Assess and follow course

—6. Investigate problem and its etiology ← —Exp—

—7. Watch for/prevent complications of problem —Exp—

—8. If indicated, institute and monitor treatment

—Emergency management —Choose another problem

—Next frame—

Fig. 3. Having picked a specific problem for planning on the previous display, one is now provided with an organized planning and problem-solving structure. Normally one would progress through all the steps if time allowed, but we shall, in this instance, proceed directly to Step 6.

/4.224 Doe, June ...... 11. Anemia: Chronic, Hypochromic, Microc Microcytic. Plan /Initial Plan 11. Anemia: Chronic, Hypochromic, Microcytic.

——————— Investigate Problem and Its Etiology ———————

—1. Check for previous patterns of similar problems
—2. Characterize problem further (if appropriate) —Exp—
—3. Check problem list for possible etiologic relationships
—4. Check current treatment for possible etiologic relationships
—5. Investigate causes of initial concern (if appropriate) —Def— ←
—6. Investigate other causes (if appropriate)
—7. Screen for predisposing problems (if appropriate)

—Next frame—

—Prob list— —Cur RX— —Outst test—

Fig. 4. It will be noted that Steps 1, 2, and 4 in the structure above take into account the unique background of the patient so that textbook knowledge is not applied out of context. It will be noted that choices are available at the bottom of this display to get immediately the patient's problem list, the current treatments, and the outstanding tests. Having completed those steps, it may be necessary to proceed to Step 5 if an answer has not already appeared. If one touches —Def—, a display appears which explains the criteria for saying a cause is of initial concern. If one touches the choice directly, he proceeds to the next display.

/5.4 Doe, June .... 11. Anemia: Chronic, Hypochromic, Microc Initial Plan 11. Anemia: Chronic, Hypochromic, Microcytic. Investigate for Cause

——— Hypochromic, Microcytic Anemia: Causes of Initial Concern ———

—Lead poisoning ←
—Iron deficiency
—Pyridoxine-responsive anemia
—Chronic disease
—Crude liver factor responsive anemia
—Beta thalassemia minor

—Begin/continue/resume other investigation —Exp—

—Next frame—

Fig. 5. This display contains the cause of initial concern for the problem hypochromic, microcytic anemia. The problem statement is carried at the top of the display in the center of the double line. If one chooses one of these (lead poisoning, for example) he is led to a display that allows him to screen for this problem to see if it is likely, or he can go directly to the displays for ruling it out (see Fig. 6).

/4.229 Doe, June ...... 11. Anemia: Chronic, Hypochromic, Microc
Anemia: Chronic, Hypochromic, Microcytic, Investigate for Cause /Lead Poisoning

——————— Investigate for Cause ———————

—1. Screen for this cause ←
—2. R/O this cause

or

—Suspend investigation for this cause due to "negative screen" —Exp—
—Discontinue investigation due to "unacceptable risk or cost" —Exp—
—Report this cause as "ruled out" (not present) —Exp—
—Report this cause as present —Exp—

Fig. 6. The choices of the above display are self-explanatory.

/01.5 Doe, June .... 11. Anemia: Chronic, Hypochromic, Microc
Hypochromic, Microcytic. Investigate for Cause /Lead Poisoning Screen for Cause

— Hypochromic, Microcytic Anemia: Cluster to Screen for Cause —

Cause: lead poisoning

—Findings that suggest this cause—	—Sensitiv.—	—Conf. rate—
Pica with residence in older housing	—	—
Prolonged exposure to lead compounds	—	—
Basophilic stippling of RBCs	—	—

—Order screening procedures
—Contingency plan —Recommendations
—Prob list— —Cur RX— → —Next frame—

Fig. 7. Having received the information in the above display, one may wish to now rule out lead poisoning. This is accomplished by touching "next frame."

/4.229 Doe, June .... 11. Anemia: Chronic, Hypochromic, Microc
G Screen for Cause Investigate for Cause Investigate for Cause/Lead Poisoning

——————————— Investigate for Cause ———————————

—1. Screen for this cause
—2. R/O this cause ←

or

—Suspend investigation for this cause due to "negative screen" —Exp—
—Discontinue investigation due to "unacceptable risk or cost" —Exp—
—Report this cause as "ruled out" (not present) —Exp—
—Report this cause as present —Exp—

Fig. 8. By touching "R/O this cause," the display in Fig. 9 is brought into view.

/1.1052 Doe, June ..... 11. Anemia: Chronic, Hypochromic, Microc
Use Investigate for Cause /Lead Poisoning R/O Cause /Lead Poisoning R/O Cause

——————————— R/O Lead Poisoning ———————————

1. Do —24 hr urine lead

or

—blood lead

2. If urine lead concentration is approximately 300 micrograms/liter, then lead poisoning is present.
If urine lead concentration is > 80 micrograms/liter in children or > 100 micrograms/liter in adults,
then lead poisoning may be present; repeat determination indicated.
3. If blood lead concentration is > 80 micrograms/milliliter in an adult, then lead poisoning is present.
If blood lead concentration is 60–80 micrograms/100 milliliters in an adult.
Then lead poisoning may be present; repeat determination indicated.
If blood lead concentration is > 60 micrograms/100 milliliters in a child, then lead poisioning is present.
If blood lead concentration is > 40 micrograms/100 milliliters in a child, then lead poisoning may be present; repeat determination indicated.

—Outst test— —Test indx— —Next frame—

Fig. 9. By touching "blood lead," for example, on the above display, the order is immediately generated in the laboratory. When they get a result, they put it directly in their terminal.

/1.1052-B Doe, June ... 11. Anemia: Chronic, Hypochromic, Microc Use Investigate for Cause /Lead Poisoning R/O Cause /Lead Poisoning R/O Cause

---

R/O Lead Poisoning

4. If urine lead and blood lead concentrations are equivocal or negative but suggestive clinical findings are present,
   Then do calcium EDTA test.
5. If > 1000 micrograms of lead are excreted per liter of urine after 8 hours of EDTA administration, then lead poisoning is present.
   If > 500 micrograms of lead are excreted per liter of urine after 8 hours of EDTA administration, then an excessive lead burden is indicated and lead poisoning may be present.
   If < 400 micrograms of lead are excreted per liter of urine after 8 hours of EDTA administration, then lead poisoning is not present.

—Contingency plan —Recommendations

References

1. Flink, Edmund: Heavy metal poisoning, pp. 57–62 in Textbook of Medicine (14th ed.), edited by Paul B. Beeson and Walsh McDermott, 1975.
2. Levine, W. G.: Heavy metals, pp. 947–957 in The Pharmaceutical Basis of Therapeutics (4th ed.), edited by Goodman & Gilman, 1970.
3. MCHV normal chemistry values, MCHV Chemistry Laboratory.
4. Poskanzep, David C., and Bennett, Ivan L.: Heavy metals, pp. 667–671 in Harrison's Principles of Internal Medicine (7th ed.), edited by Maxwell M. Wintrobe, et al., 1974.
5. Weissman, Norman: Laboratory Aids in Toxicological Problems, p. 19–20. (Copyright: Bio-Science Laboratories, 1973).

—Next frame—

Fig. 9. Continued

/4.224 Doe, June .... 11. Anemia: Chronic, Hypochromic. Microc Use Investigate for Cause /Lead Poisoning R/O Cause /Lead Poisoning R/O Cause

——————— Investigate Problem and Its Etiology ———————

—1. Check for previous patterns of similar problems
—2. Characterize problem further (if appropriate) —Exp—
—3. Check problem list for possible etiologic relationships
—4. Check current treatment for possible etiologic relationships
—5. Investigate causes of initial concern (if appropriate) —Def—
—6. Investigate other causes (if appropriate) ←
—7. Screen for predisposing problems (if appropriate)

—Next frame—

Fig. 10. It will be noted that the above display is equivalent to Fig. 4. One would return to this display and make choice number 6, "Investigate other causes," if up to this point in the workup no satisfactory resolution of the diagnostic effort had been achieved. One proceeds from choice 6 directly to the next display, and from there one systematically considers the possibilities as was done with "Causes of initial concern."

/8.1 Doe, June .... 11. Anemia: Chronic, Hypochromic, Microc Use /Lead Poisoning R/O Cause /Lead Poisoning R/O Cause Investigate for Cause

————— Hypochromic, Microcytic Anemia: Other Causes —————

—Copper deficiency (mostly infants) —Hemoglobin E
—Sideroblastic anemia
—Familial hypochromic, microcytic anemia
—Congenital atransferrinemia
—Hemoglobin koln
—Hemoglobin lepore
—Hemoglobin H
—Begin/continue/resume other investigation —Exp—

—Next frame—

Fig. 11

Having done the above, we then must have a system that allows us to modify that guidance in terms of the specific findings on a given individual patient. Statistically, a particular test can mean a specific thing most of the time in a given problem when found in an isolated manner in a specific patient. But the patient's other problems frequently modify that interpretation and one must always reconsider one's textbook plan in the context of the patient's other problems. One cannot do that without having detailed flowsheets, a complete problem list, and legible plans and progress-notes at one's fingertips, together with statistical information from the literature as one weaves a fabric of care from those two sources. One cannot overestimate the power of enlarging one's context for thought in decision-making, and this is precisely what a complete list of problems does. If problem no. 1 is terminal carcinoma of the lung and problem no. 3 is hypertension, then the plan for problem no. 3 may be to do no diagnostic studies—laboratory or otherwise. If the therapy for one problem interferes with the laboratory test for another, then we should not persist in blindly doing the latter, making misleading interpretations of iatrogenic problems. Much of the present problem of overutilization has its roots in our lack of the right information tools to deal easily with individual problems in the largest possible context.

Workers in the laboratory and radiology department should not be blindly doing tests in ignorance in the name of achieving some mythical objectivity. Knowing the microbiological problem in a patient determines, for example, what media and stain you use; what tools you use determines what you can find. In some misdirected efforts to avoid bias, you may miss the answer itself.

When dealing with the total problem list and clearly stated goals of the patient, common sense (as opposed to sophisticated technical knowledge) leads to immediate satisfactory decisions, as in the example of the carcinoma above. It is when we focus in depth on a single problem that technical knowledge must be mobilized, and even here the mind coupled to the proper computer tool can out-perform the unaided mind. If one follows the superstructure carefully and writes careful plans and designates frequencies of determinations as well as qualitative interpretations, then one has set up the basis for writing precise progress notes. Because of this use of choices to generate records and orders, the language can be common among the providers and meaningful statistical data can be acquired from many cases in the everyday practice of medicine.

When the pathologist writes the detailed displays for ruling out various causes or a problem and tries to designate which tests are best for following the course of a condition or monitoring therapy, he will find that his point of view might be quite different from that of another specialist, say an internist or radiologist, who may write detailed guidance for the same problem from another vantage point. It is important at this point for the pathologist and the other specialists to get together and work out the best possible plan among them, and the value of the computer is that they can make those efforts immediately available to all providers for use, study, and updating. In other words, our efforts can be rapidly cumulative. In contrast, in a conference one can suggest or recommend personal points of view about problems, but one cannot assure rigorous implementation of that recommendation on large numbers of patients so that it can be corrected where wrong and generalized to many other patients when correct. Looked upon from this point of view, the computerized problem-oriented record becomes not only useful but an essential tool for coupling the best knowledge to everyday practice of medicine and for the growth of medical science.

IV. PROGRESS NOTES: As any provider uses the computerized problem-oriented system to care for his patients, he and the patient generate the progress notes. The experts should be using these progress notes as the basis for their analysis of the validity of the original guidance in the displays for planning. The rigorous and meticulous practice of medicine has much to offer through the insights that everyday events provide and the new questions they raise. In this sense, practicing medicine is research at all times, and, with careful analysis, some new detail of understanding is always revealed. To the extent that this understanding can be harvested and used to make the original displays for guidance in the Data Base and the problem formulation and the Plans even more precise, the pathologist should do his share in this process.

B. *Leadership in the solving of the coordination problem:* Except for medical administrators and trustees who do not have intimate knowledge of the day-to-day data in medicine, there is no technically competent group who looks at the whole to bring about maximum coordination of effort. Each department chief is lucky if he can control in a coordinated manner the activities of his own department. Thus, all the fruits of a true synergism among departments are not realized, and the consequences to patients are enormous as they meet their health and financial obligations. The computerized problem-oriented medical record system provides a framework for coordination of such efforts. The framework will not do the job itself. Leadership is required. Historically, the pathologist has been the innovator and the conscience of medical institutions. Somewhere along the line they lost a great deal of that spirit and capacity. Indeed, failure at this leadership level dwarfs any gains or losses at a purely technical level. You know there is trouble if a pathologist in an institution speaks one philosophy and his laboratory's actions speak another. If, for example, he writes and talks about the uselessness of sputum cultures, but his laboratory processes many of them anyway, or if the frequency of certain chemical tests is out of proportion to the real need, then this lack of control of the environment has far greater significance than the number of tests available or the sophistication of the equipment or the staff. Unfortunately, discovery of a new test has a place in a journal and a bibliography, and doing tests produces income, whereas there is a poor personal reward system and even economic penalties for insisting upon the proper disciplined use of the laboratories and radiology services. For these reasons the overall system must be developed so that crucial parts are not subject to the confusion or selfish pressures of the moment. Furthermore, an overall approach eliminates the false boundary between inpatient and outpatient care. The clinical pathologist, by being isolated from much of the thinking and planning for patients and their specific problems in outpatient care, particularly physicians' offices, misses the opportunity to help with problems in their early stages and the perspective on diseases that the broader context would provide. We must remind ourselves that the hospital, particularly a teaching hospital, can become just an intellectual harvest on delayed maintenance. Computer systems with cable TV connections across a community can change all this. Focused, coordinated effort can be achieved.

Consideration of an overall computer system is a good place for leadership to be exercised. The buying of computer systems for hospitals without a total information architecture in mind can be disastrous. Development and implementation of such an architecture is not the job for a casually appointed committee that no one intends to listen to anyway, but rather a job for the best informed and strongest leaders in the institution. Leaders of today in this area have one enormous advantage over leaders of just a few years ago. That advantage is the presence of communication tools of enormous power in the very areas in which we now most need power and control of information.

In conclusion, then, we can say that no one ever took complete care of a patient for all his problems throughout his lifetime. Division of labor has been necessary and inevitable. Precise communication among the laborers with an organized framework of problem solving is one of our greatest needs. The tools are now available to meet that need. From an evolutionary point of view, it is the linkages among minds that can now develop with very fruitful results. That does not mean that we cannot continue to probe artificial intelligence and decision-making in limited areas as seen in the "great clinician" and single mind, but we must recognize if no individual has been doing in a natural way the total synthesis that is now required over time, then our goal should be to build a system whereby such synthesis can be accomplished. And as we build, we must keep reminding ourselves of the power of broadening the context of our thoughts and the power of tools for extending our feeble human memories for overcoming our parochial and frequently illogical thinking under the pressures of rapid decision making.

The Computerized Problem-Oriented Medical Record (CPOMR) represents a coordinated team effort of computer and medical people in the PROMIS Laboratory. Publications dealing with details of various aspects of this effort can be obtained by writing to the PROMIS Laboratory.

References

1. Weed, L. L., A new paradigm for medical education. In *Recent Trends in Medical Education*, E. F. Purcell, Ed. (Macy Foundation, New York, N.Y. 1976) pp. 55–93.
2. Weed, L. L., CPCs as educational instruments. *N. Engl. J. Med.* 285, 115 (1971).

The Role of the Laboratory in Monitoring Patient Care

Joseph E. Rall, M.D., Ph.D.

National Institutes of Health, Bethesda, Md. 20014

"Monitoring" is an interesting word, coming as it does from the Latin *monere*, meaning "to warn," and its relationship to the clinical chemist is apt. It seems appropriate in these days of automated clinical chemistry to point out that in a surprising number of instances the laboratory warns the clinician. I would like to discuss this warning function in three parts.

First, I would like to discuss the role of the laboratory in warning in the sense of indicating to the clinician an unexpected and abnormal value for some test.

Secondly, I would like to point out how the laboratory can be of value in multiple determinations of some substance so as to aid the clinician to tailor treatment more precisely to that particular patient's needs.

Finally, I should like to discuss the old problem of the precision of the determinations done by the clinical chemist and the distribution of normal values. Any signal that is serving as a warning should be clearly distinguishable from the background noise.

Let me begin with serum calcium as an example. In this era of automation, clinicians have a tendency to get 12 determinations when they really only want one. One result is that an extraordinary number of individuals apparently have an abnormally high value for serum calcium. In the past, almost any endocrinologist could go to a hospital calculous-disease clinic and, by doing serum calcium determinations on all patients, discover several with hyperparathyroidism. Still, it was generally considered a relatively rare disease. It has now become increasingly diagnosed, as apparently asymptomatic people have serum calcium values measured and show hypercalcemia. However, Marx and Aurbach and their collaborators at the NIH *(1)* have recently encountered a rather interesting problem with respect to this. They have now accumulated 20 patients with above-normal serum calcium concentrations, almost all of them picked up by chance. Nine of these individuals subsequently had parathyroid surgery, which in all instances was ineffective; the hypercalcemia continued. What these workers have now discovered is a new syndrome, which they call familial hypercalcemic hypocalciuria and which has the following characteristics. There is hypercalcemia—frequently substantial, as much as 130 to 140 mg/liter—there is hypocalciuria, and there is hypermagnesemia. This familial disease apparently is inherited as an autosomal dominant, and as far as these investigators can tell there are no complications of the condition—i.e., patients do not have kidney stones; they do not have bone disease; they do not have duodenal ulcers; and they apparently live normal lifetimes, with no signs or symptoms that can be related to this hypercalcemia. Extrapolating, one would estimate that several hundred patients during the last four or five years have been operated on in the U.S. for hyperparathyroidism when they in fact have familial hypercalcemic hypocalciuria. Stated another way, they were subjected to major surgery that was destined not to be effective.

Let me take another example. The first case of a congenitally high concentration of thyroxine-binding globulin (TBG) was found *(2)* when a professor of physiology volunteered to have his blood drawn to serve as a control for some study of blood iodine. He was found to have a remarkably increased serum thyroxine, but a careful study showed that he had no thyroid dysfunction. The increased serum thyroxine was due to a congenital increase in thyroxine-binding globulin *(2)*, a familial X-linked trait. This syndrome is of considerable scientific interest, because it is mainly through a family study that it was found that the gene controlling TBG synthesis is located on the X chromosome. Additionally, it is of interest because it is almost the only condition known in which there is overproduction of a protein. Clinically, it is of importance because of a tendency of some physicians to treat the results of laboratory tests rather than patients. Hence, the occasional patient with an increased TBG will find himself treated for hyperthyroidism.

The measurement of urinary sugar has produced some interesting results over the years. Almost all patients entering a hospital have a urinalysis, and this includes a urine sugar measurement. Those people in the past 15 or so years who show a positive test for sugar in their urine usually have had their blood sugar measured, or a glucose tolerance test done; frequently they are found to have diabetes in various degrees. In the asymptomatic, obese, middle-aged diabetic thus detected, there is considerable question on a statistical basis as to whether or not treatment is of any value. The reason I say this is because the results of the University Group Diabetes Study *(3)* strongly suggest that untreated, middle-aged, asymptomatic diabetics have fewer cardiovascular accidents and a lower mortality rate than do those given oral anti-diabetic drugs.

One can conclude from all this that physicians should be more alert to making a diagnosis and prescribing proper therapy than they now are. I think this is probably true, and I think the clinical chemist comes into it in his interactions with physicians in the hospital. That is to say, I believe that most clinicians who discover an abnormality in a test result that they don't anticipate should check with the clinical chemist to see if there was anything wrong with the analyses that day, or complain in one way or another. And this is where the clinical chemist can point out several things that may be involved.

In the first place—and perhaps this is the most likely possibility—there may be disease that has been unrecognized because it is in its early stages. Clearly, in many instances, particularly in the case of cancer, this is the time one most wishes to see the patient, *before* signs and symptoms have appeared.

However, there may be no disease at all and the abnormal result is the result of some process other than one commonly expected to have caused the abnormal chemistry. This is the case, for example, in hypercalcemic hypocalciuria; it also is the case in both the congenital increases and decreases of serum thyroxine-binding globulins; and there are many other examples.

So I think the clinical chemist has a responsibility to point out that asymptomatic individuals with abnormal laboratory results may well have something other than what the clinician, by reflex, might think to be the diagnosis. I believe this is a fundamental argument for a closer connection between the clinical chemist and the practicing physician, and it clearly gives rise to considerable concern about the commercial laboratories that do clinical chemistry for physicians scattered throughout the country. In some instances these laboratories offer an extensive consulting service, so that a physician may get telephoned advice about any given test. In medicine this would seem to be the most economic operation, but in the long run this kind of economy may be illusory. Hence I would suggest that it is important that clinical chemists be available and ready to discuss the results of any of their tests with clinicians at any time.

Now let me turn to the use of the laboratory for following cases during treatment. This is an old subject. It has been appreciated for a long time, for example, that treatment with dicumarol requires periodic determination of the prothombin time. The reason for this of course is the variability in response, both among patients and in any given patient as a function of what other drugs he is taking and a variety of unidentified factors. However, recent findings suggest that the laboratory has a larger role than this rather special one in the case of dicumarol. I refer to the extensive studies done by Elliot Vesell and other pharmacologists who have shown that there are major genetic variations in how individuals metabolize drugs *(4, 5)*. If one looks at the metabolic handling of any of a variety of drugs—antipyrine is frequently the example chosen—one finds that individuals vary widely in the rate at which the drug is metabolized. In contrast, if identical twins are examined, the same drug is found to be metabolized at an almost identical rate by the two subjects. Evidently the metabolism of very many drugs differs quite markedly among individuals, and in some instances a clear genetic pattern is evident *(5)*. It may differ among populations even more than it differs among the same ethnic group, but even among members of the same ethnic group most drugs show a rather wide variation in the rates at which they are metabolized.

Lately we have learned that many drugs are metabolized by microsomal hydroxylating enzymes or mixed-function oxidases and that these enzymic activities can be induced by administering drugs that are metabolized by this system *(6)*. Furthermore, there is an element of nonspecificity in the induction, so that treating an individual with phenytoin or phenobarbital, or exposure to chlordane, for example, will cause an increased rate of metabolism of most of the barbiturates, thyroxine, benzypyrine, etc., as well as an increase in the metabolism of the drug that induced the enzyme.

But these so-called detoxifying or drug-metabolizing enzymes don't always detoxify *(7)*. Benzo(a)pyrene has been particularly well studied. This substance is apparently quite unreactive in the test tube and doesn't react at all with DNA, which is surprising, because it is a quite potent carcinogen. However, if benzo(a)pyrene is exposed to drug-metabolizing enzymes, a variety of extremely reactive, very short-lived compounds—including diols, dihydrodiols, and epoxides—is formed, and it is these compounds that are the proximate carcinogens. It has also been shown that when the drug-metabolizing enzymes are induced to greater activity, substances such as benzo(a)pyrene become more carcinogenic *(8)*.

From the standpoint of the chemist this variation in drug metabolism suggests that clinicians will increasingly want to know the concentrations of drugs in the serum after administration of any of a variety of drugs. It may not be possible to be sure that the patient has a therapeutic concentration unless such determinations are done, in view of *(a)* the genetic variation among individuals and *(b)* possible unrecognized activation of drug-metabolizing enzymes.

Finally, it has been increasingly appreciated in recent years that patient compliance with a prescribed drug regimen is a problem. When I was in medical school it was assumed that if one prescribed a drug, the patient took it. Now, rather careful studies[1] have shown that the probability that a patient will take a drug exactly as the physician has prescribed is hardly better than 50/50. It has also been noted that in elderly patients who get a large number of medications, this probability is considerably less; relatively few of these patients over the age of 70 who are on the average taking 10 to 20 doses of various medications a day are in fact taking what the physician wants them to take. Hence, it is important for the physician who has prescribed furosemide for some patient's edema, without evident benefit, to know whether or not the patient actually has been taking the prescribed drug.

For all these reasons I envision clinical chemists getting more and more involved in the estimation of a whole variety of clinically important drugs in the blood. A brief survey of several major

[1] *Ed. note:* A brochure on this subject is available from Bruce H. Medd, M.D., Director of Professional Services, Roche Laboratories, Division of Hoffmann-La Roche Inc., Nutley, N.J. 07110.

laboratories shows that the Massachusetts General Hospital lists nine tests that are available for determination of drugs in blood *(9)*. The NIH also lists nine; the Diagnostic Service of the Mayo Clinic lists 32 different drugs that they are prepared to measure *(10, 11)*. As for commercial laboratories, BioScience Laboratories lists 33 and National Health Laboratories lists 26 drug assays *(12, 13)*. Evidently the demand for this service is increasing, there are solid clinical reasons behind this demand, and this is going to be an increasingly prominent service for clinical laboratories to provide.

Now I should like to say a few words about the utility of the laboratory in following the progress of patients during treatment. Here I refer not to determination of drug concentrations but rather to tests with which the patient's response to treatment is assessed.

One of the earliest laboratory tests for monitoring the results of treatment (or indeed of progress or spontaneous resolution of disease) is measurement of the erythrocyte sedimentation rate, still one of the most sensitive and reliable indicators of disease activity in subjects with rheumatoid arthritis and an important guide for treatment. In systemic lupus erythematosus, measurement of anti-DNA antibodies and C_{1q} complement now serves a similar function *(14)*. As lupus progresses, anti-DNA antibodies increase and the C_{1q} component of complement decreases.

Many different enzymes that escape into the blood stream when certain organs are injured have been used for some years as indexes for diagnosis and for following the progress of the disease. Amylase (EC 3.2.1.1), lactate dehydrogenase (EC 1.1.1.27), aspartate aminotransferase (EC 2.6.1.1), and creatine kinase (EC 2.7.3.2) are a few examples.

Perhaps the most dramatic example of the utility of a test in monitoring treatment was appreciated almost two decades ago when it was found that choriocarcinoma responded particularly well to methotrexate. As nearly as one can tell, there are several reasons for this. One is that choriocarcinoma, because it arises from the fetal contribution to the placenta, is in a certain sense not as intimately a part of the mother's genetic material as is a normal cancer. Quite clearly, another factor is that most choriocarcinomas secrete large quantities of choriogonadotropin, which for many decades has been relatively easy to measure in the urine.

Hence, when these patients were treated, the progress of their disease could be assessed by measuring the urinary gonadotropins. When the values declined slowly it was sometimes necessary to increase the dose of methotrexate, even at the risk of serious toxicity. Conversely, when no choriogonadotropin was measurable, the patients could have a respite from treatment. Furthermore, the earliest regrowth of tumor was presaged by increased urinary choriogonadotropin, a warning that treatment should be instituted. The results of this treatment have been remarkable *(15)*. If not cured, many patients have been at least rendered disease-free for five or 10 years.

A more recent example of a hormone as an index to tumor is calcitonin. This polypeptide, secreted by the "clear cells" of the thyroid, causes deposition of bone with consequent decrease in the serum calcium. The role of this material in normal human physiology is not clear; its concentration in normal humans is probably less than 250 ng/liter of serum, and it may be entirely absent *(16, 17)*. Medullary carcinoma of the thyroid is derived from the clear cells, which synthesize calcitonin. In the vast majority of these cases there is a clearly demonstrable increase in calcitonin in serum, although in absolute terms this large increase may only amount to 1 μg/liter, or about $2 \cdot 10^{-10}$ mol/liter. The exquisitely sensitive and specific radioimmunoassay can measure accurately such small concentrations. Not only is the assay of calcitonin useful diagnostically, but it also appears to be an indicator of the completeness of removal of the tumor *(16)*. Here, then, is a specific and sensitive tumor marker that is of real clinical utility in monitoring the progress of a disease, albeit a rare one.

I think several factors make it important to the clinician to have some evidence of the patient's status and how he responds to treatment. This is particularly true in cancer chemotherapy, where the drugs used not only are incompletely satisfactory but also very toxic. On the one hand, one wishes to stop treatment as soon as the tumor is destroyed. But frequently a physician cannot tell this from roentgenograms, from CAT scans, or from any of the other noninvasive procedures available to him. In the absence of sensitive and specific chemical tests, most procedures only reveal the presence of a tumor growth when it is already relatively far advanced.

Many tests have been devised to indicate the presence or absence of cancer. The tests most recently investigated in detail are those for carcinoembryonic antigen and alpha-fetoprotein. Both were considered to be fairly specific for cancer and both considered to be absent from the serum of normal adults. However, more sensitive analytical methods, especially radioimmunoassay, have shown both of these proteins to be present in the serum of normal men and women. Much experience has been accumulated with results of analyses for carcinoembryonic antigen and alpha-fetoprotein in a wide variety of individuals with different cancers *(18–20)*. Although the average values for patients with cancer and normal individuals differ highly significantly, there is nevertheless a discouragingly large overlap between the normal population and the population with cancer. In addition, people who are ill with a variety of other conditions such as inflammatory bowel disease and pneumonia may show increased concentrations of one or both of these proteins.

Edgington et al. *(21)* fractionated the carcinoembryonic antigen; one fraction, called "CEA–S," appears to be considerably more specific. But even this test was positive in less than half of patients with cancer of the gastrointestinal tract. On the other hand, false positives were low: 0 in 150 benign lung and gastrointestinal diseases, two positive among 75 smokers *(22)*. This test and the alpha-fetoprotein test have yet to be carefully evaluated as to their value in following the courses of patients with cancer. Other markers of cancer have been the kappa form of casein and the complement concentrations *(23, 20)*. More recently, the ribonuclease (EC 3.1.4.22) activity of plasma has been investigated, particularly with reference to ovarian carcinoma *(24)*. All of these can be shown to be increased, on the average, in individuals with cancer, but in any given case they are by no means diagnostic of cancer.

The problem with all these tests is manifold. In the first place, they are not specific for cancer. This is probably too much to expect because a thyroid cancer differs from an islet cell tumor of the pancreas, which differs from a carcinoma of the lung, etc. Secondly, all of these proteins are present in normal individuals, so that "abnormality" is just an above-normal concentration. Finally, in all these tests the concentrations of most of these substances vary greatly in normal individuals and it is by no means clear what physiologic or pathologic alterations other than cancer affect them. Hence, measurement of carcinoembryonic antigens, alpha-fetoprotein, kappa-casein, complement, or plasma ribonuclease are all as-yet-unsatisfactory tests for use in managing treatment in patients with cancer.

The detailed statistics also become of importance. Consider a test that is positive in 80% of patients with a certain type of cancer but gives 2% false positives. This would appear to be a good screening test. However, if the incidence of this cancer is 1 in 1000 asymptomatic subjects, when 10 000 have been tested, 208 will be positive. Of these, 200 will be normal, eight will actually have cancer, and two patients with cancer have been missed. The usefulness of this test now will depend on whether further diagnostic measures to establish the presence of cancer are easy or difficult, expensive or cheap.

Some years ago Bartter and Liddle *(25, 26)* noted that some people with cancer, especially cancer of the lung, appeared to show endocrine abnormalities. We now know that the tumors themselves may produce abnormal amounts of normal hormones. When I say "normal hormones," it should be taken with a certain reservation, because most of the hormones produced ectopically by tumors have biological activity, but in some instances in which the hormones have been reasonably well characterized they appear to differ chemically in one respect or another from the normal hormone. Nonetheless, they may be active enough so that some patients with carcinoma of the lung develop (e.g.) Cushing's syndrome, or hyperparathyroidism, or a syndrome of excessive antidiuretic hormone. It was hoped then that, with the availability of highly sensitive radioimmunoassays for polypeptide hormones, analyses for these hormones might be useful in detecting lung cancer in particular and in following the course of the disease.

But several things have recently been learned. First, only a minority of patients has increased serum concentrations of any of the pituitary proteins: depending on the series and the cut-off points, some 10 to 20% of patients *(20, 27)*. Patients with other carcinomas generally show lower values than do patients with lung cancer. In the rare patient with a tumor that produces large quantities of one of these hormones, measuring it may be useful. Furthermore, in some individuals the cancer produces choriogonadotropin *(28)*. This hormone, or at least its beta subunit (which is unique for that hormone), should not be produced by any tissue except placental tissue, so that if any beta unit of choriogonadotropin is found in the blood of a non-pregnant individual, this is abnormal. It was hoped that with a zero baseline, this analysis of serum for beta-choriogonadotropin might be a useful cancer marker. Unfortunately, in a very large series no more than one-sixth of a random group of patients with cancer showed a significant concentration of it, and although in some instances its concentration appeared to correlate with the tumor burden, in other instances it did not *(28)*. It appears, therefore, that none of the present tests for cancer fulfills the criteria of specificity, of sensitivity, and of correlating with tumor burden. Perhaps the search for such tests is doomed to failure, but the experience with choriogonadotropin in choriocarcinoma and with calcitonin in medullary carcinoma of the thyroid—in which all of these criteria have been fulfilled—encourages continued search for analytes that *will* meet these requirements.

The recent interest in immunologic defects in patients with cancer has led to extensive studies to see if any immunologic factors are altered in cancer patients. As I noted above, complement concentrations—more specifically CH_{50}, C_3, C_4, and C_{1q}—have all been shown to be usually increased in patients with cancer and decreased when the disease is ameliorated *(23)*. But the results, significant only statistically, are of little help in the case of any individual patients. An even more comprehensive study in a particularly suitable disease, Hodgkin's disease, was undertaken by Good and collaborators, and they recently reported *(30)* that although they found many immune abnormalities in patients with Hodgkin's disease, there were no consistent diagnostic findings that correlated particularly well with the progress of the disease.

Let me now take up a final point concerning the role of the laboratory in the monitoring of patients, and this has to do with the statistical nature of laboratory results. I have examined published normal values from several of the major laboratories in the United States, and in almost all instances I see normal values reported as ranging from x to y, with no indication of whether the normal range so noted represents 67% of the normal population, 80%, 95%, or what. The only exception to this is that the Mayo Clinic lists under immunoglobulins IgG, IgA, and IgM the 95% confidence limits for these values *(11)*. Numerous laboratories and hospitals and industrial laboratories point out factors that may influence laboratory results, such as which tests require the patient to be fasting before the test, which drugs may interfere with the test, any sex- or age-related difference, any circadian differences, the effect of pregnancy, the effect of smoking, and other influences that are important for a clinician who is trying to ascertain the significance of any given laboratory value. These are all of great importance, but not all such influences have been carefully evaluated. If blood glucose decreases sharply after 72 hours of fasting, this has for decades been considered to be one of the most reliable means for the diagnosis of islet cell tumors of the pancreas *(31)*. The precise degree of hypoglycemia required to establish the diagnosis has been a bit vague. Recently, Merimee and Tyson tried to establish the lower normal limit for blood sugar during 72 hours of fasting *(32)*. Surprisingly, they found that normal women showed a continual fall in plasma glucose during the fast, and values between 200 and 300 mg/liter were found in a few normal women *(32)*. Values for normal men, on the other hand, were never less than 500 mg/liter.

I think this illustrates the problem the clinician faces when good, statistically valid estimates of normal values for laboratory tests are not available.

In conclusion, I have outlined the problems the clinician faces when he is confronted with an unexpectedly abnormal value for a laboratory test. To help with this problem, the clinical chemist must be knowledgeable and available for consultation. I have then reviewed some of the tests used to monitor progress of disease, and suggest that clinical chemists increasingly will be required to develop procedures to measure important drugs in the blood. I have described some of the tests used to measure a patient's response to treatment. In cancer therapy, in particular, specific and sensitive tests are required. With only a few exceptions involving a few rare diseases, such tests are currently inadequate. Finally, I have commented on the paucity of data that have been subjected to elementary statistical analysis, to determine the normal range of values for laboratory tests provided by most laboratories.

References

1. Marx, S. J., Spiegel, A. S., Brown, E. M., et al., Divalent cation metabolism in familial hypocalciuria hypercalcemia vs. in typical primary hyperparathyroidism. *Am. J. Med.*, 1978, in press.
2. Beierwaltes, W. H., and Robbins, J., Familial increase in the thyroxine-binding sites in serum alpha globulin. *J. Clin. Invest.* **38,** 1683 (1959).
3. Meinert, C. L., Knatterud, G. L., Prout, T. E., and Klimt, C. R., A study of the effects of hypoglycemic agents on vascular complications in patients with adult-onset diabetes. I. Design, methods, and baseline results. *Diabetes* **19,** Suppl. 2, 747 (1970).
4. Vesell, E. S., and Page, J. G., Genetic control of the phenobarbital-induced shortening of plasma antipyrine half-lives in man. *J. Clin. Invest.* **48,** 2202 (1969).
5. Evans, D. A. P., Manley, K. A. and McKusick, V. A., Genetic control of isoniazid metabolism in man. *Br. Med. J.* **ii,** 485 (1960).
6. Conney, A. H., and Burns, J. J., Factors influencing drug metabolism. *Adv. Pharmacol.* **1,** 31 (1962).
7. Miller, J. A., Carcinogenesis by chemicals: An overview. *Cancer Res.* **30,** 559 (1970).
8. Jerina, D. M., and Daly, J. W., Arene oxides: A new aspect of drug Metabolism. *Science* **185,** 573 (1974).
9. Normal Laboratory Values. *N. Engl. J. Med.* **290,** 39 (1974).
10. *Clinical Pathology & Blood Bank Guide,* Clinical Center, Nat. Inst. Health, July 1, 1977.
11. Mayo Medical Laboratories, Rochester, Minn., Jan, 1, 1977.
12. *The Bio-Science Handbook. Specialized Diagnostic Tests,* 10th ed., Bio-Science Labs., Van Nuys, Calif., June 1973.
13. National Health Laboratories, 653 N. Glebe Rd., Arlington, Va. 22203.
14. Schur, P. H., and Sandson, J., Immunologic factors and clinical activity in lupus erythematosus. *N. Engl. J. Med.* **278,** 533 (1968).
15. Li, M. C., Hertz, R., and Spencer, D. B., Effect of methotrexate therapy upon choriocarcinoma and chorioadenoma. *Proc. Soc. Exp. Biol. Med.* **93,** 361 (1956).
16. Tashjian, A. H., Wolfe, H. J., and Voelkel, E. F., Human calcitonin. Immunologic assay, cytologic localization and studies on medullary thyroid carcinoma. *Am. J. Med.* **56,** 840 (1974).
17. Deftos, L. J., Bury, A. E., Habener, J. F., et al., Immunoassay for human calcitonin. II. Clinical studies. *Metabolism* **20,** 1129 (1971).
18. Zamcheck, N. Z., Moore, T. L., Dhar, P. and Kupchik, H., Im-

munologic diagnosis and prognosis of human digestive-tract cancer: Carcinoembryonic antigens. *N. Engl. J. Med.* **286,** 83 (1972).

19. Nishioka, Y., Okamoto, K., Shigeta, K., et al., Observations of α-fetoprotein in the diagnosis of diseases of the digestive tract. *Digestion* **8,** 396 (1973).
20. Franchimont, P., Zangerle, P. F., Nogarede, J., et al., Simultaneous assays of cancer-associated antigens in various neoplastic disorders. *Cancer* **38,** 2287 (1976).
21. Edgington, T. S., Astarita, R. W., and Plow, E. F., Association of CEA–S with neoplasia of the GI tract. *N. Engl. J. Med.* **293,** 103 (1975).
22. Herberman, R. B., Immunological tests in diagnosis of cancer. *Cancer* **37,** 549 (1976).
23. Verhaegen, H., DeCock, W., DeCree, J., and Vergruggen, F., Increase of serum complement levels in cancer patients with progressing tumors. *Cancer* **38,** 1608 (1976).
24. Sheid, B., Lu, T., Pedrinan, L., and Nelson, J. H., Plasma ribonuclease. A marker for the detection of ovarian cancer. *Cancer* **39,** 2204 (1977).
25. Schwartz, W. B., Bennett, W., Curelop, S., and Bartter, F. C., A syndrome of renal sodium loss and hyponatremia probably resulting from inappropriate secretion of antidiuretic hormone. *Am. J. Med.* **23,** 529 (1957).
26. Liddle, G. W., Nicholson, E. W., Island, D. P., et al., Clinical and laboratory studies of ectopic humoral syndromes. *Rec. Prog. Hor. Res.* **25,** 283 (1969).
27. Rosen, S. W., and Weintraub, B. D., Ectopic production of the isolated alpha subunit of the glycoprotein hormones. A quantitative marker in certain cases of cancer. *N. Engl. J. Med.* **290,** 1441 (1974).
28. Muggia, F. M., Rosen, S. W., Weintraub, B. D., and Hansen, H. H., Ectopic placental proteins in nontrophoblastic tumors. *Cancer* **36,** 1327 (1975).
29. Gailani, S., Chu, T. M., Nussbaum, A., et al., Human chorionic gonadotrophins (hCG) in non-trophoblastic neoplasms. *Cancer* **38,** 1684 (1976).
30. Case, D. C., Hansen, J. A., Corrales, E., et al., Comparison of multiple *in vivo* and *in vitro* parameters in untreated patients with Hodgkin's disease. *Cancer* **38,** 1807 (1976).
31. Conn, J. W., and Seltzer, H. S., Spontaneous hypoglycemia. *Am. J. Med.* **19,** 460 (1955).
32. Merimee, T. J. and Tyson, J. E., Stabilization of plasma glucose during fasting. *N. Engl. J. Med.* **291,** 1275 (1974).

Selection of Appropriate Laboratory Tests

Robert S. Galen, M.D., M.P.H.

Department of Clinical Pathology, Columbia University College of Physicians and Surgeons, New York, N.Y. 10032; and Department of Laboratories, Overlook Hospital, Summit, N.J. 07901

Today, more than 10% of our nation's health-care costs are related to laboratory testing. In addition, laboratory costs are increasing at a greater rate than health care costs in general. Does this high utilization indicate better patient care through modern medical science? Will the projected increase in laboratory expenditures to $20 billion dollars by 1980 represent further progress toward better health? The answer is, not necessarily.

Why do physicians request laboratory tests? Some results—of blood gas, glucose, and electrolyte determinations—are needed to monitor and guide treatment of critically ill patients. Other results—as from creatinine and thyroxine tests—help us to detect early renal or thyroid disease. Still other tests appear to be done primarily because they are available. In addition to seeking confirmation of a clinical impression, what other factors motivate physicians to request laboratory services? Are tests sometimes requested to show the patient or professional associates we are doing something—even if an abnormal result will never lead to any change in diagnosis or therapy? Are requests occasionally based on a desire to do a complete work-up, for the benefit of the chart, rather than of the patient? How often are requests made on the basis of a recent journal article that does not include a critical analysis of the predictive value and efficiency of the proposed test? Too often, clinical laboratory studies are neither carefully planned nor thoughtfully executed. Apparently, laboratory tests are not always selected with care, and results are not always used to benefit the patient. If unnecessary tests are performed in a particular clinical setting, then the laboratory is providing little or no new information to the clinician. This is costly and time consuming for the laboratory and wasteful of health-care resources. In addition, redundant or useless tests complicate the diagnostic process and adversely affect medical decision making.

In his book *Management*, Peter Drucker stresses the importance of feedback information if one is to obtain reliable results. He recognizes the critical role played by the introduction of the routine autopsy in medical practice 150 years ago. The autopsy provided reliable and important feedback information to the physician, thereby improving his diagnostic and therapeutic skills. Drucker attributes the rise of modern medical practice to the introduction of the autopsy *(1)*. Laboratory tests play the same important role as the autopsy, that of providing critical feedback information to the physician. It is difficult, if not impossible, to control and correct performance unless feedback is provided. The feedback information required, however, must be timely, relevant, and operational. The importance of timeliness is obvious both in theory and in practice. Every physician wants his laboratory results as quickly as possible. The importance of relevance and of an operational effect is equally obvious in theory, but not so in practice. This review presents a system for evaluating some aspects of the operational effect of a laboratory test. The sensitivity, specificity, and predictive value of a test define the operational characteristics of the test, characteristics that are not readily perceived unless quantitated.

Laboratory tests are an extension of physical diagnosis. They "amplify" the physician's senses so that he can "see," "hear," or "feel" at cellular, molecular, and atomic levels. For example, a physician cannot smell, feel, or hear oxygen, but with a p_{O_2} electrode he can detect small changes in p_{O_2}. As our understanding of disease mechanisms increases, we begin to define a "disease" by a specific molecular or functional alteration that is measurable *only* by a laboratory test. Thus "disease" may be defined by a particular test and can only be diagnosed and monitored by that test. If denied access to tests that define disease or monitor the progress or therapy of disease, a physician will become extremely angry. It is as if he had been blindfolded, had plugs placed in his ears, and heavy gloves placed on his hands, thus limiting the use of his senses in physical diagnosis. If he were then told that he could remove these impediments only when the laboratory was open, obviously the physician would prefer that the laboratory be open and ready for tests at all times. When we deny a physician a laboratory test, we are in effect cutting off his senses. If, for example, we deny a nephrologist ready access to creatinine measurements at any time of the day and night, we might just as well deny him physical access to his patients, because the only efficient way a nephrologist can detect deterioration of renal function in patients requiring dialysis or transplantation is to monitor creatinine clearance. Granted that the physician is frustrated if he is blindfolded, does that mean that any physician should be able to order any test he wants whenever he wants? The answer is no, because the laboratory is simply an extension of physical diagnosis. Can a physician examine any patient at any time and perform any physical diagnostic test he wishes? Of course not. Obviously, the patient would call a halt if he felt abused. He would question the physician if the physical examination failed to cover areas of the body or organ systems producing his complaints. Finally, the physician realizes that he has finite resources. He has only so much time for each patient and can perform only a limited number of examinations. He therefore selects those questions and procedures that training and experience have taught him are most effective in resolving diagnostic and therapeutic problems.

However, in the use of laboratory tests the physician does not sense any limitation. He considers the laboratory—like air, water, and telephones—to be an infinite resource. A simple, effortless, painless stroke of the pen is all that is required to command its use. It is not the physician's time that is used. It is not the physician's money that is consumed. It is not the physician's assistants who do the work. A physician without blindfolds, however, may still not "see" correctly unless his vision is directed to the proper place. Similarly, a physician with total freedom to order any laboratory test may not order the correct test. In both examples "correct vision" comes from training, experience, and scientific experiment. It is essential that the examination be directed to the problem under study. Being "off target" is the main reason why most laboratory tests lack sensitivity and specificity. It is essential to have a detailed understanding of the biochemical pathways in a particular disease *before* a chemical assay is established as a routine diagnostic test. Without such knowledge it is not possible to select the correct biochemical measurement. Most tests currently in use are *not* biochemically specific for the problem under study, and therefore they lack clinical sensitivity and specificity.

As an example, let us take a close look at the glucose tolerance test. This test is widely used, but it may be one of the most useless tests performed. Why? Because we do not know enough about diabetes. Diabetes is much more than glucose intolerance, but we fall back on the glucose tolerance test because currently we have no way to measure those other factors. The glucose tolerance test is way off target as far as the complications of diabetes are concerned. The glucose tolerance test tells us nothing about whether a patient is prone to develop acute ketoacidosis, nonketotic diabetic coma, peripheral neuropathy, renal disease, cardiovascular disease, blindness, or impotence. In short, the test lacks specificity and sensitivity for the complications of diabetes. Therefore it has no predictive value for the major complications of diabetes. The only thing determined by the glucose tolerance test is glucose intolerance. But occult glucose intolerance does not predict clinical diabetes with a high degree of accuracy—and, more significantly, it fails to predict the future occurrence of diabetic complications. For the patient it is the complications, not the glucose intolerance, that characterize the disease.

In thinking about the selection of appropriate laboratory tests, it is necessary to keep in mind the multiple functions of laboratory testing:

1. Provide a correct diagnosis in a patient known to be sick.
2. Provide a prognosis in a patient with a known disease.
3. Provide an indication as to whether or not a disease is present in its early or subclinical stages in an otherwise "healthy" person.
4. Provide data for monitoring the concentrations of therapeutic drugs, or the effects of drugs, or both.
5. Provide data that may indicate whether or not disease might develop at some future time, that is, delineation of risk factors.

Regardless of the clinical setting, the physician is faced with an array of laboratory tests to choose from to help solve a problem. These tests, as well as the rest of his diagnostic armamentarium, including the history, physical examination, electrocardiogram, x-ray, etc., can be subjected to a variety of statistical analyses to facilitate test selection and sequence. In addition, procedures can be evaluated in a particular clinical setting to determine their contribution to the diagnostic process or redundancy. How do most physicians select laboratory tests?

Ginsberg (2), after extensive discussions with physicians, summarized the following factors as representing the major considerations encountered in actual clinical practice:

1. The risks of complications associated with a diagnostic test or a treatment, including the possibility of death.
2. The dollar costs of the tests and treatments.
3. The diagnostic value of a test, i.e., the amount of information the test will contribute toward identifying a specific disease entity (or a specific treatment action).
4. Patient characteristics such as sex, age, economic status, and general health.
5. The possible spontaneous changes in the state of the patient after an action that requires a non-zero amount of time or when no action is taken (e.g., if the patient has a particular disease and no action is taken within a specified period, the patient may die).
6. The patient's feelings about the desirability of the various possible outcomes, both medical and economic.

Ginsberg described two particularly useful rules which were used to choose combinations of tests: only tests that require a nontrivial amount of time to perform need be considered as candidates for being combined with other tests; and if collection of the specimen must be repeated in order to perform other tests while one single-specimen test is in progress, all the tests should be combined as a single act. The following observations were made relative to the above judgements. When a state can result in a serious complication in a time period that is relatively short as compared to the time required to perform the relevant tests, these tests should be combined. Tests associated with disease states that have one or more symptoms or signs in common should be combined (organ profiles). Tests that are benign, relatively inexpensive, and require relatively long times to perform are good candidates for combination *(3)*.

In addition, laboratory technology has forced test combinations on the physician as a consequence of the introduction of profiling instruments, primarily in chemistry and hematology. Much of the physician's judgement in test selection has been co-opted by laboratory protocols requiring admission profiles, pre-op profiles, prenatal profiles, etc., removing the physician from the initial selection process. There are other considerations that affect the selection of laboratory tests, and although they are important and may dominate certain environments, they will not be dealt with in any detail: teaching, research, medical malpractice, and remuneration. For the most part, tests are compared relative to time, cost, and information content. The last, information content or diagnostic accuracy, has been dealt with by many investigators, who have taken several different approaches.

Lusted *(4)* described a method for minimizing the number of tests at any stage in the examination of the patient by applying Boolean algebra. Barnoon and Wolfe *(5)* developed a criterion that is a measure of the expected information content of a test, drawing upon information theory. The measure of expected information content of a test, $EI(T_j)$, indicates the improvement in the diagnosis that is expected to be achieved from the administration of diagnostic test T_j. One assumes that a physician would want to select a test that will improve the prior diagnosis and this would be the test with the maximum $EI(T_j)$. It should be noted that this approach assumes that reaching a diagnosis is an independent final objective, and disregards considerations of medical decision analysis, such as alpha and beta errors and associated costs, relative to ultimate patient outcome. The approach of expected information content really evaluates a test in a clinical vacuum, but does indicate clearly the relative amount of information that can be expected to be gained from each of the tests under consideration.

The two most popular models for evaluating tests are the conditional probability model and discriminant function analysis. These will be discussed subsequently.

The Predictive Value of Laboratory Diagnosis

Sensitivity, specificity, predictive value, and efficiency define a laboratory test's diagnostic accuracy *(6, 7)*. "Sensitivity" is the frequency of positive test results in patients with a particular disease, whereas "specificity" is the frequency of negative test results in patients without that disease. The predictive value of a positive test result indicates the frequency of diseased patients in all patients with positive test results; that of a negative test result indicates the frequency of non-diseased patients in all patients with negative test results. The efficiency of a test indicates the percent of patients correctly classified (diseased and non-diseased) by the test (Table 1).

In most clinical settings, we are most interested in the predictive value of the positive result. These identify patients who will be presumptive positives for a particular disease, who undergo detailed diagnostic work-ups and treatment. Of this group, those patients who are found not to have the disease will be defined as false positives. In the discussion that follows, unless otherwise indicated, predictive value will be used to refer to the

Table 1. **Predictive Value Model**

	No. with positive test result	*No. with negative test result*	*Totals*
No. with disease	TP	FN	TP + FN
No. without disease	FP	TN	FP + TN
Totals	TP + FP	FN + TN	TP + FP + TN + FN

Definitions:

TP = true positives: number of diseased patients correctly classified by test.

FP = false positives: number of non-diseased patients misclassified by the test.

FN = false negatives: number of diseased patients misclassified by the test.

TN = true negatives: number of non-diseased patients correctly classified by the test.

Sensitivity = positivity in disease, expressed in percent TP/(TP + FN) × 100

Specificity = negativity in health, or absence of a particular disease, expressed in percent = TN/(FP + TN) × 100

Predictive value of positive test = percentage of patients with positive test results that are disease = TP/(TP + FP) × 100

Predictive value of negative test = percentage of patients with negative test results that are non-diseased = TN/(TN + FN) × 100

Efficiency of test = percentage of patients correctly classified as diseased and non-diseased = (TP + TN)/(TP + FP + FN + TN) × 100 predictive value of the positive test result. Predictive value changes markedly when there is a change in the prevalence of the disease in the population under study. For example, we could know from a test's sensitivity and specificity that it was positive in 95% of diseased patients and negative in 95% of non-diseased patients. Table 2 demonstrates the change in predictive value that occurs for this test with changing prevalence of disease. It can readily be seen that a particular test has a higher predictive value when the disease occurs with a higher prevalence. This explains why a good diagnostic test frequently fails as a screening test, when the drop in prevalence is quite marked. When clinical judgement is used in ordering laboratory tests, the patient suspected of having a particular disease is placed in a new population with a high prevalence or probability of disease. Therefore, the test performs much better.

For example, if the test were a serum enzyme determination, we could know from its sensitivity and specificity that it was positive in 95% of myocardial infarction patients and negative in 95% of non-myocardial infarction patients. We can then compare this enzyme test in the out-patient clinic and the Coronary Care Unit (CCU). The prevalence of myocardial infarction in the clinic in all patients with "chest pain" is 5%, whereas in the CCU, it is 50% in patients admitted to rule out myocardial infarction. When used in the clinic, the enzyme test would have a predictive value of 50%, i.e., half of the patients with positive results would have myocardial infarcts, the remaining 50% would not and would therefore be false positives. This same enzyme test, when used in the CCU, would have a predictive value of 95%, i.e., 95% of the patients with positive results

Table 2. **Predictive Value as a Function of Disease Prevalence**[a]

Prevalence of disease, %	*Predictive value,* %
1	16.1
2	27.9
5	50.0
10	67.9
15	77.0
20	82.6
25	86.4
50	95.0

[a]For a laboratory test with 95% Sensitivity and 95% Specificity

would have infarcts, with only 5% of positive results being false positives (Table 3).

Figure 1 illustrates how sensitivity and specificity are altered by the selection of an "upper limit of normal" or "cut-off point" for a particular test. If we studied patients admitted to the coronary care unit with the diagnosis of "myocardial infarction," divided the population into myocardial infarcts and non-myocardial infarction, based on history and electrocardiographic findings, and then expressed the enzyme results for each group as a frequency distribution, we would find the classical overlapping distributions seen in Figure 1. It is impossible to select a cut-off point for enzyme activity that would completely discriminate between the two groups. At any cut-off point, one must sacrifice sensitivity for specificity, and vice versa. Herein lies the major flaw of laboratory diagnosis, in that the tests are not sensitive and specific at the same time. Looking at multiple laboratory tests improves the predictive value somewhat, but the trade-off between sensitivity and specificity always remains. If we wanted to select a cut-off point for the test illustrated in Figure 1, we would choose any point in the overlapping region. Table 4 lists several "cut-off points," with the sensitivity, specificity, and predictive value associated with each. Rather than use the word "cut-off point" or "upper limit of normal," we refer to this point of reference as the referent value. For any laboratory test studied in any disease, the predictive value of the

Table 3. **Predictive Value of an Enzyme Assay in High- and Low-Prevalence Situations (Myocardial Infarction, MI)**

Low prevalence				*High prevalence*			
Test	+	−	*Total*	*Test*	+	−	*Total*
MI	48	2	50	MI	475	25	500
Non-MI	47	903	950	Non-MI	25	475	500
Total	95	905	1000	Total	500	500	1000

Clinic: prevalence of MI = 50/1000 = 5%
Sensitivity = 48/50 = 95%
Specificity = 903/950 = 95%
Predictive value = 48/95 = 50%

Coronary Care Unit: prevalence of MI = 500/1000 = 50%
Sensitivity = 475/500 = 95%
Specificity = 475/500 = 95%
Predictive value = 475/500 = 95%

test in diagnosing the disease can be calculated at any referent value.

The relationships described above can be expressed algebraically by letting the disease prevalence equal p, the test sensitivity equal a, and the test specificity equal b. The predictive value is

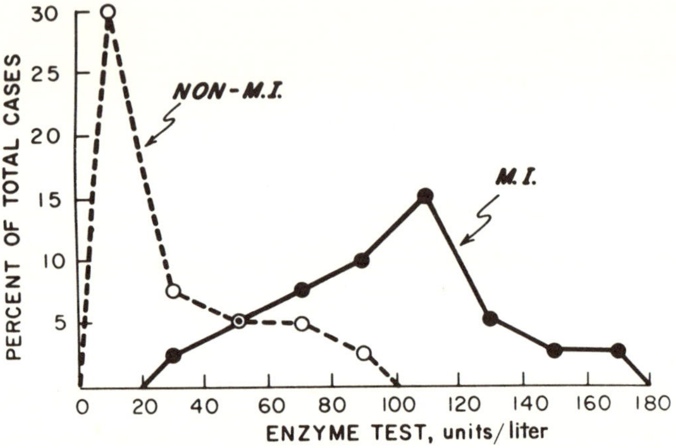

Fig. 1. Selection of an upper limit of normal on the basis of the sensitivity and specificity of a laboratory test.

Prevalence of myocardial infarction = 50%. Mean enzyme activity, MI = 97 U/liter. Mean enzyme activity, non-MI = 27 U/liter. If the upper limit of normal = 20 U/liter, the sensitivity = 100%; specificity = 60%. If the upper limit of normal = 100 U/liter, the sensitivity = 50%; specificity = 100%.

Table 4. **Evaluation of an Enzyme Test**

Referent value	*Sensitivity,* %	*Specificity,* %	*Predictive value of a positive test,* %	*Predictive value of a negative test,* %	*Efficiency,* %
20	100.0	60.0	71.4	100.0	80.0
40	95.0	75.0	79.2	93.8	85.0
50	90.0	80.0	81.8	88.9	85.0
60	85.0	85.0	85.0	85.0	85.0
80	70.0	95.0	93.3	76.0	82.5
100	50.0	100.0	100.0	66.7	75.0
Chance test[a]	50.0	50.0	50.0	50.0	50.0
Perfect test[a]	100.0	100.0	100.0	100.0	100.0

Disease prevalence = 50%
[a] Hypothetical situation.
Referent values and associated predictive values and efficiencies for the test illustrated in Figure 1.

shown calculated in Table 5. Converting back to our starting terminology,

$$\text{predictive value} = \frac{(\text{prevalence}) \times (\text{sensitivity})}{(\text{prev.}) = (\text{sens.}) + (1 - \text{prev.}) \times (1 - \text{spec.})}$$

This formula is frequently referred to as Bayes' formula or Bayes' theorem and was published posthumously in 1763 *(8)*. Before we evaluate a test result for a randomly selected patient, we know he has a probability of being diseased, the *a priori* probability, which is an estimate of the disease prevalence. After we evaluate the test result, we know he has a probability of being diseased, the *a posteriori* probability, which is the predictive value of the positive test result. If the laboratory test is any good at all, the predictive value will be greater than the disease prevalence.

Clinicians frequently compare the predictive value of laboratory tests to flipping a coin. This analogy is not as straightforward as it might appear. The most common error made is to equate a 50% predictive value with a flipping coin *(9)*. It is possible to achieve a 50% predictive value with a coin, but only if disease prevalence equals 50%. Flipping a coin has a sensitivity of 50% and a specificity of 50%. The predictive value, therefore, depends entirely on and is equal to disease prevalence. Any time sensitivity plus specificity equals 100% the value of the test being considered is equal to chance alone. Referring to Table 5, if the percentage of patients with a positive test result in the diseased group (pa/p) equals the percentage of patients with a positive test result in the nondiseased group $(1 - p)\ (1 - b)/(1 - p)$, a random selection is suspected. Since $(pa/p) = a$, the test sensitivity, and $(1 - p)\ (1 - b)/(1 - p)$ is equal to $1 - b$, (i.e., 1 - specificity), we see that if the sum of sensitivity and specificity is 100% we have a

Table 5. **Algebraic Expression of Predictive Value Table**

	No. with positive test result	*No. with negative test result*	*Totals*
Number with disease	pa	p(1 − a)	p
Number without disease	(1 − p) (1 − b)	(1 − p)b	(1 − p)
Totals	pa+(1 − p) (1 − b)	p(1 − a) + (1 − p)b	1

Sensitivity = a; specificity = b; prevalence = p.
Predictive value of positive = pa/[pa + (1 − p) (1 − b)]

test the result of which will be random. Furthermore, the sensitivity *(a)* is equal to the total percent [*pa* + (1 − *p)* (1 − *b)*]. In addition, the predictive value is equal to the prevalence. Table 6 presents a hypothetical example of a test that has a sensitivity of 30% and a specificity of 70%. We let the disease prevalence equal 100/100 000. In this case we see that the sum of sensitivity and specificity is 100% (30% + 70%). Sensitivity (30%) is equal to 1 minus specificity or 100% minus 70%.

The sensitivity (30%) is equal to the total percent positive (30 000/100 000), or 30%. We are not looking at a laboratory test, but rather at a random selection process. The predictive value of a positive result is (30/30 000) or 0.1%. The predictive value of a negative result is (69 930/70 000) or 99.9%. The efficiency of this "test" is (69 960/100 000) or 70%. Note that the predictive value (30/30 000) equals the prevalence (100/100 000). Therefore, the *a posteriori* probability of disease, given a test result, equals the *a priori* probability of disease without any test result. Obviously, this is not a useful test! Using a variety of statistical approaches, we can compare test results to those of chance, and determine if a test is significantly better than or equal to chance. The dangers in developing laboratory tests and only reporting clinical trials in diseased patients is obvious. When such a report describes a test with 90% sensitivity, there is no way of knowing whether the specificity is better than 10%.

Since sensitivity plus specificity must be greater than 100%—if a test result is to be better than chance—the next obvious question is, how *much* greater than 100%? Since clinical judgment and educated guesses are usually better than chance alone, it would be reasonable to expect a good laboratory test to be better

Table 6. **Hypothetical Example of Chance Test**

	No. with positive test result	*No. with negative test result*	*Totals*
No. with disease	30	70	100
No. without disease	29 970	69 930	99 900
Totals	30 000	70 000	100 000

Sensitivity = 30%; specificity = 70%; prevalence = 100/100 000 = 0.1%.
Sensitivity + specificity = 100%.
Predictive value = 30/30 000 = 0.1%.

than a simple history or physical examination. How good can a history be? Dr. Sheila Sherlock and associates measured the efficiency of casual estimations of blood ethanol concentrations. They recommended routine adoption of blood alcohol determinations for investigating liver disease because the test was simple and cheap. Sherlock and associates had data on 42 patients in whom they made a diagnosis of alcoholic liver disease on the basis of biopsy. Thirty-two (76%) of their patients admitted to heavy drinking, whereas only 17 (40%) had increases in their blood alcohol. Thus, the history had a sensitivity of 76% vs. 40% for the laboratory test. The specificity of both tests can be considered to be nearly 100% in this population *(10)*. Dr. S. Barley, however, found this to be a fascinating example of the power of the machine over the human mind, since Sherlock's data showed the history to be more sensitive than the laboratory test. Surely, Barley suggested, a laboratory test, no matter how cheap and simple, is not as cheap and simple as asking a question. Barley considered it a pity that the evidence of history taking was labeled mere "clinical suspicion." What, he asked, could be more diagnostic than the actual admission of heavy drinking *(11)*.

If a laboratory test is to be more useful than a history and physical examination, then the test should have greater sensitivity and specificity than the history or physical. What is the sensitivity and specificity of clinical data? Table 7 presents some of these data, together with findings for some electrocardiographic, radiological, and biochemical tests, ranked in order of increasing sensitivity plus specificity. As we will see later, there is no simple answer to the question of how great the sensitivity and specificity have to be for a laboratory test to be clinically useful?

In most clinical situations, we must contend with overlapping frequency distributions that can not be bisected to yield a test with 100% diagnostic accuracy (efficiency and predictive value). Diagnostic errors will be made regardless of the selection of the "cut-off" or referent value for the test. Cost-effective laboratory testing requires that we consider the errors of medical decision-making and utilize procedures devised to minimize the costs associated with these errors. Table 8 lists the alternative terminology for the two types of errors made in diagnosis.

Which is preferable, a test with the greatest sensitivity, the greatest specificity (highest predictive value for a positive result), or the greatest efficiency? It is not possible to have all at the same time. The greater the sensitivity, the less the specificity.

The greatest sensitivity (preferably 100%) is desired in the following situations: The disease is serious and should not be

Table 7. **Sensitivity and Specificity of Selected History, Physical, EKG, Radiological, and Laboratory Studies, Ranked According to Sum of Sensitivity plus Specificity.**

Test or function	*Sensitivity,* %	*Specificity,* %	*Sum,* %
Physical exam. for hepatomegaly *(12)*[a]	50	47	97
Blood alcohol, to detect alcoholic liver disease *(10)*	40	99	139
Plasma CEA, to detect colorectal carcinoma *(13)*	72	80	152
Hepatic scintigraphy vs. histologic exam., to detect hepatic pathology *(14)*	90	63	153
EKG criteria for left ventricular enlargement *(15)*	58	97	155
Clinical exam. and mammography to detect breast carcinoma *(16)*	67	98	165
Plasma glucose test, in biopsy-proven lactase deficiency *(17)*	76	96	172
History of heavy drinking, to detect alcoholic liver disease *(10)*	76	99	175
LD isoenzymes, to detect myocardial infarction *(7)*	78	98	176
CK isoenzymes, to detect myocardial infarction *(7)*	100	85	185
History of diarrhea after lactose ingestion, in biopsy-proven lactase deficiency *(17)*	100	88	188

[a] Numbers in parentheses are references from which the data in the table were derived.

Table 8. **Two Types of Errors in Making a Diagnosis: Terminology**

False positive	*False negative*
Type II error	Type I error
Beta error	Alpha error
Error of commission	Error of omission
Some proportion of well patients diagnosed as sick	Some proportion of sick patients diagnosed as well

misdiagnosed, and the disease is treatable. Moreover, false-positive results should not lead to serious psychological or economic trauma to the patient. An example of such a disease state is pheochromocytoma. This disease can be fatal if missed, but if diagnosed it is nearly 100% curable. In this case we want a test with the greatest sensitivity possible. If we make a falsely positive diagnosis we not only can repeat the test but can perform a different test on urine before attempting surgery. Other examples include hypothyroidism, venereal disease, and other treatable infectious diseases.

The greatest specificity (preferably 100%) is desired in the following situations: The disease is serious but is not treatable or curable. False-positive results can lead to serious psychologic or economic trauma to the patient. The knowledge that the disease is absent has psychologic or public health value, or both. An example of such a disease state is multiple sclerosis. This disease is serious, but not curable. In this case we want a test with the highest specificity possible. We do not want any false-positive results, and must be prepared to sacrifice test sensitivity. If a case of multiple sclerosis is missed, there is no need for undue concern, because the patient eventually will return. On the other hand, a false-positive diagnosis can cause serious psychological harm. Other examples include most occult cancers and any other serious but untreatable disease. As an example, a test for occult cancer of the lung must have a specificity of 100%. The only treatment available at present is lobectomy or radiation, and treatment based on a false positive would have serious consequences.

The greatest efficiency is desired in the following situations: the disease is serious but treatable, and false-positive results and false-negative results produce essentially equally costly or serious errors. An example of such a disease state is acute myocardial infarction. This disease may be fatal, but it is treatable. If a case is missed, irreparable harm may result, but equal harm can be caused if the diagnosis is made when an infarct is not present. Other examples include diabetes mellitus, lupus erythematosus, and some forms of leukemia and lymphoma.

Using the predictive value model, how does one go about selecting an appropriate test? There are at least 10 rates that can be calculated from the basic predictive value table (Table 1). We have already discussed sensitivity, specificity, predictive value (positive and negative) and efficiency. Some of the more popular rates are:

• presumptive positive—the percentage of the total tested that have positive test results: (TP + FP/total tested)
• detection rate—the percentage of the total tested that are true positive: TP/total tested
• combined error—the percentage of the total tested that are either falsely positive or falsely negative: (FP + FN/total tested).
• error ratio—the ratio of the total number of false diagnoses to the number of true-positive diagnoses: (FP + FN/TP. The best discriminating test or referent value has the smallest error ratio *(18)*.
• Youden index—equal to the (sensitivity + specificity) - 1. The index therefore can range from 0 to 1. An index of 0 means that the test is no better than chance. A perfect test scores 1, since sensitivity plus specificity can never exceed 2 *(19)*.
• likelihood ratio—the ratio of sensitivity/ (1 − specificity). Tests with high likelihood ratios are better discriminators than tests with low ratios *(4)*.

Which of the above rates is best in selecting a laboratory test or determining a referent value? The answer is, unfortunately, none of them. All of the above parameters are based on the assumption that alpha and beta errors (i.e., false negatives and false positives) are equally costly in the decision making process. But what if alpha and beta errors *were* equally costly, which referent value would be best? Murphy and Abbey *(20)* propose that the smallest error is then made, on the average, if a referent value at the intersection of the two distribution curves is selected. When alpha and beta errors carry different risks, it is necessary to consider the relative cost of each type of error. The total cost of errors then can be minimized in selecting a referent value by factoring disease prevalence to reflect the relative risk of alpha and beta errors *(20)*. Once we decide on the parameter to maximize, we can evaluate some laboratory tests, comparing their sensitivity and specificity.

Two situations are found in the clinical pathology literature that involve test comparisons and that require additional statistical methods:

1. Two different tests are evaluated in the same patient population (correlated data).
2. Two different tests are evaluated in two different patient populations (independent data).

The same question is asked in both situations. Which test is better? To obtain an answer, we must be certain that any difference in sensitivity and specificity noted between the two tests is real. We want to be certain that the difference is statistically significant, and that it is not caused by chance alone (sampling error).

Statistical methods such as the *t*-test or Chi Square analysis allow us to make this comparison. We evaluate two types of tests in this way: qualitative and quantitative. For qualitative tests we merely compare different test sensitivity and specificity values. For quantitative tests, it is necessary to compare test sensitivities at equal specificities and test specificities at equal sensitivities. It is then possible to determine which test is more sensitive or specific, or both *(7)*.

In dealing with cost-benefit relations more objectively, we can use the receiver-operating-characteristic (ROC) curve. This requires a graph in which sensitivity is plotted vs. (1 − specificity) (Figure 2). The resulting concave plot is known as an "ROC curve." If health costs are calculated for alpha and beta errors, then they can be factored into an equation along with disease prevalence, as follows: AC_{FN} is the additional cost associated with false negatives and AC_{FN} is the additional cost associated with false positives; $P(D-)$ is the *a priori* probability of no disease and $P(D+)$ is the *a priori* probability of disease (prevalence). The optimal "operating position" or test cut-off point occurs when the slope of the ROC curve (the likelihood ratio) equals the following expression:

$$\frac{AC_{FP}\, P(D-)}{AC_{FN}\, P(D+)}$$

If the cost data are lacking, the ROC curve can be approached simply as a way to minimize errors. Errors are minimized when the slope equals $P(D-)/P(D+)$ *(21)*.

In summary, three relationships are possible between sensitivity and specificity for any test: the sensitivity may be greater than the specificity, the sensitivity may be equal to the specificity, or the sensitivity may be less than the specificity. Let us review

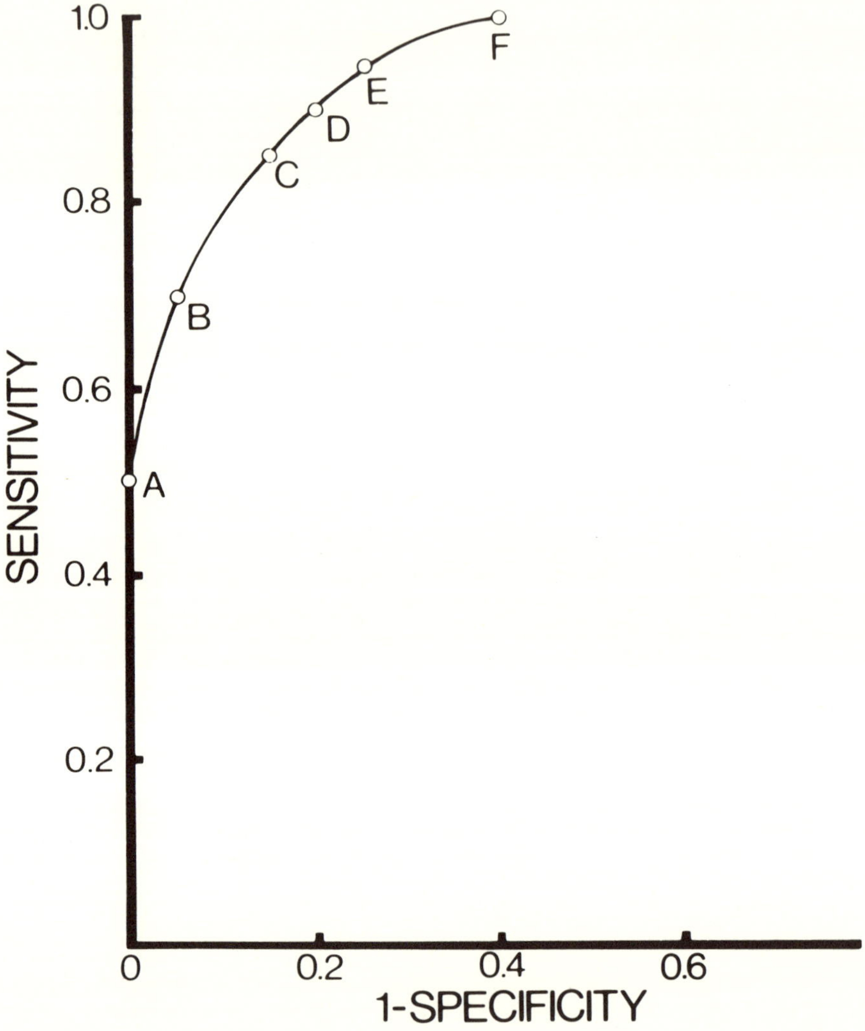

Fig. 2. Receiver-operating-characteristic curve for hypothetical enzyme test data illustrated in Figure 1.

At point *A*, the sensitivity is 50% and the specificity is 100%. At point *F*, the sensitivity is 100% and the specificity is 60%. Points *A* through *F*, the operating positions, correspond to the referent values in Table 4.

what happens to the predictive value and efficiency under each of these circumstances in the face of fixed as well as increasing disease prevalence:

At a given prevalence, an increment in specificity results in a greater increase in the predictive value of a positive test than the same increment in sensitivity. As prevalence increases, the predictive value of a positive test increases regardless of the relation between sensitivity and specificity.

At a given prevalence, an increment in sensitivity results in a greater increase in the predictive value of a negative test than the same increment in specificity. As prevalence increases, the predictive value of a negative test decreases regardless of the relationship between sensitivity and specificity.

At a given prevalence (up to 50%) an increment in specificity results in a greater increase in the efficiency of a test than does the same increment in sensitivity. As prevalence increases, the efficiency of a test increases only if sensitivity exceeds specificity. When sensitivity is less than specificity, the efficiency of a test decreases with increasing prevalence. When sensitivity equals specificity, test efficiency is independent of prevalence and equal to sensitivity (or specificity). The value of test efficiency in all other cases falls between the value of test sensitivity and specificity, regardless of the relationship between sensitivity and specificity and disease prevalence.

Combination Testing

So far, we have discussed the predictive value of a single test result. But in fact the result of a single test rarely is used as the final arbiter of a medical decision. Therefore, let us explore some hypothetical and some actual examples of multiple-test situations to illustrate what happens when we apply more than one test to a clinical problem.

Two independent tests in a screening or diagnostic situation can be used in three different ways:

1. Test A is applied first, and all those with a positive result are retested with test B (series approach: (++) = +).
2. Test B is applied first and all those with a positive result are retested with test A (series approach: (++) = +).

3. Tests A and B can be used together and all those with positive results for either or both tests are considered to be positives (parallel approach: (+ −) = +, (+ +) = +, (− +) = +).

Which approach or sequence is best? This depends on the testing situation and the sensitivity and specificity of the individual tests and their combinations. For the sake of this discussion, we examine the hypothetical data presented in Table 9.

The sensitivity of test A is (190 + 760)/1000, or 95%. The sensitivity of test B is (40 + 760)/1000, or 80%. The sensitivity of the series combination (A and B positive) = 760/1000, or 76%, but the sensitivity of the parallel combination (A or B positive) is (190 + 40 + 760)/1000, or 99%. With parallel testing, the combined sensitivity is greater than the individual sensitivities of the contributing tests.

Similarly, the specificity of test A is (4850 + 84 250)/99 000 = 90%. The specificity of test B is (9800 + 84 250/99 000 = 95%. The specificity of the series combination is (9800 + 4850 + 84 250)/99 000 = 99.9%, since A+B−, A−B+, and A−B− are all interpreted as negative results in series testing. The specificity of the parallel combination is 84 250/99 000, or 85.1%, since only A−B− is considered a negative response.

Parallel testing results in the highest sensitivity but the lowest specificity, whereas series testing results in the lowest sensitivity but highest specificity (Table 10).

The series approach can be taken as either A then B, or B then A. Let us use our hypothetical tests to see how test sequence affects the final outcome in screening for disease. We let disease prevalence be 1% in a population of 100 000. Test A is applied first and all those with a positive result are retested with test B (Tables 11 and 12).

Table 9. **Hypothetical Data for Two Tests, A and B**

	Test results				
	A+B−	*A−B+*	*A+B+*	*A−B−*	*Totals*
Disease	190	40	760	10	1000
Nondisease	9800	4850	100	84 250	99 000
Totals	9990	4890	860	84 260	100 000

We see that on initial testing there are 10 850 positive test results using test A alone. These are, however, presumptive positives and will be retested with test B (Table 12).

Now let us take the same tests and the same prevalence, but this time test B is applied first and all those with a positive result are retested with test A (Tables 13 and 14).

We see that on initial testing there are 5750 positive test results

Table 10. **Combination Testing for (Hypothetical) Data**

	Test(s)	*Sensitivity, %*	*Specificity, %*
Single	A	95.0	90.0
Single	B	80.0	95.0
Series	A and B	76.0	99.9
Parallel	A or B	99.0	85.1

Table 11. **Series Approach: A then B**

	Test A positive	*Test A negative*	*Totals*
Disease	950	50	1000 (sensitivity, 95%)
Nondisease	9 900	89 100	99 000 (specificity, 90%)
Totals	10 850	89 150	100 000

Table 12. **Series Approach: A then B**

	Test B positive	*Test B negative*	*Totals*
Disease	760	190	950 (sensitivity, 80%)
Nondisease	495	9405	9900 (specificity, 95%)
Totals	1255	9595	10 850

The predictive value of a positive result is 60.6% (760/1255).

Table 13. **Series Approach: B then A**

	Test B positive	*Test B negative*	*Totals*
Disease	800	200	1000 (sensitivity, 80%)
Nondisease	4950	94 050	99 000 (specificity, 95%)
Totals	5750	94 250	100 000

on using test B alone. These are, however, presumptive positives and will be retested with test A (Table 14).

The test analyses show the most effective way to use laboratory tests in series. The first sequence (A then B) resulted in the retesting of 10 850 patients. There were then 1255 patients diagnosed as positive with a predictive value of 60.6%. When the sequence was reversed to B then A, only 5750 patients needed retesting. There were 1255 positives with a predictive value of 60.6%. Both sequences, using the series approach, produced 1255 positives with 760 true positives in the group. However, the latter sequence (B then A) is considered more effective since fewer patients required retesting to get to the same end result. This analysis assumes that both tests cost about the same to perform in the laboratory. If two tests are going to be used in series, the optimal sequence can be determined by using the predictive value model as described here.

What about parallel testing in this same situation? In this case tests A and B are applied simultaneously. If either A or B or both is positive, this is considered a positive test. A negative test requires A *and* B to be negative (Table 15).

Table 14. **Series Approach: B then A**

	Test A positive	*Test A negative*	*Totals*
Disease	760	40	800 (sensitivity, 95%)
Nondisease	495	4455	4950 (specificity, 90%)
Totals	1255	4495	5750

The predictive value of a positive result is 60.6% (760/1255).

Table 15. **Parallel Approach: A or B positive = positive test**

	A or B positive	*A and B negative*	*Totals*
Disease	990	10	1000 (sensitivity, 99%)
Nondisease	14 750	84 250	99 000 (sensitivity, 85.1%)
Totals	15 740	84 260	100 000

The predictive value of a positive result is 6.3% (990/15,740).

For tests run in parallel (A and B determined simultaneously), but considered positive if *either* component is positive, and negative only if *both* are negative, the sensitivity is higher and the specificity is lower than in comparable series testing. The sensitivity is increased because some diseased patients are positive for one test, but not for the other. Similarly, there are more false-positive results in nondisease patients. The parallel approach outlined above has the highest sensitivity. It detects the greatest number of diseased patients in the population. However it also produces the greatest number of false positives. In this case the predictive value is so low (6.3%) that this test might be unacceptable for any disease. Of all the approaches, the parallel approach requires the most laboratory work, because both tests are performed on all patients in the population.

The above relationships can be demonstrated with actual data from a study of enzymes and isoenzymes in the diagnosis of acute myocardial infarction *(22)*. Using the final clinical diagnosis as the ultimate criterion, we calculated the sensitivity and specificity of single and multiple enzyme determinations. Positive results were defined for creatine kinase (CK; EC 2.7.3.2) and aspartate aminotransferase (AST; EC 2.6.1.1) on day one and lactate dehydrogenase (LD; EC 1.1.1.27) on day two, the laboratory's usual upper limit being used for each test. The results are summarized in Table 16. It can be seen that interpreting results in a series fashion always decreases sensitivity and increases specificity relative to single tests or smaller test combinations. The overall effect of the series approach is that more cases will be

Table 16. **Sensitivity and Specificity of Enzyme Measurements, Singly and in Combination, in Diagnosing Myocardial Infarction in the Coronary Care Unit**

Test[a]	*Sensitivity, %*	*Specificity, %*
CK[b]	95.7	57.1
AST	91.4	73.5
LD	87.2	90.5
AST and LD	82.2	92.9
CK and LD	82.2	92.9
CK and AST	87.2	83.7
CK, AST, and LD	77.7	95.2

[a] All combination results interpreted in series fashion as listed.
[b] Abbreviations: CK, creatine kinase (EC 2.7.3.2); AST, aspartate aminotransferase (EC 2.6.1.1); and LD, lactate dehydrogenase (EC 1.1.1.27).

missed, but there will be a higher predictive value for the positive result (i.e., fewer false positives).

Since it is widely held that the clinical judgment of experienced laboratorians brings to problem solving some techniques that are not reducible to computer programming, we submitted serial enzyme data for CK, AST, and LD on 96 consecutive patient admissions to the CCU to five professors of clinical pathology. The pathologists were assigned the task of placing each case into one of two categories, either myocardial infarction or not *(7)*. It is quite remarkable, but their average sensitivity of 77.6% and their average specificity of 94.8% was no different from that obtained from our simple computer analysis, interpreting the same data in a series fashion.

If we consider biochemical profiles or batteries of laboratory tests, interpreted in a parallel fashion, we see that they have extremely high sensitivity but low specificity. Because of the high sensitivity, these profiles, when negative, have a very high negative predictive value. For that reason, they are used by clinicians to exclude or rule-out a variety of diagnoses with a high degree of probability. Later on in the diagnostic work-up, series testing is performed, which is highly predictive of the presence of a particular disease because of its high specificity. As the clinician follows his own algorithm for working up a patient, the probability continues to increase as he receives more information. Ultimately, in the series of diagnostic tests and procedures the physician needs a test with nearly 100% specificity if he is to make the diagnosis with complete certainty.

Tests may be administered in a series, in sequence, but interpreted as if they were performed in parallel, thereby maximizing sensitivity. An analysis of the six-sequential-stool guaiac protocol for screening for colonic cancer is an example of this approach. As expected, the sensitivity increases and specificity decreases as the number of tests in the sequence increases.

Cancer detection is compared to the cost of screening in Table 17. We see that the marginal cost (the extra cost required to find extra cases) increases most strikingly. The marginal cost must not be overlooked when thinking about combination testing or profiling, because it is an important consideration that is obscured by the average cost. In this study the marginal cost of the sixth test is 20 000-fold the average cost. Marginal cost rises less rapidly with sequential testing when the test involved has a lower sensitivity. In addition, decreased disease prevalence, as would be expected, increases the marginal cost strikingly *(23)*.

It is essential that laboratorians and clinicians be able to use the predictive value model for evaluating the diagnostic accuracy of laboratory tests as well as other procedures. This will facilitate rational test selection and anticipation of outcomes. Since false-positive results can be predicted, the average and marginal cost per case can be calculated. Cut-off points (referent values) for laboratory tests can be selected to regulate the rate of false positives to keep them within medically or economically derived guidelines. Furthermore, multiple laboratory tests for a single disease can be evaluated and a decision reached as to whether adding a test to a profile or series increases the predictive value enough to justify the extra cost.

Multivariate Analysis

In our discussion of series testing, we discussed applying two independent tests and proceeded to illustrate our points with a hypothetical example. The application of sensitivity and specificity percentages in series testing can only be valid if the tests (A and B) are independent of each other. Otherwise the first test would serve to preselect the population for the second test. Therefore, if the tests are not independent, the sensitivity and specificity percentages are no longer valid in the selected sub-population. If the tests are not independent, then it is not valid to use the available sensitivity and specificity to predict the

Table 17. **Cancer Detection and Screening Costs with Sequential Guaiac Tests**

No. of tests	*Cancer detection*			*Screening costs ($)*		
	No. of cases	*Incremental gain*	*Total*[a]	*Incremental*	*Marginal*[b]	*Average*[c]
1	65.9469	65.9469	77 511	77 511	1 175	1175
2	71.4424	5.4956	107 690	30 179	5 492	1507
3	71.9004	0.4580	130 199	22 509	49 150	1810
4	71.9385	0.0382	148 116	17 917	469 534	2059
5	71.9417	0.0032	163 141	15 024	4 724 695	2268
6	71.9420	0.0003	176 331	13 190	47 107 214	2451

[a] Calculated by addition of cost of guaiac stool testing on 10 000 people and cost of barium-enema examination of all those with positive tests. Prev = 72/10 000

[b] Calculated by dividing incremental cost by incremental gain for cancer detected.

[c] Calculated by dividing total cost by no. of true positive results.

Reproduced from Neuhauser and Lewicki *(23)*, with permission.

outcome accurately. However, such an analysis would indicate the general effects to be derived from a series approach. Similarly, the performance of a repeat test and the use of the same sensitivity and specificity as applied to the unselected population will provide only an approximation of what would happen. An accurate analysis requires that the second sample (selected subpopulation) redistribute itself entirely to resemble the first population. This, of course, is unlikely. Therefore, the calculation of predictive values based on repeat testing with the same test, or a combination of tests in series, is fraught with statistical difficulties. It does however, demonstrate the general beneficial effects of repeat testing, but does not result in a precise predictive value percent that can be affixed to the diagnostic process. The calculated predictive value is always higher than the actual predictive value of the testing situation. Furthermore, the degree of error depends on the degree to which the two tests are correlated.

Having therefore disqualified simple predictive value analysis of multiple dependent variables, we have no choice but to proceed to more sophisticated models of multivariate analysis. Direct and stepwise discriminant function analysis is quite suitable for this purpose *(24, 25)*.

Discriminant analysis begins with the desire to distinguish between two or more groups of cases, such as myocardial infarction and non-myocardial infarction. To distinguish between the groups, we collect data on discriminating variables such as enzymes and isoenzymes for which the results for the two groups are expected to differ. Discriminant analysis attempts to do this by forming one or more linear combinations of the discriminating variables. These "discriminant functions" are of the form:

$$D_i = d_{i1} Z_1 + d_{i2} Z_2 + \ldots + d_{ip} Z_p$$

where D_i is the score on discriminant function i, the d's are weighting coefficients, and the Z's are the standardized values of the p discriminating variables used in the analysis. Ideally, the discriminant scores (D's) for the cases within a particular group will be fairly similar. Regardless, the functions are formed in such a way as to maximize the separation of the groups.

When faced with the problem of test selection in a given clinical situation, discriminant analysis can be used to define test redundancy. The efficiency of all the tests in diagnostic discrimination can be evaluated against the efficiency when selected tests are eliminated from the analysis. When we have more discriminating variables than necessary to achieve satisfactory discrimination, as is frequently the case with laboratory testing, the most useful variables can be selected by stepwise discriminant analysis. With this procedure, the single best discriminating variable is selected. A second discriminating variable is selected as the one best able to improve the value of the discrimination criterion in combination with the first variable. Subsequent variables are selected according to their ability to contribute to further discrimination. At each step, tests already selected may be removed if they are found to reduce discrimination when combined with more recently selected tests. Either all tests will be selected or the stepwise procedure will determine that the remaining tests do not contribute to further discrimination. These tests that do not contribute to the discrimination are redundant, and are eliminated. The stepwise procedure results in the selection of an optimal set of tests for a given diagnostic discrimination.

Conceptually, it should be noted that discriminant function classifies cases without reference to probabilities of misclassification (alpha and beta errors). As pointed out by Lusted, "The larger the value of the linear discriminant, the more certain we are an individual belongs to population 2; the smaller the value, the more certain he belongs to population 1. But there are intermediate values which leave us doubtful about how to assign the patient. This brings us back to consider not classification but the probability that, given t, the patient belongs to population 2" *(4)*. Lusted related this probability by using Bayes' theorem, incorporating the likelihood ratio and the ratio of prior probabilities (prior odds) *(4)*. There are also some statistical limitations to linear discriminant analysis, since the underlying assumptions of normal distribution of each variable, equal variances, and the existence of a linear relationship are made. However, non-parametric discriminant analysis programs are available if linear discriminant function cannot be applied to the data.

The data referred to earlier from our myocardial infarction study were subjected to discriminant function analysis. Laboratory tests included CK, LD, AST, ALT,[1] HBD, CK-MB and LD isoenzymes. Table 18 shows the results of discriminant analysis of the classical triad of cardiac-related enzymes (CK, AST, LD) performed on admission, day 1 and day 2 in 46 myocardial infarction and 48 non-myocardial patients admitted to the CCU. These enzyme tests were not normally distributed. In calculation of the discriminant function, a non-normal distribution provides a function that may not be optimum. A log transformation of the data, while normalizing the distribution, successfully increased the discriminant power of these cardiac enzymes. Table 19 shows the results of discriminant analysis of data on CK-MB and LD isoenzymes obtained on admission, day 1, and day 2 in these same patients. Results of these isoenzyme tests were considered in a binary mode. CK-MB was positive if it was greater than 3%

Table 18. **Discriminant Function Analysis of Cardiac Enzymes: CK, AST, LD Measured on Admission, Day 1, and Day 2 in MI (46) and Non-MI (48) Patients Admitted to the CCU**[a]

Tests:	*CK, AST, LD*	*Log transformation*
Sensitivity	67.4%	80.4%
Specificity	91.7%	93.7%
Efficiency	79.8%	87.2%

[a] Abbreviations as in Table 16; also: MI (myocardial infarction) and CCU (Coronary Care Unit). No. of patients in parentheses.

Table 19. **Discriminant Function Analysis of Data for Isoenzymes (CK and LD) as Compared to Isoenzymes (CK and LD) Plus Total Enzyme Activities, Measured on Admission, Day 1 and 2 in MI (46 Patients) and Non-MI (48 Patients) Patients Admitted to the CCU**[a]

	CK-MB, LD isoenzyme	*CK-MB, LD isoenzymes, CK, LD, AST, ALT, HBD*
	Percent	
Tests:		
Sensitivity	97.8	95.7
Specificity	87.5	93.7
Efficiency	92.6	94.7

[a] Abbreviations as in Tables 16 and 18; also: HBD, hydroxybutyrate dehydrogenase, and ALT, alanine aminotransferase (EC 2.6.1.2).

[1] Additional nonstandard abbreviations used here are: ALT, alanine aminotransferase (EC 2.6.1.2); HBD, hydroxybutyrate dehydrogenase; CK-MB, the MB isoenzyme of creatine kinase.

of total CK activity. LD isoenzymes were considered positive if LD_1 was greater than LD_2 ("flipped" LD pattern). Isoenzyme data alone correctly classified 92.6% (87/94) of the patients. The redundancy of additional test data is shown in the same table, where in addition to the isoenzyme results, the log transformation of CK, LD, AST, ALT, and HBD enzyme activity were included in the discriminant analysis. Because each test was performed on admission, day 1, and day 2, we have entered 15 additional results for each patient into the analysis. The discrimination was increased to 94.7% (89/94), a net gain of only two patients.

As indicated earlier, rather than selecting the tests for analysis, one can enter all results into a stepwise discriminant program for test selection. There are several different stepwise programs, in each of which a different selection criterion is used. Our data were analyzed by the method of Rao et al. *(24)*, in which the variable is selected that yields the largest increase in the generalized distance as measured by Rao's V, which amounts to the greatest overall separation of the groups. The tests selected by the program were CK-MB (admission and days 1, 2), LD isoenzymes (1, 2), LD (2), and HBD (2). This combination resulted in exactly the same discrimination as when all tests were analyzed by the program.

Discriminant function analysis would seem to offer a logical approach to test selection for multichannel analyzers. Once the disease entities were identified, the optimum tests could be selected by means of discriminant function, to avoid redundancy and thereby increase the scope and information content of the test profile. Certainly the proliferation of multichannel analyzers and biochemical profiling warrants this type of investigation. Frequently, we hear laboratory directors absolve themselves of responsibility for test selection "because the clinicians do all of the ordering." The configuration of test profiles on multichannel analyzers, however, is clearly the responsibility of the laboratory director (not of the manufacturer!) and provides an opportunity for the laboratory to have an impact on better patient care through a logical approach to test selection.

Decision Analysis

Decision analysis is a problem-solving technique that is finding its way slowly into the literature of clinical and laboratory medicine. Although we will not explore decision analysis here, it is essential that we consider its impact on test selection. The interested reader is referred to Raiffa's *(26)* excellent theoretical introduction to the subject as well as to several articles on clinical decision making *(27–30)*.

Figure 3 depicts a simplified version of a "decision tree" for a patient with coronary-type chest pain and describes the possible courses of action and consequences. It is possible to calculate the probabilities and utilities (values) detailed in the decision tree. The decision to be made (square node) is whether or not to treat the patient for acute myocardial infarction. In either case, there is some chance (circular nodes) that the patient either has or has

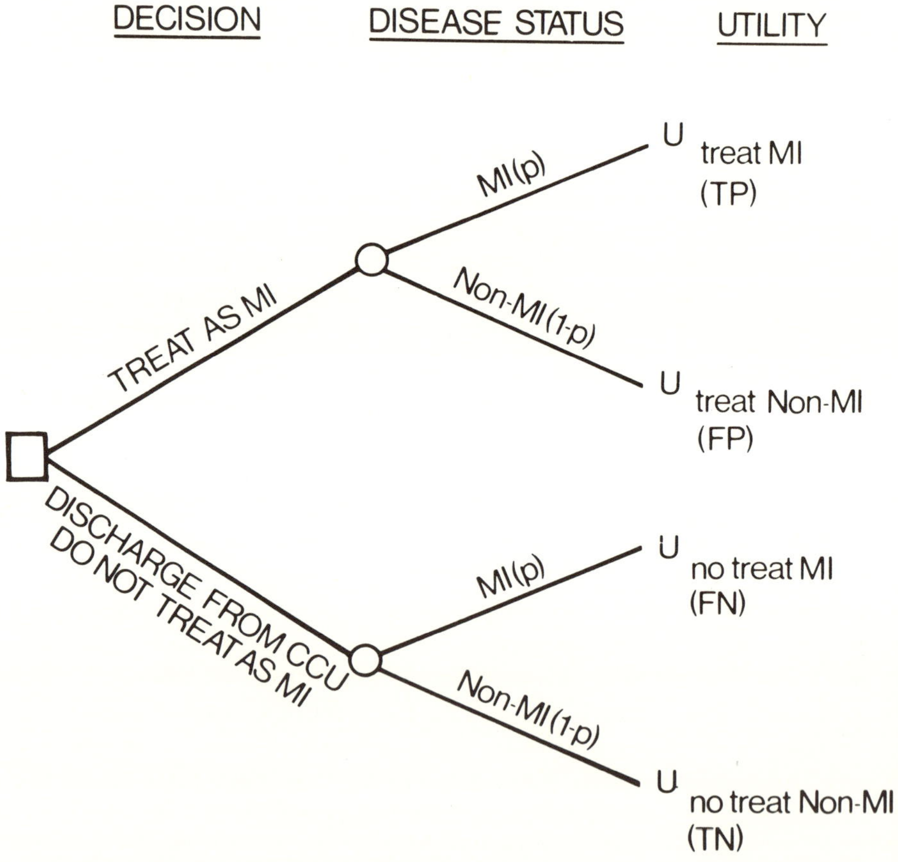

Fig. 3. Decision tree, showing two choices for patients with coronary type chest pain admitted to the Coronary Care Unit to rule out myocardial infarction.

not suffered an infarction. Each of the outcomes has a certain value, and this relative value is denoted as the utility of the outcome. Ginsberg *(3)* suggests defining the components of a consequence as follows: *(a)* out-of-pocket dollar costs; *(b)* days in bed with no pain; *(c)* days in bed with mild pain; *(d)* days in bed with severe pain; *(e)* days to diagnosis or definite action; *(f)* permanent complications; *(g)* temporary complications; *(h)* short-term severe pain; *(i)* long-term mild pain; *(j)* days of restricted activity; and *(k)* death. These components must then be translated into a common scale, such as dollars.

The "expected value" or worth of treating and not treating can be calculated by assigning values to the probabilities and utilities. The solution of the decision tree is presented very well by Pauker and Kassirer *(30):*

$$EV_{\text{treat}} = (p)U_{\text{treat dis}} + (1 - p)U_{\text{treat no dis}}$$
$$EV_{\text{no treat}} = (p)U_{\text{no treat dis}} + (1 - p)U_{\text{no treat no disease}}$$

The likelihood of achieving the best outcome or consequence is maximized by choosing the option with the higher EV.

The net benefit (B) and net cost (C) are derived as follows:

$$B = U_{TP} - U_{FN}$$
$$C = U_{TN} - U_{FP}$$

The two courses of action (treating and not treating) are equivalent when $EV_{\text{treat}} = EV_{\text{no treat}}$ Solving these equations for p, we calculate a probability at which the clinician should be indifferent to either course of action.

$$p = \frac{U_{TN} - U_{FP}}{(U_{TP} - U_{FN} + U_{TN} - U_{FP})}$$

It follows that if the probability of disease is greater than the probability at the indifference point, treatment is preferred. Similarly, test results that increase the probability of detecting disease (predictive value of positive result) can be evaluated relative to the indifference point and selected on the basis of this model.

Pauker and Kassirer *(30)* denote the indifference point probability *(p)* as (T), since it is a threshold, and substitute B and C for utilities in the above equation:

$$T = C/(B + C)$$

This is further simplified to $T = 1/[(B/C) + 1]$

In a given testing situation, various laboratory tests can be evaluated, and the test or test combination that has a predictive value above the threshold value (T), would be selected, knowing that the expected value of treatment exceeded the expected value of no treatment. One now could truly answer the question, "How good should a laboratory test be?" Regardless of sensitivity and specificity, if T cannot be exceeded in a given clinical setting, the testing scheme can be considered useless in terms of diagnosis and treatment.

Sisson et al. *(31)* demonstrate how the acquisition of laboratory data may make patient-care worse, rather than better. This can only be appreciated by developing a decision tree. The authors propose that a patient population would have a prevalence of pancreatic cancer of 12/1000 if they were older than 40 years and had experienced mid-abdominal pain of one to three weeks duration. They propose a hypothetical risk-free test for carcinoma of the pancreas with 80% sensitivity and 95% specificity. In addition, because the test detected the tumor early, half of the cases were curable after total pancreatectomy. The risk of pancreatectomy carries a 10% mortality. The decision tree (Figure 4) is based on these assumptions and the possible outcomes are noted. The use of the new test increases the death rate (12.5 vs. 12) and among the group who die are included five patients with benign disease. In addition, 44 patients without cancer must suffer the consequences of pancreatic insufficiency. The hazard of obtaining new information in this hypothetical clinical situation is clear and stresses the need for the use of decision analysis in the test selection process.

Summary

In summary, there are four levels at which laboratory tests and procedures can be evaluated:

I. *Analytical analysis of laboratory test:*
precision, accuracy, technical sensitivity, technical specificity.

II. *Diagnostic analysis of laboratory test:*
diagnostic sensitivity, diagnostic specificity, Youden index, likelihood ratio, etc.

III. *Operational analysis of laboratory test:*

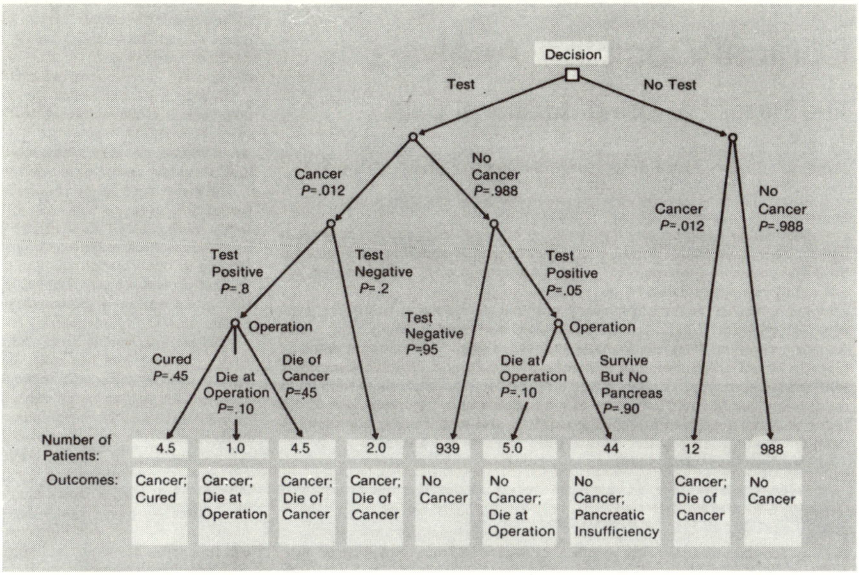

Fig. 4. Beginning with a 1000-patient base at the decision node *(square)*, probabilities at each branch give the number of patients arriving at each probability node *(circles)*.

The number of patients, recorded at the bottom for each outcome, totals 1000 for each decision: test or no test. Reproduced from Sisson, Schoomaker and Ross *(31)*, with permission.

predictive value of positive result, predictive value of negative result, efficiency, discriminant function, etc.

IV. *Medicial decision-making analysis of laboratory test:*
threshold probability, cost/benefit analysis, solving the decision tree.

Test selection can be done at any level, without knowledge of the test's evaluation and performance at the remaining levels. Alternatively, the development of new laboratory tests can proceed from level one to level four, or vice versa. Unfortunately, the former is usually the case and most of the tests in use today have never been evaluated at the medical decision-making level (level IV). Recent efforts at developing a glucose Reference Method represent a disproportionate amount of time and energy expended at level one, and typify our backwards approach to laboratory medicine. In thinking about the development of new diagnostic tests, we should begin at level four to characterize the properties and specifications that the test must meet. As an example, an in vitro test for the diagnosis of pulmonary embolism could be characterized in this fashion, with criteria specified at each of the lower levels.

Returning to the question of how good a laboratory test should be, we see that the answer must come from an analysis of the benefit/cost equation (level IV). Figure 5 is a plot of the net benefit and cost of treatment vs. the threshold probability. Because the threshold probability defines how certain one must be of the diagnosis before proceeding with treatment, it serves as a minimum probability that should be exceeded by the predictive value of the test. When the benefit/cost ratio is low, a test with a very high predictive value is required to exceed the threshold probability. On the other hand when the benefit/cost ratio is high, even a test with a low predictive value would be helpful to the physician in making the decision to treat the patient. Within this framework, several clinical situations could be evaluated and problems requiring the development of highly predictive laboratory tests (low benefit/cost ratios) could be identified. In laboratory medicine too much emphasis has been placed on the

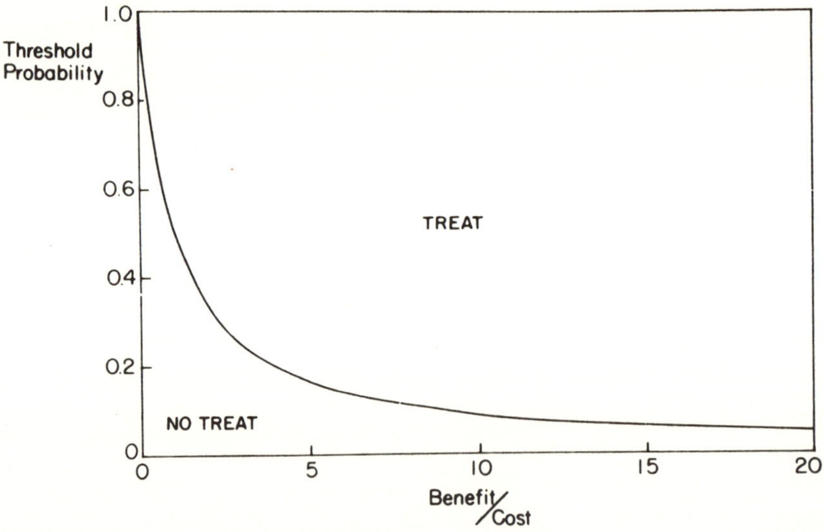

Fig. 5. Relations between benefit and cost of a treatment and the threshold probability.

For a given benefit/cost ratio, treatment is preferred when the probability of disease in an individual patient exceeds the threshold value. Reproduced from Pauker and Kassirer *(30)*, with permission.

"laboratory" and not enough on the "medicine." How important is the coefficient of variation when the benefit/cost ratio is high? Tests can not be developed or selected appropriately in a therapeutic vacuum.

The laboratory is not alone to blame, however. Physicians should select laboratory services with an understanding of the inherent statistical limitations of all procedures, including history and physical diagnosis. Few medical schools provide formal education in this area, and there is little applied research on the clinical value of our multibillion dollar annual expenditure on laboratory tests. Unless physicians regard the ordering of laboratory tests as a serious responsibility, worthy of the same thoughtful consideration as a history and physical examination, the privilege to select such services freely may not remain unrestricted in the future.

References

1. Drucker, P. F. *Management*, Harper and Row, New York, N.Y., 1974, p 268.
2. Ginsberg, A. S., The diagnostic process viewed as a decision problem. In *Computer Diagnosis and Diagnostic Methods,* J. A. Jacquez, Ed. Charles C Thomas, Springfield, Ill., 1972, pp 203–240.
3. Ginsberg, A. S., *Decision Analysis in Clinical Patient Management with an Application to the Pleural Effusion Syndrome*, The Rand Corp., R–751–RC/NLM, Santa Monica, Calif., 1971, p 125.
4. Lusted, L. B., *Introduction to Medical Decision Making,* Charles C Thomas, Springfield, Ill., 1968, p 271.
5. Barnoon, S., and Wolfe, H., *Measuring the Effectiveness of Medical Decisions*, Charles C Thomas, Springfield, Ill., 1972, p 229.
6. Vecchio, T. J., Predictive value of a single diagnostic test in unselected populations. *N. Engl. J. Med.* **274**, 1171 (1966).
7. Galen, R. S., and Gambino, S. R., *Beyond Normality: The Predictive Value and Efficiency of Medical Diagnoses*, John Wiley & Sons, New York, N.Y., 1975, p 237.
8. Bayes, Rev. Thomas, An essay toward solving a problem in the doctrine of chance. *Philosoph. Trans. Roy. Soc.* **53,** 370 (1763).
9. Galen, R. S., Interpretation of laboratory tests. *N. Engl. J. Med.* **292,** 433 (1975).
10. Hamlyn, A. N., Sherlock, S., Brown, A. J., and Baron, D. N., Casual blood-ethanol estimations in patients with chronic liver disease. *Lancet* **ii,** 345, (1975).

11. Barley, S. L., Blood-ethanol in liver disease. *Lancet* **ii,** 770 (1975).
12. Blendis, L. M., McNeilly, W. J., Sheppard, L., et al. Observer variation in the clinical and radiological assessment of hepatosplenomegaly. *Br. Med. J.* **i,** 727 (1970).
13. Hansen, H. J., Snyder, J. J., Miller, E., et al., Carcinoembryonic antigen (CEA) assay. *Human Pathol.* **5,** 139 (1974).
14. Drum, D. E., and Christacopoulos, J. S., Hepatic scintigraphy in clinical decision making. *J. Nuc. Med.* **13,** 908 (1972).
15. Jelliffe, R. W., Quantitative aspects of clinical judgment. *Am. J. Med.* **55,** 431 (1973).
16. Shapiro, S., Strax, P., Venet, L., Periodic breast cancer screening in reducing mortality from breast cancer. *J. Am. Med. Assoc.* **215,** 1777 (1971).
17. Newcomer, A. D., McGill, D. B., Thomas, P. J., Hoffmann, A. F., Prospective comparison of indirect methods for detecting lactase deficiency. *N. Engl. J. Med.* **293,** 1232 (1975).
18. Simonson, E., *Differentiation between Normal and Abnormal in Electrocardiography,* Mosby, St. Louis, Mo., 1961, p 328.
19. Youden, W. J., Index for rating diagnostic tests. *Cancer* **3,** 32 (1950).
20. Murphy, E. A., and Abbey, H., The normal range—a common misuse. *J. Chronic Dis.* **20,** 79 (1967).
21. McNeil, B. J., Keeler, E., and Adelstein, S. J., Primer on certain elements of medical decision making. *N. Engl. J. Med.* **293,** 211 (1975).
22. Galen, R. S., Reiffel, J. A., and Gambino, S. R., Diagnosis of acute myocardial infarction: Relative efficiency of serum enzyme and isoenzyme measurements. *J. Am. Med. Assoc.* **232,** 145 (1975).
23. Neuhauser, D., and Lewicki, A. M., What do we gain from the sixth stool guaiac? *N. Engl. J. Med.* **293,** 226 (1975).
24. Dixon, W. J., Ed. *BMD Biomedical Computer Programs,* University of California Press, Berkeley and Los Angeles, Calif., 1973.
25. Anderson, T. W. *Introduction to Multivariate Statistical Analysis,* John Wiley & Sons, New York, N.Y., 1958.
26. Raiffa, H., *Decision Analysis: Introductory Lectures on Choices Under Uncertainty.* Addison-Wesley, Reading, Mass., 1968.
27. Schwartz, W. B., Gorry, G. A., Kassirer, J. P., et al., Decision analysis and clinical judgment. *Am. J. Med.* **55,** 459 (1973).
28. Gorry, G. A., Kassirer, J. P., Essig, A., et. al., Decision analysis as the basis for computer-aided management of acute renal failure. *Am. J. Med.* **55,** 473 (1973).
29. McNeil, B. J., Varady, P. D., Burrows, B. A., and Adelstein, S. J., Measures of clinical efficacy: Cost effectiveness calculations in the diagnosis and treatment of hypertensive renovascular disease. *N. Engl. J. Med.* **293,** 216 (1975).

30. Pauker, S. G., and Kassirer, J. P., Therapeutic decision making: A cost-benefit analysis. *N. Engl. J. Med.* **293,** 229 (1975).
31. Sisson, J. C., Schoomaker, E. B., and Ross, J. C., Clinical decision analysis: The hazard of using additional data. *J. Am. Med. Assoc.* **236,** 1259 (1976).

PANEL DISCUSSION

(Ed. note: The discussions in this book are edited versions of the answers of the speakers to questions submitted to the session chairmen by persons in the audience.)

Dr. Young: While we are waiting to get together the questions from the audience, maybe Dr. Whitehead will say a few words.

Dr. Whitehead: At the Queen Elizabeth Hospital in Birmingham, we have now over two million dollars in computer hardware, and probably an equivalent amount in software. But at this stage, we only have the ability to report cumulatively all patient information for all laboratory disciplines and also to collect drug-therapy information, and we can relate these.

Otherwise, we are now almost at a full stop in what we can do.

I feel that the kind of computer system suggested here is going to be much the same.

Let me say that we do have groups of people in our own country working with small computer systems, doing very similar things with, let us say, thyroid disease or with gastric surgery and gastric complaints and decision making on these, but they are isolated and they are the super specialists of whom we have talked.

The second comment I would like to make is that patients come to Sheila Sherlock who are reporting for care mostly because of alcohol consumption. She is very concerned about the heavy drinker who doesn't come into care. Alcohol measurement can in this situation be very useful.

Dr. Young: There are many questions. The first one and perhaps the very one likely to produce a thoroughly critical response is: Is there logical substance in the clinical notion that I believe a test result only if it conforms to my clinical expectation.

Dr. Galen: That is from the point of view of a clinician?

Dr. Young: Yes.

Dr. Galen: If that in fact is the framework in which a test is ordered, then there is no point in ordering the test. What you are dealing with is a unity concept. It is like saying that you are going to do a study to validate use of an EKG to study myocardial infarctions and then setting up your criteria for myocardial infarction by studying all patients that have EKG abnormalities. If that is the viewpoint, I see no point in ordering a laboratory test.

Dr. Young: Dr. Rall?

Dr. Rall: That is perhaps a slightly extreme version of what I would say, because it is quite clear (and it depends on the particular laboratory) that some test results seem perfectly ridiculous to a clinician, and this is where I think the clinical chemist comes in, a personal contact to let the clinician know what happened that day—were normals run, were they okay—and you repeat the test.

Clearly, if a test value is twice normal and you have no reason to expect the patient has any disease that this test would suggest (usually they don't suggest a single disease), then one is highly suspicious of it. Nonetheless, unexpected things occasionally are picked up and obviously it is ridiculous to ignore such a result. Just as obviously, it is ridiculous to make a diagnosis on the basis of that single erroneous test.

Dr. Young: Dr. Sackett?

Dr. Sackett: I suspect this notion may have arisen from the frustrated clinician who is receiving unsatisfactory laboratory results.

Dr. Young: Dr. Weed?

Dr. Weed: We always should look at information from another point of view. If it is all collected and all computerized and can be looked at from population points of view, there is much value and much information that wasn't gotten from the point of view of that particular physician. It may not be useful to him, but it may be very useful to others and in a much, much larger context.

Dr. Young: What are the strategies for discarding poorly sensitive and specific laboratory tests; and, coupled with that, how many tests has your laboratory stopped doing?

Dr. Galen: I was once asked to write a short communication to pathologists called "How Does One Get Rid of an Obsolete

Test?"—and we all know that there *are* obsolete tests. I think I will try to pick benign ones here, so we don't have a palace revolt.

I think the thymol turbidity is an obsolete test. I think cephalin flocculation is an obsolete test. I think that the PBI has been replaced by thyroxine measurement—and, hopefully, I sense agreement on this and that we no longer have to discuss why or the statistics behind it.

The next question is how one implements those changes, and I am sure you are all familiar with the rational approaches to the answer.

One first tries to make the new test available and hopes that physicians will order the new test and see its superior performance. That seldom works.

The next thing then is simply to actually use some consensus method of a laboratory committee that includes the chiefs or interested people from the departments of medicine, surgery, pediatrics, or ob-gyn, and try to rationally explain to them how an aspartate aminotransferase assay is more predictive than a cephalin flocculation test and how you would like to replace it in the laboratory. If by committee consensus they approve, then you just stop doing the old test and any orders that come in for it are converted to orders for the new test. Moreover, you have newsletters, for example, with which to spread the word.

As a last resort, what I would do is say that the PBI is obsolete; it will not be done; we are now doing thyroxine assays instead. And not only won't it be done here, but we are not even going to send it out to a specialist laboratory to have it done for you. That is taking a very hard line in saying I know when a test is obsolete, I know there is a better test to be done, and I want to talk to you about it face to face, because I can convince you that a thyroxine assay is better than a PBI, for example.

No one can force us to do a test in my laboratory, but he certainly can bring pressure to bear for it to be sent out. What I usually do then is pick a reference laboratory that is geographically as far away from my hospital as possible, maybe in another country, and then say to the requesting physician, "Yes, I really understand that there is an application for this test and I understand your position, but they don't make reagents for it any more; I am going to send it to this laboratory in New

Zealand and I hope the results come back in time for it to be clinically useful." Usually, in time, requests for the test just fade away.

Dr. Young: Dr. Weed?

Dr. Weed: I think this is one specific way the whole new computer system can be useful. In all these detailed plans, in all these steps, the test would disappear; with this National Library of Displays, it wouldn't appear. But so that you can always have freedom to do as you please, on the bottom of these screens is a function tape called "typing" and you can hit it, you can request thyroxine assays for flat feet if you want to, but they are done at the end of the day. That's the first step. Let them type it in, because they *know* now that they are going outside the routine. You don't say it is right or wrong. Instead, you go to the physician after a while and say, "Would you defend this for me, because we want to keep those displays up-to-date?" Also, you read their plan and progress notes—and what will you find? Half the time they don't even interpret in their progress notes what they planned to do in their plan, or they ignore it. When the physician is queried about the logic of this, it shakes him up. If they are using it and using it wrong, you ask for a logical or literature-citation defense. After the second or third time, they don't want to do it anymore.

You see, people don't mind being told they are not thorough; they don't mind being told they are not reliable; but when you tell them they can't think straight, it gets to them.

Dr. Young: Now another follow-up, and it is obviously a great burning issue with the experienced laboratorians. All the questions are coming to the same point. You have now got your system perfected. What do you do when it breaks down?

Dr. Weed: You mean the technical system?

Dr. Young: Yes. Say the computer system on which everybody totally depends doesn't work today.

Dr. Weed: Well, first of all, what I would like to have you think of is the human mind, which is a computer system. My hand here is a printer, and teaching is writing programs for it. My mind is debugging it.

Now, people are "down" most of the time, whereas our computer system (we have many computers and we have backup) went for four years with less than 0.04% down time, because we can control that. For instance, if the power goes off in the whole institution on the two systems, the third can operate off the auxiliary power.

Say the speed goes from 40 μs down to 25 μs or so. Some people will say "This thing is too slow!" But I have been in some pretty distinguished hospitals where we didn't get the diabetics chart one out of three times.

The point is that we are already in deep trouble, so the computer makes our job a lot easier.

Dr. Young: Perhaps unsolicited results or screening testing would be of more value if performed *after* diagnosis and (or) treatment rather than as part of the initial studies. This could eliminate the noise or red-herring effect while simultaneously suggesting any other problems that may be concurrent with the major problem or a result of the major problem and (or) its treatment, at a time when the physician is better able to concentrate on secondary problems.

Dr. Sackett: Yes. I think what you are suggesting is the separation of the diagnostic process; that is, to try to deal with that immediate feature that is responsible for bringing the clinician and the patient together.

But looking for disorders unrelated to the chief complaint can result in treatment that is either easier or more efficacious because the condition was found early rather than late. It is certainly the sort of thing that many clinicians initiate themselves, and obviously it is going to be the topic for much of the later discussion here. I think that in considering it one raises a whole other series of issues, not the least of which is whether or not therapy in these situations really is worthwhile, because obviously if you make an early diagnosis and then are unable to apply efficacious therapy, you increase the duration of illness, not of health.

Second, one must recognize the increasing concern about the problems of labeling individuals as diseased. We have recently found some quite disturbing data about what happens when you tell an individual he has hypertension. His absenteeism in the next year will be about three- to fourfold what it was the year before, and that increase in absenteeism is associated with psychosocial changes. Those changes occur regardless of whether you treat him or not, so they cannot be labeled a side effect of the drugs, and they are augmented by health education, not relieved, and are quite closely related to compliance. As an example of the non-compliant hypertensive individual, steelworkers in my town lost three weeks of additional work and income the year after the diagnosis of hypertension. This is not target-organ damage; it is the effect of their treating themselves as if they are much more fragile, staying away from work substantially longer when they have intercurrent infections and illnesses.

So I think there is a whole other series of issues raised in the case-finding circumstance such as this, but I also think most clinicians clearly separate, at least in their own mind, the case finding from the diagnostic activities.

Dr. Young: Any comments?

Dr. Weed: I think we have got to go back, starting with small children, to educate about health, because we have their records from the first grade on. They should recognize they are unique, they should come to discover from their own personal data a lot of the things we are now finding mathematically and so forth. If this were done I think responses such as this that you are finding with the hypertensives might not happen.

All sorts of weird things are going to happen if we suddenly throw things like this at the public. I am not surprised about the hypertensives, but the real moral of that story is: stop telling people things, stop doing things that may adversely affect the way they feel they have to deal with these realities. It is a tough problem. I wouldn't [want to] draw the wrong conclusions.

Dr. Young: Next question: How is it possible to determine the prevalence of a disease in a general population and be sure it is a valid number?

Dr. Galen: That is a question we are frequently asked relative to the predictive value model. As it turns out, if you start playing with these numbers, you realize that the *exact* prevalence is not that important. You are looking at a number such that if you are off by 10-fold you are okay, so what you want to know is the frequency of the disease. One per 100? One per 1000? Are you talking about a genetic thing like PKU, for which the frequency is 1 per 15 000, or neuroblastoma, which is 1 per 40 000? Or are you talking about diabetes, incidence 2 or 3%, where a ballpark figure is all you need to enter into that little fourfold table and evaluate a laboratory test? What we tend to do is to give the test the benefit of the doubt; we go ahead and ask ourselves the most probably frequency of disease we could imagine in my community for disease X, and we add to it by a factor of 10. For instance, if I am worried about using the carcinoembryonic antigen test, and they want to do it on admissions or patients' visits to gastrointestinal clinic, I ask myself if I could possibly imagine the frequency of cancer of the bowel being more than 1%. If the answer is no, I will use 1% and then apply the sensitivity and specificity of the assay, and if the predictive value is still 3 or 4% I am satisfied not to probe any further; I am satisfied that the test is useless.

If I do that analysis and get an intriguing predictive value, then I start to dig deeper. Public health departments frequently have data; if you have a hospital-discharge computer-summary system, you can get your disease prevalence figures on your own population, working backwards from your discharge diagnoses. But the exact prevalence is a lot less important than it is to appreciate the *effect* of prevalence in the model, so that if you are off by a factor of 10, it doesn't matter. Data on sensitivity and specificity, by the way, are available in the literature. If you flip through clinical chemistry journals, they are there. It may not be spelled out that this a new assay for a disease and this is the sensitivity and specificity. Instead, a paper will tell you somebody looked at 100 cases and 90 were positive, and at 1000 healthy controls and 90 were negative. The numbers are there.

You then apply the prevalence in your own locale. That has to be really a rough estimate, from which you can calculate the predictive value and make a decision very easily.

Dr. Rall: This rather depends on why you want to know the prevalence. One might want to know it for a whole variety of different reasons, and it can turn out to be a very complicated thing to find out the prevalence of a particular disease, because it is very hard to get an unbiased or unweighted sample, and normal values, for example, frequently are chosen from those for either hospital personnel or people with apparently unrelated diseases, and can be rather badly skewed. Or incidence or prevalence figures will be chosen from convenient but unrepresentative groups, such as union members or people coming to clinics, and these may give really quite misleading figures.

Such data frequently are exploited by health lobbyists. They go to Congress and say the incidence or prevalence (Congressmen don't know the difference) of disease X is enormous, and show them the results for a particular population.

To find the real prevalence of some abnormality, you first have to define the group for which you are going to measure, then you have to go to considerable epidemiological pains to make sure that you really are not getting some terribly biased sample.

Dr. Galen: I agree with that entirely. There is a point that I would like to stress about the application of a test. Suppose someone comes to me and says, "Everybody is screaming for this new test because it was in *Reader's Digest,* let's set it up," and I have five manufacturers already knocking on the door with kits, and so I know I could set it up. The question is: Will I encourage my clinicians to use it, and under what circumstances? Should it be used as an addition to an admission profile test? Should we use it only in symptomatic patients? Should we use it only in followup patients with no disease? In that sense the relative prevalence is very important, regardless of the absolute number per hundred thousand, defined epidemiologically. It is important to know that once a patient is triaged into the coronary care unit, the frequency of myocardial infarction is about 50%, or maybe it is 30 to 50%, but the frequency of myocardial infarction of patients with chest pain in the emergency rooms, before they have been seen by anyone, is much less than 1 or 2%. One can deal with those relative frequencies, look at a certain assay and decide not to do it on admission or not to do it at all, or whatever.

So I think the relative percentage is very useful information for the laboratory in deciding the environment in which a test would be used, but I agree with you entirely as to how virtually impossible it is to ascertain the true prevalence of any disease.

Dr. Young: The next question is: How general do you intend your statement to be that unsolicited tests are confusing? Should not specialists who recognize a potentially serious state contribute test information about that state, whether solicited or not, because we are concerned with total patient care.

Dr. Sackett: I think that goes back to the question we just discussed earlier. Is one carrying out a diagnostic process or is one doing "case finding," as clinical epidemiologists would call it? If the latter, I think additional laboratory data, and history, and physical maneuvers can be justified individually on their own merits.

Dr. Young: Dr. Galen, what do you propose as a practical approach to developing new tests?

Dr. Galen: I would propose looking at the four-layered model and ultimately looking at the practice of medicine and identifying areas where the benefit/cost ratio of treating a patient is very low. That is to say, [areas where] we need to be very certain of the diagnosis before going ahead with treatment. If we looked at everything that we do today in the hospital, we could find certain diagnostic and therapeutic situations where the physician must draw upon all the ancillary diagnostic services available, until he is certain, or certain enough—which is referred to as the "threshold probability"—that he is willing to incur the risks of treatment over the risks of no treatment. Working back in those clinical areas, we can then really pinpoint the tests that need to be developed. Say we want a test for pulmonary embolism, we want a test for tendency to clot, we want a test for X, Y, and Z. Then, working the numbers and the levels back, one can demonstrate for the clinical pathologists and clinical chemists that we need a test for a given situation, characterize what type of test it should be and just how precisely it needs to be performed, and whether it should be qualitative or quantitative. But it really relates to the benefit/cost ratio; even though a lot of the numbers look fancy, and there are all sorts of curves, it turns out that most good physicians are intuitively following the principles and the lines of medical decision-making. They *know* that when the benefit/cost ratio is high, they do not need to have a high degree of certainty before proceeding to treat.

This is why most of the treatments of infections have begun long before we get reports back from the bacteriology laboratory. Before the results of sensitivity testing are reported back, the physician has already chosen his antibiotic, begun therapy, because there is very little in the way of risk for him in that situation.

On the other hand, if the physician is going to treat a fungal infection with use of a very potent antimicrobial, he is going to wait for a high degree of evidence and confidence in his diagnosis because of the risk involved in the treatment. So I think doctors intuitively know when they are uncomfortable about beginning treatment, and they get x-rays, EKGs, and the laboratory data, and gather a lot of information. That raises their probability of making a correct diagnosis above the threshold probability; in effect, the clinician can then show it would be better to treat rather than not to.

It is important for us in the laboratory to recognize that ultimately we are trying to support the physician's decision—the decision he must make and must live with—and that he himself needs a certain amount of information before he can feel comfortable with that decision. There are areas in medicine where doctors are very uncomfortable about making decisions as to treatment, and that is because we don't have good tests for those areas.

Dr. Rall: I think what you say is logical and that you undoubtedly are right. I would only point to the unexpected and ad hoc nature of most discoveries. From a practical standpoint, for example, the appreciation of the potential sensitivity of radioimmunoassay and its development has sort of permeated tests, not only in the clinical laboratory, but certainly in the research laboratory, and in a perfectly extraordinary way. My point is that frequently you are going to find something else than what you are looking for.

Dr. Galen: I think to some extent we want to separate methodological approaches from analytes. The spin-off of stumbling across a new way to measure anything, something old or something new, is different from saying that what are we not now testing for we should be testing for, and then let the laboratory and everybody else worry about the methodology, or come back to us and say "We understand what you are saying, we have no way of measuring the analyte you have told us you need to have measured," and leave it there. Rather, I think we have got to communicate with the laboratory and tell them where we need more information, as opposed to areas where the physician is very secure in making diagnoses, where the benefits of treatment so outweigh the costs that he doesn't even need the laboratory test to begin with in order to proceed with treatment.

We are still chugging along working on new glucose methods. There comes to a point where if you recognize how the test fits into the medical framework, you don't *need* another method and you don't *need* to be more precise. I think that to some extent we in the laboratory have really gone astray relative to how the numbers are finally used by physicians. Anyone who has ever been involved in emergency situations where a laboratory has to streamline itself will know how few tests one actually can do and still not compromise patient care. Those that have shared that experience, coming from our environment of high testing, are amazed at how well medical care can be delivered with very few laboratory tests, and what those few critical tests are that really make a difference. That is because there really is such a thing as clinical judgment, and it tends to work very well most of the time.

Dr. Young: I just happen to have a slide showing the traditional evolution of a new laboratory test. It starts with a patient-care physician who has a need. He develops it himself until it finally is a more or less useful working test. The laboratory's job is then to adapt it from the type of test that can be done one at a time to a factory basis, so that instead of doing it once you can equally well do it a hundred times.

Dr. Weed: I think that probably one of the most important areas now for the clinical pathologist is to start looking critically at the tests they *already* have.

In our place, if some doctors come to me or go to the pathologist and demand certain things, by knowing that individual's performance and his past habits, you will interpret that demand in one way; another you will interpret in a completely different way. Well, how do you know what they think and what their habits are? You look at what they are already doing.

For instance, I go to a meeting and I hear a clinical pathologist give a very convincing paper on why you shouldn't do sputum examinations; it is a waste of time. Then I go into his own institution and find that every other request coming to the laboratory is for a sputum examination. Now, you would say, "What is he talking about new things for? They have no control over the present." Such new things I have often looked upon as diversionary tactics. Somebody wants to talk about something else, not because he is interested, but because he wants to get you off the subject you are already on.

Now, there is no reward in the system for cleaning up the present mess. For instance, suppose you receive a lot of sputum-culture requests and you think they should be stopped.

You are not going to write a paper and get the Nobel Prize for getting 100% agreement on what you believe and what goes on in your own institution. No one would publish it and you are not going to get a prize for it.

Furthermore, if you started to practice that, the x-rays and the laboratory tests would be halved and then the hospital administrator would go beserk, because that is where the hospital makes its money, so that we have a negative feedback here. But I think that is where all the progress is going to be in the next decade.

You are like the farmer in Vermont who said to the Department of Agriculture, "Don't bring me any more brochures; I am not farming half as well as I know how to right now."

Dr. Young: Let me put a couple of questions once more to Dr. Weed. How does your program accommodate for the variation in normal values from laboratory to laboratory because of use of different methodologies? And what implications does the problem-oriented approach have for clinical chemists and their needs for uniformity of methods and coefficients of variation across the country? In brief, how do you accommodate the differences from place to place?

Dr. Weed: First of all, we visualize that if there were a National Library of Displays, they would have what we call a generic set, and we would all agree that on coronary cases we would follow this enzyme. Then in given localities there would be differences; for instance, they might have a different bill for the drug or test, or they don't look for histoplasmosis in this laboratory as they look for it some place else. There would be the instant set of displays, so you would have the generic set, which would then be modified to fit your situation.

Now, some people think of those programs discussed this morning as inflexible and difficult, but they can be changed by a secretary in seconds. That is the revolution in information technology. You sit in front of the terminal, and you hit a thing at the bottom called "frame editor," and you sit at the typewriter and you can just change it, and it can be re-edited throughout the system—so there is no problem there. These are questions for a National Library to start dealing with. People can always type in alterations so that they have total freedom, but never again freedom without accountability. Do anything you want, but you will have to explain it to somebody.

Now, concerning the remarks that were made at the beginning about costs. Aren't they so great that it would be beyond our present situation economically? One of the reasons we are now in such economic trouble medically is because there is no synergism in the system. We have an uncoupled system without synergism. You are doing the same things over and over again.

For instance, they think they have Legionnaire's Disease in Ohio. Because they don't have a good record system and they aren't computerized, it isn't coupled to the right authorities, they have to send a group from Atlanta, put them up in hotels, to ask their own questions. Then another group has to come from here. Then they hire an infectious disease expert. And so on. We are just stumbling over ourselves, so I would say we can't afford *not* to automate, and then people will say, "But these things cost so much!" Read the latest issue of *Scientific American* on what is happening to technology. We can store and move information at a rate you wouldn't believe, and you know as well as I do that if we once say we have to do it we ultimately will. When airlines decided to automate did they live through a year of hell? Yes. Did they consider stopping? No, because they knew they *had* to continue.

If we once decide philosophically that we must become coupled, that there must be synergism, we will start, not go around complaining about how much the terminal will cost. According to David Sarnoff, the first television set cost two hundred million dollars to develop. Now they cost two hundred. The physicists understand this and they get a cyclotron for a million, but in medicine we swither around. If philosophically it is needed then *start,* get working, make it work, two years, 10 years, do whatever we need, have a little bit of driving spirit, and it really works. In 10 years, the Japanese will have it miniaturized in a wrist watch. You know it.

Dr. Young: At this point, we must end, even though there are many other questions, both unasked and unanswered.

II

SCREENING FOR DISEASE

Cost Effectiveness of Multiphase Health Testing

Morris F. Collen, M.D.

Medical Methods Research, Kaiser Foundation Research Institute; and The Permanente Medical Group, 3700 Broadway, Oakland, Calif. 94611

A cost analysis of multiphasic health testing is presented with data on the cost per positive test for each of its clinical laboratory tests, showing the effects of age, sex and prevalence of the abnormality in the population being tested. A cost effectiveness study of health checkups by the systemized multiphasic mode versus the traditional approach showed that the former decreased clinical laboratory costs by 45 percent.

Screening for disease is an important component of preventive medicine; and in primary health care, screening for disease is generally provided to patients by means of health checkups. In contrast to a medical diagnostic workup for specific patient problems or complaints, the goal of such a health checkup is to provide health care and to detect early asymptomatic disease. Thus a medical workup is primarily problem- and symptom-oriented, whereas a health checkup is essentially examination- and test-oriented. The selection of tests is critical for each group of health examinees.

The traditional method for a patient to obtain a health checkup is to see a primary-care physician, who takes a medical history from the patient, provides a physical examination, and then arranges for those diagnostic tests and procedures that the physician judges are necessary to complete the health evaluation. The physician then makes a determination as to whether the patient is well or sick and recommends appropriate followup care.

There has evolved over the past 25 years a systematized approach to providing the laboratory testing portion of a health checkup called "multiphasic health testing"; it involves automated laboratory procedures and specially trained allied-health personnel to collect data on patients' medical histories, clinical laboratory tests, x-ray procedures, and other physiological measurements in a programmed sequence *(1–3)*. A "multiphasic health checkup" is a health examination provided by using multiphasic health testing followed by a physical examination and a physician's decision if the patient is well or sick, with recommendations for appropriate followup health counseling or sick care, or both.

Here, I will present a cost-effectiveness evaluation of automated multiphasic health testing, with emphasis on its clinical laboratory test phases.

The evaluation of the costs and effectiveness of automated laboratory testing is not simple. A government-sponsored committee organized to develop guidelines for automated multiphasic testing defined "cost analysis" as the determination of costs within an operational system, which may be presented as total cost of operation, cost per test, cost per positive test, etc. *(4)*. They defined "cost effectiveness analysis" as a comparison of the economic efficiency of alternative systems directed at the same objective. This latter form of analysis determines whether or not the output of one system requires fewer resources than another to accomplish what is desired, or, alternatively, whether the same dedication of resources to one system or another can provide greater output.

Cost Analysis Per Positive Test

Table 1 presents the costs incurred to identify a clinically important positive test in the clinical laboratory component of a multiphasic testing program, and indicates which tests would be most cost effective for a specified population. These data are based upon a previously reported analysis of 44 663 examinations performed in 1968 *(5, 6)*. The costs listed in Table 1 include *all* costs, direct (personnel, fringe benefits, amortized equipment, and supplies) and indirect (overhead, plant maintenance, regional administration, etc.).

The total costs per patient for all clinical laboratory tests on blood and urine were $4.49 (including a blood-chemistry panel of eight tests) in 1968 and $6.26 (including a blood chemistry panel of 12 tests) in 1973. During this five-year period, although total costs have increased—primarily because of higher personnel costs—the unit cost per patient has been kept relatively constant by gradually increasing the volume of patients processed from about 100 to 180 per day.

The frequency of positive tests depends on the normal limits established for each test. The methodology for setting screening threshold values for the positive tests herein reported has been described elsewhere and will be referred to below for each test *(7–10)*.

Hemoglobin and Leukocyte Count

These two blood tests were measured by automated equipment at a cost of $0.42 per test. Table 1 shows that 3.1% of all men had a positive test (hemoglobin <130 g/liter), at a cost of $13.55 per positive case, and 10.3% of all women had a positive test (<120 g/liter), at a cost of $4.10 per positive case. Only 1.9% of young men had a low value for hemoglobin, in contrast to

Table 1. **Cost per Positive Laboratory Test, According to Test and Age Group**[a]

Test		*All ages*		*<40 y*		*40–59 y*		*≥60 y*	
	Unit cost ($)	*% Positive*	*Cost/ positive case ($)*	*% Positive*	*Cost/ positive case ($)*	*% Positive*	*Cost/ positive case ($)*	*% Positive*	*Cost/ positive case ($)*
Hemoglobin (men)	0.42	3.1	13.55	1.9	22.10	3.1	13.55	5.6	7.50
Hemoglobin (women)	0.42	10.3	4.10	2.6	3.30	10.1	4.10	5.5	7.60
White-cell count	0.42	2.2	19.10	2.6	16.15	2.2	19.10	1.7	24.70
Serum glucose (1-h)	0.75	5.7	13.15	4.6	16.30	5.6	13.40	8.3	9.05
Serum cholesterol	0.29	2.4	12.15	1.7	17.05	2.7	10.75	3.0	9.70
Serum uric acid	0.29	4.5	6.40	3.4	8.50	4.8	6.05	6.0	4.80
Serum albumin	0.29	0.3	96.70	0.4	72.50	0.3	96.70	0.4	72.50
Serum total protein	0.29	4.0	7.25	3.5	8.30	4.4	6.60	3.9	7.45
Serum calcium	0.29	1.3	22.30	1.4	26.70	1.3	23.30	1.5	19.30
Serum creatinine	0.29	1.4	20.70	0.8	36.25	1.4	20.70	2.7	10.70
Serum aspartate aminotransferase	0.29	4.2	6.90	3.6	8.05	4.6	6.30	4.5	6.45
VDRL	0.16	1.5	10.65	0.8	20.00	1.9	8.40	2.3	6.45
Urine culture (men)	0.20	0.4	50.00	0.1	200.00	0.3	66.70	1.0	20.00
Urine culture (women)	0.20	3.3	6.05	3.3	6.05	3.2	6.25	4.0	5.00
Urine glucose	0.18	8.2	2.20	7.6	2.40	8.9	2.00	7.3	2.50
Urine protein	0.18	6.4	2.80	7.2	2.50	5.6	3.20	7.0	2.60

[a] Modified from Collen et al. *(6)*

12.6% of young women. Over the age of 60, both men and women had a similar prevalence (5.5 and 5.6%) of low hemoglobin values (at a cost of $7.50 and $7.60 per positive case). In 5.1% of examinees an abnormally high hemoglobin value was found (that is, >150 g/liter in women and >167 g/liter in men).

In 2.2% of all examinees, a leukocyte count of less than 4.0×10^6 liter or greater than 12.0×10^6 liter was found, at a cost per positive case of $19.10. Counts of less than 4.0×10^6 liter were reported in 0.7% of all adults, and greater than 12.0×10^6/liter in 1.5%.

Serum Chemical Assays

Eight chemical determinations were performed on serum by an eight-channel AutoAnalyzer (Technicon Instruments Corp., Tarrytown, N.Y.) at a cost of $2.78 (including a glucose challenge). Serum glucose determination, including the 75-g glucose challenge dose, cost $0.75 per person; the remaining cost was distributed equally so as to arrive at a unit cost of $0.29 for each of the other chemistry tests. Normal limits for each test had previously been established for this population group according to age and sex.

Serum glucose was tested 1 h after a 75-g glucose load was given orally. The upper normal limit for men and women was similar, but directly related to age (up to 29 y, 205 mg/dl; 30-39, 215; 40–49, 225; ≥50, 235). In addition, a correction was made whenever a patient had eaten a meal within 3 h before ingesting the challenge dose of glucose *(10)*. In 5.7% of all examinees (at a cost per positive case of $13.15), the value at 1-h was above normal limits. In such positive cases, a 2-h serum glucose determination was also routinely performed, with a designated upper normal limit of 150 mg/dl (8.3 mmol/liter). The proportion of examinees with positive tests at both 1 and 2 h was 3.5%.

Serum cholesterol had age- and sex-related normal limits *(10)*. In 2.4% of all examinees an abnormally high value was found (at a unit cost per positive test of $12.15). The frequency of positive tests increased with age.

For serum uric acid the upper normal limits for women were 6.3 mg/dl (375 μmol/l) for under age 45 y and 7.2 for 45 y and over. For men, any value greater than 8.4 mg/dl (500 μmol/l) was considered to be a positive test *(10)*. An increased value for uric acid was reported in 4.5% of all examinees, at a unit cost of $6.40 per positive test.

Serum albumin was abnormally low (less than 28 g/liter in women, less than 30 g/liter in men 40 years of age and over, and less than 32 g/liter in men younger than 40) in only 0.3% of persons tested, at a unit cost per positive test of $96.70.

Serum total protein was reported as abnormally low or high (normal limits, 60 to 80 g/liter in women, 59 to 81 g/liter in men over the age of 40, and 60 to 83 g/liter in men under 40 years of age) in 4.0% of adults, at a cost per positive test of $7.25.

The serum calcium concentration was considered to be abnormally high when >105 mg/liter (2.6 mmol/liter) in women, and 108 mg/liter (2.7 mmol/liter) for men *(10)*. In 1.3% of adults, these values were exceeded. The cost per positive test was then $22.30.

Serum creatinine was considered elevated [over 13 mg/liter (115 μmol/liter) in women and over 16 mg/liter (1.40 μmol/liter) in men] in 1.4% of all persons, at a cost of $20.70 per positive case. The prevalence of positive tests was only 0.8% in young adults, but increased with age to 2.7% in the oldest age group. The cost per positive case was $10.70.

Serum aspartate aminotransferase (SGOT; EC 2.6.1.1) was found to be above normal (that is, 41 U/liter) in 4.2% of persons, at a cost of $6.90 per positive case.

Other Tests

The Venereal Disease Research Laboratories (VDRL) test for syphilis was performed at a unit cost of $0.16. A positive test was reported in 1.5% of all examinees, at a cost per positive test of $10.65.

Urinary sugar and protein. Urinalysis for sugar (1 h after a glucose load) and protein was performed by means of enzyme "dip-stick" paper strips, at a cost of $0.18 each test. Positive tests for urinary sugar were reported in 8.2% (at a unit cost of $2.20 per positive) and for urinary protein in 6.4% of adults (at a unit cost of $2.80 per positive test).

Urine culture for bacilluria. Mid-stream "clean-catch" urine specimens were collected and incubated at 37 °C for 6 h in triphenyltetrazolium chloride (TTC) at a cost of $0.20 per test.

The development of a pink color in 6 h was classified as a positive test, and corresponded to 100 000 or more bacilli per cubic millimeter by colony count. A positive test (for both TTC and colony count of 100 000 or more) was reported in 3.3% of women (at a cost of $6.05 per positive case). In men, the prevalence of bacilluria was very much less, and only in men over the age of 60 did the frequency of positive tests reach 1%. Obviously, urine culture is not a cost-effective screening test for men.

It must be emphasized that the unit costs here reported were critically related to a multiphasic laboratory load of about 500 patients per week. If only 250 persons were screened weekly, the cost per patient would probably double. If 750 persons could be screened per week, the unit cost would probably decrease by a third.

It is also important to emphasize that evaluation of the cost effectiveness of this method of case finding requires an extension of this approach to determine the cost per proved "true" positive case, which requires expensive followup confirmatory and validating procedures to determine both the sensitivity and the specificity of each test. For example, it was reported that the cost per positive serum glucose test is about $13; since only about half of these cases are confirmed by their physicians to have diabetes mellitus, the cost per "true" positive for diabetes is about $25. (The cost of "false" positives is referred to below in the *Discussion*.)

Cost Effectiveness of Alternative Testing Methods

Since 1950, the Kaiser-Permanente Oakland Medical Center has provided a "multiphasic health checkup," which consists of a systemized battery of tests (now with a panel of 12 blood chemistry tests) and a self-administered history (automated multiphasic health testing), followed up by a 15-min scheduled visit for a physical examination by a physician in the medical department. Also available, is a "traditional" health evaluation provided by the same medical department physician, who, during a 30-min scheduled visit, takes a history and physical examination and arranges for supplemental diagnostic tests and procedures as indicated. After the physician sees the patient, in either the traditional or multiphasic initial examination visit, he may refer the patient to appropriate specialty clinics and for followup clinical laboratory tests, x-rays, electrocardiograms, or other special diagnostic procedures as necessary to "rule-in" or "rule-out" a diagnosis.

A study was conducted in 1972 comparing similarly selected patients receiving traditional and multiphasic health checkups; their health status was determined by chart review and they were classified as well, worried-well, asymptomatic-sick, or sick; and the following data then adjusted so as to be comparable by age, sex, and health status *(11–16)*. The same physicians provided the examinations and arranged followup care for both groups, and so the quality of care was assumed to be similar.

Table 2 shows the use and cost of services for the initial and followup examination visit by traditional medical care (TMC) as compared to the multiphasic health testing (MHT) mode. Multiphasic health testing replaced the individually selected tests which were ordered in the traditional mode.

The costs shown are costs to the Health Plan for the services provided to the Plan members, and do not represent fees or charges which would have been paid by non-member patients. For example, MHT cost to Health Plan for a member was $17.46 (of which the clinical laboratory tests cost $6.26), but the charge to a non-member patient would have been $30–$40 for an MHT.

Table 2 also shows the followup-visit costs initiated by the physicians to complete the health checkups. Most patients did

Table 2. **Comparative Cost of Clinical Laboratory Services (Initial and Followup) for a Health Checkup (Adjusted for Age, Sex, and Health Status)**[a]

	Initial Examination		*Examination Followup*		*Total Costs*	
	TMC[b]	*MHT*[c]	*TMC*	*MHT*	*TMC*	*MHT*
Physicians	$28.52	$14.75	$10.82	$ 7.45	$39.34	$22.20
MHT-Total	0	17.46	—	—	0	17.46
(Lab)		(6.26)[d]				(6.26)
Clinical lab.	10.71	0	5.21	2.45	15.92	2.45
X-ray	3.60	0	1.27	1.57	4.87	1.57
ECG, etc.	0.71	0	0.58	1.12	1.29	1.12
Total	$43.54	$32.21	$17.88	$12.59	$61.42	$44.80

[a] Modified from Collen et al. *(13)*
[b] TMC = 2040 persons
[c] MHT = 1916 persons
[d] Includes prorata share of registration, supervision, and overhead

not have their health evaluation fully completed at the initial examination visit, because the detection of possible variations from normal required further diagnostic tests (clinical laboratory, radiology, electrocardiogram, etc.) and/or physician specialist consultation visits (internal medicine, ophthalmology, gynecology, dermatology, etc.) to confirm the validity of the finding or to "rule-out" or "rule-in" diagnoses being considered. The final report of the health evaluation was provided to the patient at a followup return visit; or if the entire examination was essentially "negative"—i.e., there were no clinically significant abnormalities to report to a well patient—this might be handled by a telephone call or a letter.

MHT patients used less physician time than did patients who took the traditional examination. The costs for ancillary services (clinical laboratory, radiology, electrocardiogram, and other diagnostic procedures) used for the followup-evaluation workups are also shown, and the impact of the more comprehensive initial clinical laboratory testing of MHT is clearly evident ($2.45 for the MHT group vs. $5.21 for the traditional group).

Table 2 also compares the total costs for providing health checkups by the two methods tested. The total cost for an evaluation is the sum of the resources used on the initial examination visit and on the followup-evaluation visits to complete the health evaluation of the patient.

The data show that the average cost for a health evaluation by the traditional (TMC) physician examination and followup method was $61.42. As an alternative, by first providing a multiphasic health testing (MHT) battery of tests, followed by a physician examination, the total costs for a health evaluation were decreased to $44.80 for an average savings of $16.62 per examination, or a decrease in total costs of 27%. The total costs for clinical laboratory tests by the traditional mode were $15.92, as compared to $8.71 ($6.26 + $2.45) by the MHT mode, or a decrease of 45%.

Discussion and Conclusion

The cost per positive test depends upon the yield rate of positive findings, which in turn depends upon the sensitivity and specificity of the test and the prevalence of the abnormality in the population being tested. The cost effectiveness of health checkups is also impacted by the costs of false-positive and false-negative tests, which have been discussed in detail elsewhere *(15)*. Multiphasic testing has often been alleged to increase costs of care due to generation of excess false-positive tests. Table 2 refutes this allegation. In fact, the decreased costs of care after an initial multiphasic health checkup, as compared to a traditional checkup, persist for at least 12 months *(16)*. On the other hand, increasing concern for the costs of false negatives, which are one factor in generating some medical malpractice liability, raises the question as to whether the higher followup costs (Table 2) of the traditional checkup are possibly produced by false-negative tests.

Our conclusion is that if health checkups are provided to a large population, the multiphasic health testing approach is the most cost effective from the viewpoint of the clinical laboratory and the total health care program.

Data referred to in this paper were from studies supported in part by grants from the U.S. Dept. of HEW (CH 05–8, DC 00142, HSM 110–70–407, HS 00288) and the Kaiser Foundation Research Institute.

The new medical-care delivery system project was conceived by Sidney R. Garfield, M.D., and the associated data were analyzed by James H. Duncan, B.S.I.E. Robert Feldman, M.D., supervised the Oakland multiphasic program. P. H. Kidd prepared the unit cost data for the cost per positive test study. John Mosher, C.P.A., prepared the unit cost data for the cost effectiveness study; Gary Friedman, M.D., participated in its study design; and A. B. Siegelaub, M.S., performed all the age–sex adjustments to the data.

References

1. Collen, M. F., Periodic health examinations using an automated multi-test laboratory. *J. Am. Med. Assoc.* **195,** 830 (1966).
2. Collen, M. F., The multitest laboratory in health care in the future. *Hospitals* **41,** 119, (1967).
3. Collen, M. F., Ed., *Multiphasic Health Testing Services.* John Wiley & Sons, New York, N.Y., 1977.
4. Zucker, L. W., et al., Cost and cost analysis guidelines, part IV, *Provisional Guidelines for Automated Multiphasic Health Testing Services*, **3,** DHEW Publ. No. (HSM) 72–3011, 1970. U.S. Govt. Printing Office, Washington, D.C.
5. Collen, M. F., Kidd, P. H., Feldman, R. F., and Cutler, J. L., Cost analysis of a multiphasic screening program. *N. Engl. J. Med.* **280,** 1043 (1969).

6. Collen, M. F., Feldman, R., Siegelaub, A. B., and Crawford, D., Dollar cost per positive test for automated multiphasic screening. *N. Engl. J. Med.* **283,** 459 (1970).
7. Cutler, J. L., Collen, M. F., Siegelaub, A. B., and Feldman, R., Determination of normal values for blood chemistry tests. *Bull. Coll. Am. Pathol.*, 319–326 (October 1969).
8. Collen, M. F., Siegelaub, A. B., Cutler, J. L., and Goldberg, R., Aspects of normal values in medicine. *Ann. N.Y. Acad. Sci.* **161,** 572 (1969).
9. Friedman, G. D., and Collen, M. F., Derivation of normal values. Chap. 19 in *Multiphasic Health Testing Services,* M. F. Collen, Ed., John Wiley and Sons, New York, N.Y., 1977.
10. Bassis, M. L., and Collen, M. F., Normal chemistry values in an automated multiphasic screening program. In *Automation in Analytical Chemistry,* Technicon Symposia 1966, **1,** N. B. Scova, et al., Eds., Mediad Inc., White Plains, N.Y., 1967, p. 309.
11. Collen, M. F., and Garfield, S. R., *New Medical Care Delivery System, Final Report,* January 1975, P.B. 246 630, National Technical Information Service (NTIS), Springfield, Va.
12. Collen, M. F., *A Case Study of Multiphasic Health Testing.* Committee on Technology and Health Care, U.S. Natl. Acad. Sciences, Washington, D.C., 1977.
13. Collen, M. F., Garfield, S. R., Richart, R. H., Duncan, J. H., and Feldman, R., Cost analyses of alternative health examination modes. *Arch. Int. Med.* **137,** 73 (1977).
14. Collen, M. F., Cost analysis of the Kaiser Foundation's systemized health evaluation, pp. 707–714. In *Preventive Medicine USA.* PRODIST, New York, N.Y., 1976.
15. Collen, M. F., Systems evaluation of MHTS. Chap. 17 in *Multiphasic Health Testing Services.* John Wiley & Sons, New York, N.Y., 1977.
16. Garfield, S. R., Collen, M. F., Richart, R. H., Feldman, R., Soghikian, K., and Duncan, J. H., Evaluation of a new ambulatory medical care delivery system. *N. Engl. J. Med.* **294,** 426 (1976).

Cost Effectiveness of Laboratory Testing in Genetic Diseases

Harvey L. Levy, M.D.,* Ann J. E. Bennett, ** and Elizabeth D. K. McDevitt***

**Department of Neurology, Harvard Medical School; Metabolic Disorders Screening Program, State Laboratory Institute; Massachusetts Department of Public Health; and Joseph P. Kennedy, Jr. Laboratories at the Massachusetts General Hospital, Boston, Mass. 02114; and **New England Newborn Screening Program for Hypothyroidism and Metabolic Disorders, State Laboratory Institute, Massachusetts Department of Public Health, Jamaica Plain, Mass. 02130; and ***Metabolic Disorders Screening Program, State Laboratory Institute, Massachusetts Department of Public Health, Jamaica Plain, Mass. 02130*

This discussion of the cost effectivness of laboratory testing for genetic disease will be limited to that which is routinely performed on specimens from newborns. It is this type of testing that involves the greatest number of individuals and families, arouses the greatest controversy, and has the greatest capacity for good as well as for evil. This type of screening has also been generally referred to as the prototype for genetic testing, by advocates as well as opponents of this form of activity. Testing of individuals who present with clinical signs of genetic disease has a far more limited cost effectiveness, because such signs often indicate irreversible damage, though this form of testing may be invaluable for purposes of proper diagnosis, genetic counseling, and research. Laboratory testing for chromosomal abnormalities as well as that done for prenatal diagnosis has been covered elsewhere and will not be considered here, inasmuch as we have not had extensive involvement with these areas.

System of Routine Neonatal Screening

In most of the United States and in some other countries, screening for phenylketonuria (PKU) is routinely performed on virtually all newborn infants. A blood specimen, obtained from the heel of the infant, is impregnated into a filter paper card (the "Guthrie" specimen) and mailed to a central (usually public) laboratory, where it is tested for phenylalanine, increased phenylalanine concentration being a marker for PKU. The test for phenylalanine is usually a bacterial inhibition assay known as the "Guthrie" test (Figure 1), though it may also be a fluorometric assay that utilizes the automated analyzer principle *(1, 2)*. As of 1973, over 15 million newborn infants had received this test; perhaps as many as 20 million newborns have by now been so tested. Over 1500 infants have been discovered with PKU, a prevalence of approximately 1:13 000.

In several screening programs other genetic diseases of metabolic type are also sought in the same blood specimen that is used for PKU screening. These disorders include galac-

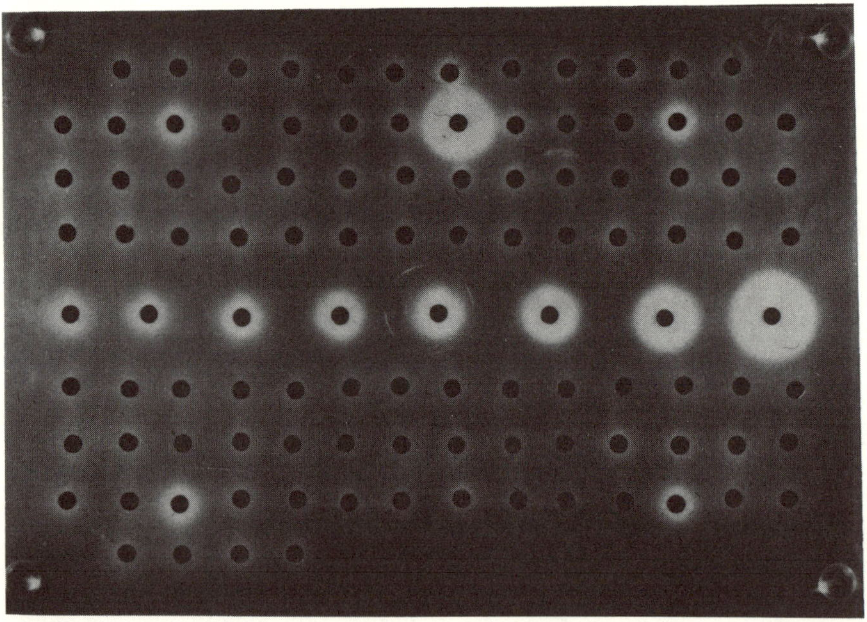

Fig. 1. "Guthrie" plate for phenylalanine.

The row of discs in the center are "control" specimens containing concentrations of phenylalanine ranging from 20 mg/liter *(far left)* to 500 mg/liter *(far right)*. The bacterial growth surrounding the disc in the middle of the second row from the top indicates a phenylalanine concentration in the specimen exceeding 200 mg/liter. This specimen was from an infant with phenylketonuria.

tosemia, maple-syrup urine disease, and the cystathionine β-synthase (4.2.1.22) deficient form of homocystinuria *(2)*. Most recently, screening for neonatal hypothyroidism has been added to several programs *(3)*. While this latter group of disorders is not strictly a "genetic disease," there are inherited forms of neonatal hypothyroidism and it is often included within the concept of "genetic screening." Aside from neonatal hypothyroidism, which has a relatively high frequency of 1:5000, these other disorders have frequencies that are far lower than that of PKU, ranging from 1:75 000 for galactosemia to about 1:200 000 for maple-syrup urine disease and homocystinuria.

In Massachusetts, one of the most comprehensive newborn screening programs currently in existence has been developed. Three separate filter-paper specimens are obtained from each neonate (Table 1). The specimen of blood from the umbilical cord is obtained in the delivery room, the blood specimen from the newborn is obtained at the time of nursery discharge or on the fourth day of life, whichever is the earlier, and the urine specimen is obtained by the parents when the infant is three to four weeks of age. These specimens are mailed to the State Laboratory Institute of the Massachusetts Department of Public Health, where the cord-blood specimen and the newborn-blood specimen are tested by several different bacterial or other assays and the urine specimen is tested by paper-chromatographic methods

Table 1. **Massachusetts Metabolic Disorders Screening Program**

birth-cord blood	Galactosemia
	Maternal PKU
	Maternal histidinemia
	Maternal homocystinuria
2–4 days-newborn blood	PKU
	Hypothyroidism (cretinism)
	MSUD
	Homocystinuria
	Galactosemias
3–4 weeks-urine	Amino acid disorders
	Organic acid disorders

(4). The results of this program are listed in Table 2. As noted, many genetic disorders have been identified. Not all are of clinical importance, but many are associated with clinical complications unless dietary (or other) therapy is begun early in infancy. The most common and most important of these complications is brain damage, with resulting mental retardation. The screening of newborn blood now includes specimens from several other New England states to form a New England Regional Newborn Screening Program.

Table 2. **Metabolic Disorders and Their Estimated Frequency among Newborns in Massachusetts**

Disorder	*Total screened*	*Total detected*	*Estimated frequency*
Hypothyroidism	69 646	15	1:5 000
Phenylketonuria (classical)	1 295 292	89	1:14 000
Phenylketonuria (atypical)	1 295 292	80	1:16 000
Phenylketonuria & PMH (maternal)	424 928	15	1:30 000
Iminoglycinuria	577 303	50	?1:11 000[a]
Cystinuria	577 303	43	1:13 000
Hartnup disorder	577 303	32	1:18 000
Histidinemia	577 303	29	1:20 000
Histidinemia (atypical)	577 303	2	1:290 000
Histidinemia (maternal)	99 575	1	1:100 000
Galactosemia	820 069	11	1:75 000
Maple-syrup urine disease (MSUD)	1 155 935	4	1:290 000
MSUD (intermediate variant)	1 155 935	2	1:600 000
Argininosuccinicacidemia	577 303	8	1:70 000
Argininosuccinicacidemia (atypical)	577 303	1	<1:500 000
Cystathioninemia	577 303	9	1:65 000
Homocystinuria	863 546	3	1:290 000
Hyperglycinemia (non-ketotic)	577 303	3	1:190 000
Hyperprolinemia	577 303	2	1:290 000
Methylmalonicacidemia	153 508	2	1:75 000
Propionicacidemia	577 303	1	<1:500 000
Hyperlysinemia	577 303	2	1:290 000
Sarcosinemia	577 303	2	1:290 000
Hyperornithinemia	577 303	1	<1:500 000
Carnosinemia	577 303	1	<1:500 000
Urocanicaciduria	577 303	1	<1:500 000
Hyperglutamicaciduria	577 303	1	<1:500 000
Fanconi syndrome	577 303	1	<1:500 000
Rickets (vitamin D dependent)	577 303	1	<1:500 000

[a] The number of cases and incidence for iminoglycinuria may be falsely high because carriers for this disorder who only have hyperglycinuria later in life may have iminoglycinuria as infants.

Cost Effectiveness of Newborn Screening

Several years ago we began to study the cost effectiveness of the newborn screening program in Massachusetts *(5)*. As we were directly involved in the screening laboratory and its findings, we knew that the costs were considerable, because we employed 16 full-time individuals, needed material and equipment, and utilized much laboratory space. However, as pediatricians involved in the diagnosis and care of infants and children with genetic disease, we have also been aware of the expenses incurred in hospitalization, outpatient care, and institutionalization of those with mental retardation. Consequently, we have sought to compare the costs of screening with the costs of failure to screen newborn infants for metabolic disorders.

The traditional method of considering costs of laboratory testing has been to include only those expenses of the testing process per se. Where the expenses in collection and delivery of specimens are inconsequential or considered within other categories, these may be ignored. In newborn screening, however, these costs are as basic to the total procedure of testing as are the costs of maintaining the laboratory. Consequently, the initial costs of screening are those of specimen collection and mailing (Table 3). Since our 1972–73 study, collection costs have increased due to increases in salaries and increased costs of postage. The average annual collection and mailing costs for the years 1973–77 have been about $163 000.

The direct costs of laboratory testing currently constitute the major expenses in genetic screening. Table 4 lists these average annual expenses for 1973–77. Of the total $283 597, 76% is for

Table 3. **Average Annual Costs of Specimen Collection and Mailing for Years 1973–77**

Specimen		*Individual cost*	*No. of specimens*	*Total cost*
Umbilical cord blood		$1.05	64 861	$ 68 104
Newborn blood		1.05	67 026	70 377
Newborn urine		0.13	54 467	7 081
	Totals	$2.23	186 354	$145 562
Repeat blood		16.50	1 065	17 572
Repeat urine		.13	1 123	156
	Total		188 542	$163 280

personnel, 17% for laboratory overhead such as Institute administration and building maintenance, and the remaining 7% is for supplies, kits, equipment, telephone, and postage. These costs do not include those for hypothyroidism screening, which involves a program and laboratory entirely separate from that devoted to screening for purely genetic diseases.

Table 5 lists cost data for the laboratory testing of each specimen within the program. The testing of umbilical cord blood specimens is the least expensive, costing about $66 000 annually or $1.02 per specimen, whereas the testing of urine specimens is

Table 4. **Average Annual Costs of Laboratory Testing, 1973–77**

Item	*Cost*
Direct expenses	
Personnel	$172 146
Fringe benefits	43 036
Supplies	9 959
Kits	1 736
Equipment	2 100
Telephone	6 000
Postage	500
Indirect costs	48 120
Total	$283 597

Table 5. **Average Annual Laboratory Costs for Testing Each of the Specimens, 1973–77**

	Cost		
Item	*UM CD blood*[a]	*NB blood*[a]	*Urine*
Direct expenses			
Personnel	$37 322	$ 62 384	$ 72 440
Fringe benefits	9 330	15 596	18 110
Supplies	2 000	3 428	4 531
Kits	500	1 236	—[b]
Equipment	500	1 100	500
Telephone	500	2 750	2 750
Postage	50	225	225
Indirect costs	16 040	16 040	16 040
Total	$66 242	$102 759	$114 596
No. of specimens	64 861	67 026	54 467
Cost/specimen	$1.02	$1.53	$2.10

[a] UM CD = umbilical cord; NB = newborn

[b] The urine specimen is collected by the parent, using a card that was originally part of the newborn blood kit. Consequently, these costs have been assigned to newborn blood testing.

the most costly, averaging almost $115 000 annually, or $2.10 per specimen. The testing of the newborn-blood specimen, certainly the most valuable part of the program, costs approximately $103 000 annually, or $1.53 per specimen (exclusive of testing for hypothyroidism).

Weighed against these costs are the monetary benefits to be gained by routine screening for genetic diseases. This is a very difficult area to assess, because the benefits to be gained are in the prevention of clinical abnormality rather than in the correction or amelioration of an existing disease state. Hence, though this represents the highest form of medicine and the goal to which all medical care should strive, its benefits must be estimated and can only be realized in future years.

Table 6 lists the average number of infants with a genetic disease of definite clinical significance detected annually in the years 1973–77. Table 7 lists the annual expenses of caring for those with PKU. These expenses have not changed substantially from those of 1972–73, and total $70 187. The personnel and other basic clinic costs have remained essentially the same. The costs of the individual clinic visits and hospitalization have increased owing to inflation, but these have been offset by a decreased number of clinic visits and hospitalizations due to greater efficiency in the care of PKU. The costs of the diet, however, increased about 50% from the 1972–73 level while the number of infants and children under dietary therapy remained the same.

Table 8 lists the average annual costs of evaluation and treatment of diseases other than PKU. These expenses of $20 470 are about 40% less than those that we had estimated for 1972–73.

Table 6. **Average Annual Number of Individuals with Genetic Disease of Potential Clinical Significance Detected in 1973–77**

Disorder	*No. of infants*
Phenylketonuria (PKU)	7
Phenylketonuria (maternal)	1
Galactosemia	1
Argininosuccinicacidemia	1
Methylmalonicacidemia	1
Cystathioninemia	1
Other (maple-syrup urine disease, hyperglycinemia)	1
Total	13

The major difference is in the lesser expenses incurred for extended hospitalization of infants and children with such disorders as maple-sugar urine disease, propionicacidemia, and homocystinuria. In retrospect, the year 1972–73 was unusual in that such large costs for these services were necessary. In addition, improved outpatient care for disorders such as these has reduced the need for extended hospitalization.

Table 7. **Annual Costs of Evaluation and Treatment for Phenylketonuria (PKU), 1973–77**

Clinic:			
Personnel		$40 776	
Laboratory tests		2 000	
Clinic visits		5 321	
Hospitalization		2 000	
		Total	$50 097
Diet:			
7 infants 0–1 yr. of age @	$548	$3 836	
7 infants 1–2 yr. of age @	507	3 549	
7 children 2–3 yr. of age @	507	3 549	
7 children 3–4 yr. of age @	654	4 578	
7 children 4–5 yr. of age @	654	4 578	
		Total	$20 090
Total annual cost of care for children with PKU			$70 187

Table 8. **Annual Costs of Evaluation and Treatment for Disorders Other Than Phenylketonuria**

Disease	*No. identified*	*No. already under care*	*Initial care & diet (Age 0–1 yr)*	*Care & diet (Age <1 yr)*	*Totals*
Galactosemia	1	5	$1 800	$ 400	$ 2 200
"Classic" MSUD[a]	0	2	—	4 576	4 576
"Intermediate" MSUD	0	1	—	1 124	1 124
Argininosuccinicacidemia	1	7	200	1 428	1 628
Hyperglycinemia, nonketotic	0	2	—	306	306
Homocystinuria (vitamin B_6 resistant)	0	3	—	1 459	1 459
Homocystinuria (vitamin B_6 responsive)	0	1	—	112	112
Methylmalonicacidemia	1	1	6 272	1 186	7 458
Cystathioninemia	1	4	210	415	625
Fanconi syndrome	0	1	—	220	220
"Other"	1	1	506	256	762
Totals	5[b]	28	$8 988	$11 482	$20 470

[a] Maple syrup urine disease

[b] Does not include the cases of maternal PKU for which there was no evaluation or treatment expense.

Table 9 summarizes the costs incurred in newborn screening for metabolic disorders and compares these costs to estimated savings to be realized by the early detection and treatment of these diseases. The total costs of screening are $537 534. Expenses that may be averted include those of institutionalization of mentally retarded individuals. In Massachusetts, costs for such institutionalization have increased by about 10% per year so that the average annual cost per patient for the years 1973–77 was about $9723. In using the same basis for calculation as we did for the previous study, we consider that four of the seven infants with PKU and one of the infants with another disorder detected annually would subsequently be institutionalized for an average of 20 years each, yielding avoided expenses of $777 840 and $194 460, respectively. We feel that these are very conservative data, because inflation would almost certainly push costs of institutionalization to much higher levels in the future years when such individuals would require this type of care. Apart from institutional care, individuals with mental retardation and other clinical abnormalities resulting from genetic disease require diagnostic evaluations and treatment (e.g., for convulsions), much of which is averted by newborn screening. These averted costs have been estimated at $7200 for PKU and $10 800 for other disorders, again a conservative estimate when the inflationary rise in hospital and medical care expenses is considered. Hence, the total estimated averted annual costs for

Table 9. **Average Annual Benefit-Cost Ratio, 1973–77**

Item		*Cost*	
Cost of specimen collection & mailing		$163 280	
Cost of laboratory testing		283 597	
Cost of PKU care		70 187	
Cost of care for other disorders		20 470	
	Total		$537 534
Averted costs:			
Institutionalization			
PKU		777 840[a]	
Other disorders		194 460[a]	
	Total		$990 300
Estimated annual savings			$452 766
Benefit-cost ratio			1.8

[a] Based on average annual cost of $9723/year for each individual institutionalized in Massachusetts.

1973–77 are $990 300, or $452 766 more than the costs of screening. The positive benefit/cost ratio of 1.8 is virtually identical to that which we had formerly estimated for the year 1972–73 *(5)*.

Regionalization of Newborn Screening

For virtually any type of laboratory service, the costs per specimen are inversely related to the number of specimens tested in a single laboratory. The screening of newborn blood specimens for metabolic disorders is no exception. This is so because certain basic personnel, overhead, and equipment expenses are necessary regardless of the number of specimens tested. The nature of the testing techniques enables additional specimens to be added to an existing program with little or no additional costs in personnel. Additional costs are only in supplies, one of the least expensive components of the program. Furthermore, testing large numbers of specimens enables a laboratory to develop expertise in interpreting test results. These factors have led to the development of a regional testing laboratory in one state to which newborn blood specimens from several nearby states are sent. Such a laboratory was first developed in Oregon and includes the testing of newborn blood specimens from Alaska and Montana as well as from Oregon *(6)*. A similar program has now been developed in New England, where the Massachusetts laboratory tests newborn blood specimens from Maine and Rhode Island as well as from Massachusetts for hypothyroidism, PKU, galactosemia, homocystinuria, and maple-syrup urine disease and specimens from Connecticut and New Hampshire specifically for hypothyroidism. In considering the regional screening in New England exclusive of that for hypothyroidism, 100 000 newborn blood specimens are tested at little more cost than would be necessary to test the 67 026 specimens from Massachusetts alone. Consequently, the laboratory costs for testing each specimen have now been reduced from $1.53 to about 1.10. The relatively large equipment costs as well as the expertise necessary in screening for neonatal hypothyroidism make regionalization even more imperative for this type of program.

Conclusion

A previous analysis of cost effectiveness for newborn screening was criticized for not discounting to net present value the dollar amounts of averted costs for institutionalization, given the fact that costs for institutionalization would only be realized in future years *(7)*. In actuality, the rising costs of institutional care due to inflation would almost certainly negate such discounting. Moreover, we have not considered averted costs due to loss of individual productivity which, if considered, would substantially increase the benefit/cost ratio *(8)*.

There would seem to us little doubt that newborn screening for metabolic disorders when performed in large laboratories testing at least 50 000 specimens per year is well justifiable on a monetary basis alone.

References

1. Guthrie, R., and Susi, A., A simple phenylalanine method for detecting phenylketonuria in large populations of newborn infants. *Pediatrics* **32,** 338–343 (1963).
2. Levy, H. L., Genetic screening. In *Advances in Human Genetics*, **4,** Harry Harris and Kurt Hirschhorn, Eds., Plenum Press, New York, 1973, pp. 1–104.
3. Dussault, J. H., et al., Preliminary report on a mass screening program for neonatal hypothyroidism. *J. Pediatrics* **86,** 670–674 (1975).
4. Levy, H. L., Shih, V. E., and Madigan, P. M., Massachusetts metabolic disorders screening program. I. Technics and results of urine screening. *Pediatrics* **49,** 825–836 (1972).
5. Van Pelt, A., and Levy, H. L., Cost-benefit analysis of newborn screening for metabolic disorders. *N. Engl. J. Med.* **291,** 1414–1416 (1974).
6. Brandon, G. R., Regionalization of public health metabolic laboratories. *Public Hlth. Lab.* **34,** 56–60 (1976).
7. Gordon, J. B., Cost benefits of newborn screening. *N. Engl. J. Med.* **292,** 1246 (1975).
8. Levy, H. L., Cost benefit of newborn screening. *N. Engl. J. Med.* **292,** 1246 (1975).

PANEL DISCUSSION

Question: Dr. Selikoff, would you comment on the risk of lung cancer to non-smokers who work in offices where there are smokers? *(Ed. note: Dr. Selikoff's paper does not appear in this book.)*

Dr. Selikoff: I don't know the answer to this. I am urging Dr. Hammond to extend his epidemiological studies because that group of a million people includes couples in which one spouse smokes and the other does not. I am asking him to look at the risks separately for women who are non-smokers married to smoking husbands as compared to women who don't smoke and are married to non-smoking husbands. If he can get some computer time, I will be glad to give you the answer in due course.

Question: Dr. Collen, how were the 12 tests in your multiphasic automated testing selected? What percentage of positive results are false-positives, i.e., "SMA-12 induced" disease? If this occurs, to what extent would you modify the cost for a true-positive lab test?

Dr. Collen: Eight tests were used in our initial study, and then in the second study we used 12 tests. The data I presented included the 12 tests. For the second study some of the original eight were omitted, as I indicated, because they were not cost effective. The 12 tests that were selected were chosen from a study of 20 tests in which the usefulness of eight were compared with that of 20. Cost per positive test was the criterion for selection of the tests for use by the department of medicine that saw these patients. Their medical staff selected the 12 tests that they believed to be most cost effective.

Determination of the true-positive test does require going back to the charts on these people several years later to establish essentially whether the positives and negatives were verified. We have not done this for all tests because it is very expensive. I mentioned a few for which we know the percentage of false positives, e.g., for glucose it is about 50%. Of course, I must hasten to point out that the term "false positive" depends on your objective. We were very interested in hyperglycemia because years later half of the patients with a high blood glucose will have frank diabetes mellitus. The clinician isn't interested in this group; he wants to do something about existing diabetes, so that's why 3% of the adult population have diabetes mellitus confirmed but twice that number have hyperglycemia. If a physician is not considering what may happen to these people in the future he calls all these hyperglycemics false positives.

So, you must again define your terms very precisely. I cannot tell you the proportion of false positives for all the 12 tests.

Dr. Selikoff: Of the people in the audience who have had a medical examination in the past year or two, how many subjected themselves to automated screening? And how many went to a doctor? I think the point is obvious. If someone suspects he is ill, he will generally visit a physician, not an automated health-examination center. Most people believe that they get better medical care that way.

Dr. Collen: I have gone to a doctor who has taken a blood specimen from me and sent that blood specimen for multiphasic screening, but your question was how many went to multiphasic screening without speaking to a physician and were satisfied thereby.

Dr. Selikoff: Automated testing is a completely different story.

Dr. Collen: I think this poll has a problem of great self-selection, but also I don't think the question is the right question. I think your answers don't apply to a relevant question. No one recommends that you have a panel of tests performed without follow-up. Twenty-five years of experience advise against this because, as was brought out by probably everyone here, normality is greatly influenced by age, sex, and other factors. You can't interpret a test just by reading the result from a machine. You must go to someone who knows medicine, knows you, and interprets it in relation to you specifically. We have done almost a million tests during 25 years, and in no instance have we provided any information to the patient directly. He must go to a physician, who will interpret that test in relation to the patient's particular problem.

So I would suggest you rephrase the question as to how many

had a health checkup using the multiphasic technique as a method of test collection and processing, as an alternative to going to a physician who ordered tests in accordance with what in his judgment was most useful to you.

Question: Could you comment on the costs of screening for neonatal hypothyroidism?

Dr. Levy: I believe that in the New England program it has been estimated that costs for screening for hypothyroidism currently run somewhere in the neighborhood of $1.50 per specimen. That is the laboratory cost, and does not include the cost of the specimen collection. I believe the laboratory costs run somewhere in the neighborhood of $200 000 a year for 125 000 specimens.

Question: A supplement to the September issue of *Pediatrics* addressed that question. The cost was 60 cents per T–4.

Dr. Levy: I thought the question was how much does it cost to screen for hypothyroidism rather than how much does a thyroxine cost. I think that the reported cost is low. I think that's low because I don't think that tends to include repeat testing. It only includes initial testing. I strongly disagree with that kind of analysis because as a matter of fact when testing for PKU about 10 or 12 years ago, it was said that it only cost about 25 cents per PKU test. If you do it in a totally irresponsible way, I think that is probably what it *does* cost, but if you include your actual cost it is more: the cost for repeat testing, the cost for a TSH assay, which is an absolutely positive indicator of hypothyroidism and is essential for confirmation of the diagnosis of hypothyroidism in a screening program.

The only way you will be able to differentiate between normality and hypothyroidism in a premature infant with a disorder such as TBG deficiency is by measuring TSH. When you do that, you find out, as for instance in the New England program where the costs are somewhere in the neighborhood of $75 000 per year, that screening for low T–4 will cost a great deal more money when you really look at it as a total package. I strongly feel that when you look at multiphasic screening, or newborn screening, you must look at the total package because that's really all that matters.

Question: In view of the latency factors of some substances, are saccharin tests valid when done on rats for short periods of time and with large doses of saccharin? Can permissible body burdens be established by short-term experiments?

Dr. Selikoff: To answer the second part of the question first: No, we have no way of measuring body burdens nor do we have to know that there *are* any body burdens; indeed, about 99.5% of all saccharin is excreted unchanged in the urine, so we could anticipate very little accumulation.

The human studies of the effects of saccharin are inconclusive. First of all, in the United States, we used very little saccharin except in 1943, during World War II. We used very little saccharin until around 1964 when soft drinks with cyclamates and saccharin and so forth were introduced. Thus we have no way of knowing now in any large number of people what saccharin does over a long time, but we may know 20 or 25 years from now.

What do you do when you don't know, and you have no human data? Our colleagues in the laboratory tell us that if you do animal testing, and carcinogenicity is shown by reliable tests—using good pathologists, enough animals, accurate diagnosis, with no worms in the liver and no stones in the bladder, especially if cancer occurs in more than one species, and there is some evidence for dose response, then we should regard the compound as potentially carcinogenic in humans.

Does that mean it will inevitably happen? No, we have very little information so far about this; however, every identified carcinogen in humans, with the exception of arsenic, where the testing has been limited so far, is also carcinogenic in animals. We don't know yet about the reverse—whether every agent that is carcinogenic in animals will also prove to be carcinogenic in humans.

What do you do in such a case? By law, the FDA has to make a judgment as to whether or not the tests were valid, but once that judgment is made they are not allowed to use judgment anymore, and if a compound appears to be carcinogenic the FDA simply cannot allow it to be used as a food additive. The FDA is not allowed to use judgment to say how much of a substance that causes cancer in animals in reliable tests may be allowed in human food. We are in a bind, and we are going to be in a worse bind in the next several years, because as the Toxic Substance Control Act gets into high gear, probably within a year's time, we are going to have discovered other agents which are carcinogenic in animals and for which there are no human data one way or the other. We as a society have to make these choices from here on in, and they are very difficult choices.

Question: Should cost effectiveness be based on the cost to determine a positive abnormal result, or should it be based instead on the possible severity of the illness uncovered?

Dr. Collen: If by severity we mean morbidity, disability, and mortality, these factors cannot be ignored. We don't screen for conditions that have no impact on an individual, so one assumes that you are only going to screen for conditions which cause either physical or organ or psychological impairment. Given those conditions, then, you look at the prevalence of the condition and at the tests, to determine whether screening is cost effective.

I don't know whether I have answered the question well, because I'm not sure what the term "severity" means.

Question: In some cases screening may cost $60, and that may seem like a high cost in relation to $1 per positive, but uncovering a serious situation where life or death is at stake may very well be worth $60.

Dr. Collen: That helps clarify the question. Consider clinical mammography, which is one of the items that I discussed earlier. It is our most expensive test. It costs us $5 per examination. However, we have a forthcoming publication on the cost/benefit of mammography vs. physical examination. It is a very expensive test per positive when we have one in 500 women over age 50 with a positive test, because at $5 per test it costs $2 500 per positive. Yet it still is cost effective over a five-year period. The question is an excellent question; and the answer is that you *do* have to consider the effect of the condition on the patient.

Dr. Levy: I am glad that question was raised, because this has bothered me for a long time about multiphasic screening. For instance, I wonder how much good we do by finding hyperglycemia in the absence of glycosuria—is it really doing the patient any good for him to know that he has a high blood glucose and will develop diabetes in 20 years, if he lives that long, in contrast to finding overt diabetes? Are we, for instance, going to treat the hyperglycemia, and if we do are we really going to prevent anything? If we find early hyperglycemia, and we don't prevent the later onset of diabetes mellitus or its complications, then indeed would we not be better off to forget screening for hyperglycemia, and simply concentrate on screening for glycosuria, which is far less expensive? It seems to me that perhaps there might be other multiphasic analyses which would fall into that category.

Dr. Collen: I certainly agree with the importance of differentiating between hyperglycemia and diabetes mellitus. Actually, however, when you determine your normal limits for blood sugar, you will find that 10 or 15% of people over age 60 have an abnormal blood sugar, at least by our definitions. But then the clinician who is interested in sick people and the care of people who have diabetes mellitus can reduce the number of abnormalities by half, to about 3% of all adults.

Now, the long-term benefits of treatment is a subject worthy of another whole symposium. The American College of Physicians held a panel discussion on what is diabetes a year ago, and they had another panel recently, and the definitions differ. There is no question but that we must define what is called diabetes mellitus, and there is no agreement on it. But however you define the condition that you are testing for, you can establish a test to identify it. If you want to do other measures to define a different kind of diabetes, then you can. We have said that essentially the upper five percentile of blood sugars are those that the clinicians want to look at.

Question: I have heard a pediatrician say that the five PKU cases identified in my state last year would all have been expected from the family history. Can screening be restricted to a smaller group? Should a history not be taken before the screening, to reduce the need for screening?

Dr. Levy: Yes. You can argue that if one simply concentrated on those families in which there is a PKU sibling, and then screened for PKU in subsequent siblings in that family, then one wouldn't have to screen the general population. You can pick up PKU's by that method, but you will only pick up 20% or at the most 25% of PKU's by that method. At least that is our experience in Massachusetts, and it has been the experience in Oregon, and I think it is the experience of most states. As a matter of fact, I looked at that question a few years ago, because that is the same argument that was raised in a conference in Europe. I think that it would be most unusual for all PKU's in one state to be anticipated from their history. It can happen of course, but I think that none out of five or none out of 10 might be detected in the adjacent state.

This is a sort of circular argument, however, because the way the first sibling was identified was by routine PKU testing, and therefore if one concentrated on those families one indeed would be able to zero in by PKU screening, but you have to have the *first* case identified in order to zero in on the second case. As a matter of fact, it turns out very often in PKU families that two or three siblings with PKU are born before the first case is identified. So, if you didn't have a newborn screening program, your yield would certainly not be five out of five. When you start a screening program you will find one or two PKU parents in already-identified families. But the other positives would occur in previously unidentified families.

So, the answer is that you cannot have an effective screening for PKU, or for that matter any genetic disease, simply by focusing only on families in which there is a family history. In the vast majority of genetic disorders transmitted in an autosomal recessive fashion the carriers are asymptomatic, and there may be no family history of that disorder. For dominantly inherited diseases the history yields much more information.

Dr. Nipper: Dr. Selikoff, this comment was apparently prompted by comments about PBB's and PCB's and mother's milk. The questioner wants to know whether those same compounds are present in commercial infant formulas and in cow's milk, and I presume he also wants to know their effects on infants.

Dr. Selikoff: They are not contained in the usual defatted dry milk formula. PBB's and PCB's are fat soluble, and when the products are defatted they are not present in the commercial mixtures that are available. So, if infants are fed the mixture they get no PCB's and no PBB's.

It is somewhat ironic that those mothers who are most conscientious, most concerned, and want the best for their babies, want natural care with breast feeding, and end up with the children who are afflicted with this chemical difficulty.

Secondly, cow's milk, in general, does not have much PCB. In Michigan it used to contain PBB, but now it does not, because the PBB-contaminated cattle were killed. The slaughter is still going on, and I think another 40 000 cows will be slaughtered. We expect that there will be no PBB's eventually in cow's milk in Michigan.

There is another ironic situation here. Cow's milk tends to lose its PCB's and PBB's before human milk does. The reason is that a cow begins to give milk at around 18 months, and therefore has then had only a year and a half in which to collect the chemicals. Women generally don't breast-feed for about 20 years, and so they have 20 years to collect the various chemicals and there are many other chemicals in breast milk besides the halogenated biphenols. Cows, if owned by a good farmer, will give milk for a long time, and a lot of it. Therefore, whatever they have taken in they tend to get rid of. It is not so with women. Women tend to breast-feed for only limited periods of time, and so whatever they have in their fat stores, in their tissues, they tend to retain and give to the next baby, and the next baby, and so forth. So, there are these constraints based upon fat solubility of biphenols.

We certainly don't know the effects on human health at this time. Dr. Mason Barr of the University of Michigan has studied some 350 children in our survey, and found a very significant difference in symptoms among Michigan children as compared to Wisconsin children. What the ultimate effect will be we simply don't know now.

Question: A 45-year-old patient goes to his physician for a check-up. He states he has no complaints whatsoever, but he is going for a check-up because he is reacting to advertisement of the benefits of a periodic health examination. What would be the frequency of finding any significant abnormalities during a multiphasic screening?

Dr. Collen: That would depend, of course, on the particular test panel he received. In that age group we find in our population that, if they have no complaints, essentially half of them are verified as having no clinically important abnormalities, about 20% are found to have some clinically important abnormalities for which they are referred for definitive medical care, and the remainder are divided between two groups. About 5% have previously unknown asymptomatic diseases—hypertension, diabetes, and so forth. The rest have problems of living, psychosomatic problems that are benefited by psychological guidance, health counseling, and so forth.

Question: How effective is the treatment of aminoacidurias? Are the children "normal" or merely not abnormal enough to require institutional care?

Dr. Levy: It varies from one inborn error of metabolism to another. In some instances, such as PKU, the treatment is quite effective. By "quite," I mean that these children clearly do not need institutional care, and as a matter of fact, they are not mentally retarded. Their IQ tests essentially place them in normal IQ levels. However, it has also been found that many children with PKU detected by inborn screening a number of years ago and treated biochemically are well-controlled, but are having such things as learning disorders in school and to some extent some hyperactive behavior, which has called for some additional care within school. So treatment of PKU is not 100% effective, but it is certainly better than institutionalization.

For maple-syrup urine disease, and others, the response to treatment varies more. Generally, we have some patients with maple-syrup urine disease who are doing quite well now within the first five years of life. Whether or not they are going to have learning disorders when they get into school is a question we are not able to answer now or for the next few years.

Galactosemics do exceedingly well, and most of the galactosemics that we have had experience with, who were detected by newborn screening, and treated from the newborn age onward are doing very well, and seem perfectly normal, with normal IQ's and no hyperactivity or other behavior problems.

Hypoglycemia of the neonate has evaded effective treatment. To my knowledge, nobody yet has an effective treatment to prevent mental retardation in that disorder.

We still have a lot to learn about treatment of inborn errors of metabolism, so I am sure that in the next 20 to 40 years we will certainly be learning about newer treatments.

Question: One of the environmental chemical problems that cause damage to people, and it has been in the news recently, has dealt with a fire-retarding chemical ["Tris"] used on infant and toddler sleepwear. Could you comment on the possible damage from the exposure of children to Tris? I understand that there is a possible link to male sterility, but it has not been reported anywhere I've see except in the lay press.

Dr. Selikoff: All of the information we now have on Tris has been related to the Ames test results with salmonella and with animal studies of a limited nature. Major economic decisions were taken on the basis of these very limited data. I am reluctant to say more about it. I would urge you to look at the original broad data yourself, their quality, their relevance, and then judge them in accordance with the very major decisions that were taken as a result.

This is an important test case because it has almost set a standard for what we will do with similar data in the future.

Question: Would you comment on what additional cost/benefit could be attributed to lab testing vs. history and physical exam alone?

Dr. Collen: We have never studied the cost/benefit in quite that way. I think others have tried to, but I don't think I can answer that question.

Question: Could you tell us how the normal ranges were established in your studies?

Dr. Collen: Yes. We developed distribution curves by age and sex, by time of day, and hours since last food. This information was published quite a few years ago. Then we usually take an upper percentile—and this is related to the clinician's objectives. For example, if he wants the upper 3% of the blood sugar, or 5%, or 8%, or 10%, we use that as the reference value. That's the usual method, though with time and experience we will provide to them and refer to them exactly that group of people that they are interested in for any particular test, such as uric acid or glucose.

Question: During the years of study of asbestos workers, were precautions or preventive measures implemented, or did the study continue with working conditions unimproved?

And secondly, one criterion for screening is that a positive and effective treatment should exist. In the asbestosis or PCP or PBB examples, are there effective treatments other than the removal of the hazard?

Dr. Selikoff: There was virtually little control of the insulation workers until around 1972. At that point, asbestos insulation virtually disappeared, and there has been very little exposure in the last five years. Exposure in the factories, despite the information that was available, remained unaltered until around 1969 or 1970. In many factories now, at least in places like Johns-

Manville, there has been virtually a revolution in the industrial hygiene control, which is now excellent in many if not all factories. At least from what the workers tell me it's like the difference between day and night, because before they often couldn't see the windows. So there has been marked improvement in many aspects of exposure in the last five years.

With regard to the very important second question that was asked about whether we can do anything when we find these things. One criticism that has always been made of the medical profession has been that they are much more interested in rat bites than what to do about the rats. Yes, there are some very important things that can be done.

You can do nothing whatsoever about some cancers. For some everyone who has one will die. In the case of lung cancer, if you know the individual has been a cigarette smoker and an asbestos worker, there is some reason to believe that it is useful to examine him every four months after 20 years of exposure—before that you are giving him excess radiation—and also probably to do a sputum cytological examination every four months, focus your effort after the 20-years period in the smokers. The non-smokers get some excess lung cancer but not very much. The smokers get about seven times more lung cancer than non-smokers. If you do that you will be able to save a few lives.

Our results are still not brilliant, and as in lung cancer in general, I doubt in the best of circumstances whether we save more than 25%; that is by getting the cancers early when they are little nodules and asymptomatic. We can save some lives, and it's a common disease, and it's worth doing.

For asbestosis, which causes a large proportion of deaths in asbestos workers you can do a great deal. You can't affect a pulmonary fibrosis because that stays there just as you see it on the X-ray for the rest of the individual's life, and in fact will probably progress. But these people don't die generally of pulmonary insufficiency due to the fibrosis. If they have fibrosis they have a lowered respiratory reserve, and they can go along fairly well but they can't walk fast. They go around the block instead of going up the hill, they take the elevator instead of walking, and so forth, but they do fairly well. What kills them are intercurrent infections such as bronchitis for which you and I would just go to bed for a day or do nothing at all; but that may be enough to kill them. Bronchial pneumonia and viral pneumonia may also be terminal events.

If the doctor knows a patient has asbestosis he will treat him vigorously with antibiotic therapy, often with hospitalization. It's a rough four, five, six, seven days, but then the patient is well and returns to his previous state. You can save the life of most people with asbestosis until they die of something else.

For gastrointestinal cancer, you can do a stool guaiac for occult blood because that is simple enough and you can save some lives. Obviously, you can't do much about cancer of the esophagus and stomach, but these are infrequent. You can save about half the people with colon cancers; about 55% will probably be saved with that type of screening. For oro-pharyngeal cancers—that is, cancers of the tongue, oro-pharynx, and larynx—you can save a fair proportion of people, but these cancers are uncommon and so you aren't saving very many people overall.

So you *can* do something as a result of screening for cancer.

III

STRATEGIES FOR LABORATORY TESTING

Organ Panels and the Relationship of the Laboratory to the Physician

John Bernard Henry, M.D., and Peter J. Howanitz, M.D.

Department of Pathology, Division of Clinical Pathology, State University of New York, Upstate Medical Center, Syracuse, N.Y. 13210

For many years, physicians have selected and critically evaluated clinical laboratory data (measurements and examinations) in the course of patient care for at least one of the following reasons:

(1) Confirmation of a clinical impression
(2) Rule out a possible diagnosis or disease
(3) Therapeutic or management guide
(4) Prognostic guide
(5) Detection of disease (screening)

In general, the guiding philosophy of the attending physician in regard to such laboratory data has been to request only what is essential or identified as a clear-cut indication or need.

In terms of chemical pathology, a "blood" (actually serum) urea nitrogen (BUN) or serum creatinine measurement may be used to confirm a clinical impression of uremia and renal failure. To rule out cirrhosis of the liver in a patient with upper gastrointestinal bleeding, one or more hepatic-function tests may be requested. The magnitude of the increase in serum creatine kinase activity 24 to 48 h after the onset of a myocardial infarction may provide valuable prognostic information to the attending physician concerning the extent of cardiac necrosis. As a guide to therapy, periodic measurements of blood glucose are valuable in adjusting the therapeutic regimen of a patient with diabetes mellitus in terms of insulin as well as diet and exercise.

Finally, a number of blood chemical measurements have demonstrated value in detecting disease in selected populations. These include measurements of serum calcium for hyperparathyroidism; fasting blood glucose for diabetes mellitus; serum protein electrophoresis for multiple myeloma (plasma cell dyscrasia); serum bilirubin, alkaline phosphatase, serum aspartate aminotransferase and protein electrophoresis for hepatic disease; and blood lipids for coronary artery disease or atherosclerosis.

In recent years, the philosophy of ordering only what is absolutely necessary or specifically indicated has been challenged, a departure from the medical school teaching that one obtains a complete history and complete physical examination, but only selected and limited laboratory data. Because medicine is an art as well as a science, scientific medicine revolves around collection of scientific data relevant to a particular patient's illness. Furthermore, in the interest of better patient care through comprehensive health screening and application of preventive medicine, a number and variety of clinical laboratory measurements and examinations can be generated for each patient in a high-risk population even though there is no present apparent illness, the aims being early detection of disease, before the onset of clinical manifestations, as well as delineation of an individual's normal values before the history or physical findings indicate disease.

The Problem-Oriented Medical System (POMS) or more specifically, the Problem-Oriented Medical Record (POMR) has become widely accepted *(1)*. Organ panels may be identified in the POMR in terms of a relevant data base related to the laboratory in addition to history and physical information. Because the POMR focuses on patient's problems rather than diagnosis, it is imperative to identify the selection of the laboratory data base in a manner that is responsive to medical problem solving. Organ panels by their very nature—i.e., selection of the most sensitive and specific measurements and examinations relative to a particular medical problem—fulfill this need.

Furthermore, it is feasible in terms of subsequent evaluation to validate the selection of measurements and examinations for a particular medical problem in light of their effectiveness in problem solving and management. In other words, if the data base in an organ panel does not appear to be sufficient in terms of sensitivity or specificity for medical problem solving, it can be rejected or modified accordingly. Periodic reassessment of the composition of organ panels in terms of the selection of measurements and examinations through correlation with medical problem-solving in the POMR should become increasingly important with increasing emphasis on cost-effective medical care and better laboratory utilization.

Normal values for clinical laboratory data generally represent data collected from population studies and, when compared to an individual's normal or reference values, have a much wider range. If detection of disease depends on finding a deviation from "normal" in clinical laboratory data, the much narrower range for an individual obviously will provide a more sensitive baseline for such detection. From our own studies, we have identified individual normal ranges for several measurements that are about one-third to one-half of the range for the population's normal values.

Finally, instrumentation has become available that makes the generation of a number and variety of clinical laboratory measurements more readily available and economical. In general, when the volume of laboratory measurements increases, the unit cost decreases. For the individual patient, this means that more laboratory data can be obtained at the same or lower cost than currently available when selected, limited (low-volume) laboratory studies are performed.

Evolution of Organ Panels

A history and physical examination serve to delineate the patient's problem and permit the attending physician to arrive at a clinical impression, but additional data—radiologic, pathologic, and laboratory—and consultation with other medical specialists may be required to clarify and document a patient's problem and eventuate in correct diagnosis and appropriate treatment. The physician can order whatever he requires to meet the unique needs of any patient.

It seems logical to take advantage of the attending physician's information derived from the patient's history and physical findings to serve as a point of departure for the selection of a battery or panel of measurements and examinations in the form of an "organ panel." For example, a battery or panel of liver-function tests has served as a useful reference in the study of patients with liver disease for many years. The hemogram or complete blood count has similarly served in evaluating patients with anemia. A natural extension of these familiar batteries or "organ panels" to encompass other organs, systems, or pathophysiologic reflections of the sequence of disease seemed logical.

In terms of data collection, "organ panels" reflect a critical, comprehensive, and current selection of clinical pathologic measurements and examinations geared to a particular patient's problem, organ, or system. "Organ panels" focus on medical problems and hence facilitate medical care from the standpoint of how or why the attending physician evaluates a patient, as well as the use of a POMR.

"Organ panels" as a form of health-care delivery also developed in our medical center partly in response to the increased use of general screening procedures that are now in operation in health centers, clinics, and hospitals throughout the United States. There are many philosophies regarding the efficacy of screening patients for a variety of diseases. One approach has been to order a group of tests that are convenient for the laboratory to perform, with no thought to the individual patient, his presenting complaint, his signs or symptoms, or his physical examination. With this philosophy a dehydrated patient with metastatic disease and a woman awaiting delivery would be screened with exactly the same 12 or 20 tests—and conceivably would have the same values! Because advances in automation have mostly been in chemical tests, most of these screening procedures have been restricted to chemistry determinations.

Not surprisingly, our medical staff have been quite receptive to the philosophy of "organ panels." From the patient's history and physical findings, a physician usually identifies a particular organ(s) as a basis of his initial impression and problem list; from this he usually orders appropriate laboratory tests to aid in establishing a diagnosis. Thus the "organ panel" follows the logic of a physician's evaluation of his patient's medical problem.

When a physician is faced with a medical problem that is not within his particular sphere of interest or training, he can depend on the "organ panels" to provide him with the most germane as well as sensitive and specific measurements and examinations appropriate for a particular ailment. This is not only didactically useful, the physician, regardless of his specialty or subspecialty, is confident and secure in the belief that the most appropriate laboratory information has been requested.

We selected the most sensitive and specific tests available for the different panels, in consultation with specialists and using basic medical science information derived and translated into patient care. Our panels also are constantly subject to revision, to keep abreast of new developments and ensure optimal patient care. Hence the pathologist participates in the selection of laboratory measurements and examinations as well as their translation and interpretation.

From our preliminary data, it appears that organ panels not only may help shorten the average hospital stay of patients, but they also lend themselves to more specific and comprehensive ambulatory patient care. This, in turn, results in great savings to the patient for hospitalization, with economy in overall delivery of medical-care services. For example, after the admission workup of a patient, additional determinations are often necessary to clarify a patient's problem. This requires re-ordering and often another (or more) blood collection on a subsequent day and more time to process specimens (i.e., separate sera from whole blood) as well as additional time (days) to accumulate the requisite data to establish a diagnosis. All this is obviated with the use of organ panels.

In addition, the charge for a test within a particular panel is now about half what the same test would cost if ordered singly, and greater reduction in costs may ensue with an increased volume and decrease in unit cost. Furthermore, organ panels obviate a certain amount of repetition and overlapping of determinations, and this also may lower a patient's laboratory expenses.

From the standpoint of clinical pathology, we have been able to better allocate our existing resources in terms of automation and better organize our technical and clerical staff, the net result being more efficient and effective service in patient care. There is increased interest and response from our professional and technical staff, who feel they are better informed about individual patients, since the name of the organ panel requested identifies the medical problem.

Finally, "organ panels" represent a departure from the past and an innovation in the delivery of health care that is responsive to patients and their attending physicians. In an academic environment it is essential that such innovations not only be cost effective but also compatible with the educational philosophy and highest standards of medical care. Although our panels' prime function is in diagnosis, under specific circumstances they may also be used as management panels, or specific management panels can be developed.

Several of our currently available diagnostic panels are shown in Figures 1, 2, and 3, with a request form shown in Figure 4. Comments that explain why certain measurements are selected as a part of a particular panel are reviewed periodically. It is anticipated that each patient would have a complete blood count (hemoglobin, hematocrit, leukocyte count, peripheral smear evaluation and differential) and a urinalysis. Hence these are not included with all "organ panels." Other panels or profiles may be anticipated in the future, e.g., "Coma Panel", "Bleeding Panel," etc.

General Health Screen Panel (Figure 1)

This panel may be used to screen each patient who is visiting our hospital for the first time and for subsequent periodic health examinations. It is designed to evaluate renal, liver, endocrine, musculoskeletal, and hematopoietic systems. If an abnormality is noted, an appropriate organ panel can be utilized. Routine screening has been criticized *(2)* as having little beneficial impact on patient care despite the significant number of abnormal laboratory results generated. This panel has been designed for the early detection of those diseases for which treatment is either easier or more effective when undertaken earlier. However, the measurement of the effectiveness of screening is judged by the usefulness in altering the clinical course of a patient. Although this is subjective, our clinical colleagues believe a screening panel is of value for selected patients.

Parathyroid Panel (Figure 1)

A prospective study of 1630 asymptomatic adult patients in a Mayo Clinic series showed 1.4% of patients with hypercalcemia diagnosed as hyperparathyroidism *(3)*. Serum calcium mea-

STATE UNIVERSITY HOSPITAL
UPSTATE MEDICAL CENTER

CLINICAL PATHOLOGY – ORGAN PANEL A

Last Reporting Date: ______________________

GENERAL HEALTH (Fasting)				
Creatinine mg/dl	0.6–1.2			
Urea Nitrogen (BUN) mg/dl	8–18			
SGO Transaminase IU/L	1 - 39			
Uric Acid mg/dl	Female 2.0–6.4 Male 2.1–7.8			
Chloride mEq/L	98–109			
CO_2 Content mM/L	24–30			
Sodium mEq/L	135–145			
Potassium mEq/L	4.0–4.8			
Calcium mEq/L	4.5-5.4			
Lactic Dehydrogenase IU/L	71-207			
Glucose mg/dl	65-115			
Cholesterol mg/dl	Age 20–30 130–230			
Triglyceride mg/dl	10–160			
CBC	See Hematology Report Sheet			
Urinalysis	See Microscopy Report Sheet			
PARATHYROID				
Calcium Total mEq/L	4.5–5.4			
Alkaline Phosphatase IU/L	Adult 4–13 Child 13–20			
Inorganic Phosphorus mg/dl	Adult 2.0–5.2 Child 4.0–7.0			

PROSTATIC				
Acid Phosphatase IU/L	0-0.8			
Creatinine ng/dl	0.6-1.2			
Calcium meq/L	4.5-5.4			
Alkaline Phosphatase IU/L	Adult 4-13			
Phosphorous mg/dl	Adult 2.0-5.2			
PANCREATIC				
Amylase, Serum Somogyi Units/dl	50-150			
Amylase Urinary units/min.	1-5			
Lipase, Serum units/ml	0.5-1.8			
Calcium mEq/L	4.5-5.4			
Creatinine Serum mg/dl	0.6-1.2			
Urine Creatinine mg/dl				
Amylase/Creatinine Clearance %	1-3.5			
THYROID				
T_4-RIA ug/dl	5.5-12.3			
Free Thyroxine ng/dl	0.9-2.3			
T_3-RIA ng/dl	92-217			
TSH-RIA µIU/L	Less than 10			

LEGEND

A. Laboratory Accident
B. Confirmed
C. Clotted
D. Moderate
E. Recollect Specimen
F. Unit Notified
G. Specimen not received
H. Hemolysis
J. Xanthochromic
K. Marked (Over 100/hpf)
L. Lactescence
M. Minimum (Less than 0-1/hpf)
N. Not done – see schedule
P. Patient not available
Q. Quantity not sufficient
R. Patient uncooperative
T. Improperly labeled
U. Test cancelled by physician
W. Improper specimen
X. Report to follow
Y. Obscured by dye

______________________ M.D.
Clinical Pathologist

Fig. 1. Cumulative Report Form for Organ Panels (A).

surement is the best diagnostic test for primary hyperparathyroidism. Serum phosphorus is included to distinguish primary hyperparathyroidism from other forms of hypercalcemia, and alkaline phosphatase is assayed as an indicator of bone involvement. We recommend that this panel be used on

STATE UNIVERSITY HOSPITAL
UPSTATE MEDICAL CENTER

CLINICAL PATHOLOGY -- ORGAN PANEL B

Last Reporting Date: ________________

	HYPOPITUITARISM PANEL (Insulin Tolerance)		
Time	Glucose mg/dl	Cortisol RIA mcg/dl	Growth Hormone ng/ml
Control 1			
Control 2			
15 Min.			
30 Min.			
45 Min.			
60 Min.			
90 Min.			
120 Min.			

MICROCYTIC ANEMIA				
Serum Iron µg/dl	50-150			
TI BC µg/dl	250-450			
Hemoglobin Electrophoresis	See Hematology Report Sheet			
CBC	See Hematology Report Sheet			

PULMONARY				
Blood pH units	Art. 7.38–7.44 Ven. 7.36–7.41			
Blood PCO_2 mmHg	Art. 35–40 Ven. 40–45			
Blood PO_2 mmHg	Art. 95–100			
Oxygen Saturation Percent	Art. 94–100 Ven. 60–85			
CO_2 Content mM/L	24–30			
Sodium mEq/L	135–145			
Potassium mEq/L	4.0–4.8			
Chloride mmol/L	98–109			
Urea Nitrogen (BUN) mg/dl	8–18			
Glucose mg/dl	65-115			
CBC	See Hematology Report Form			

MUSCLE				
Creatine Phosphokinase (CPK) IU/L	Male 55-170 Female 30-135			
Lactic Dehydrogenase IU/L	71-207			
T_4-RIA ug/dl	5.5-12.3			
Creatine Phosphokinase Isoenzymes	See Special Report			
Aldolase (S-L) units	3-8			

MALABSORPTION				
Fecal Fat gm.day	2-5			
Xylose Tolerance Urine gm/5hr	5.3-7.7			
Calcium Total mEq/L	4.5-5.4			
Carotene ug/dl	50-360			
Albumin gm/dl	Age 15-35 3.44-5.64 Age 35-55 3.22-5.10			
Total Proteins gm/dl	6.78-8.3			
Folic Acid ng/ml	5-21			
Electrophoresis	See Chemistry Report Form			

METABOLIC				
Glucose mg/dl	65-115			
Uric Acid mg/dl	Female 2.0-6.4 Male 2.1-7.8			
Cholesterol mg/dl	Age 20-30 130-230			
Triglycerides mg/dl	10-160			
T_4-RIA ug/dl	5.5-12.3			
Creatinine mg/dl	0.6-1.2			
Calcium mEq/L	4.5-5.4			
Alkaline Phosphatase IU/L	Adult 4-13 Child 13-20			
SGO Transaminase IU/L	1-39			
Lactic Dehydrogenase IU/L	71-207			
Phosphorous mg/dl	Adult 2.0-5.2 Child 4.0-7.0			

LEGEND

A. Laboratory Accident
B. Confirmed
C. Clotted
D. Moderate
E. Recollect Specimen
F. Unit Notified
G. Specimen not received
H. Hemolysis
J. Xanthochromic
K. Marked (Over 100/hpf)
L. Lactescence
M. Minimum (Less than 0-1/hpf)
N. Not done – see schedule
P. Patient not available
Q. Quantity not sufficient
R. Patient uncooperative
T. Improperly labeled
U. Test cancelled by physician
W. Improper specimen
X. Report to follow
Y. Obscured by dye

________________ M.D.
Clinical Pathologist

Fig. 2. Cumulative Report Form for Organ Panels (B).

three successive occasions to evaluate patients for primary hyperparathyroidism. If the diagnosis of primary hyperparathyroidism is still not forthcoming, we then recommend measurement of parathyrin (parathyroid hormone) by a radioimmunoassay specific for the carboxyl-terminal end of the

STATE UNIVERSITY HOSPITAL
UPSTATE MEDICAL CENTER

CLINICAL PATHOLOGY -- ORGAN PANEL C

Last Reporting Date: ______________________

CARDIAC INJURY I				
Creatine Phosphokinase (CPK) IU/L	Male 55-170 Female 30-135			
Lactic Dehydrogenase IU/L	71-207			
CARDIAC INJURY II				
Lactic Dehydrogenase Isoenzymes	See Special Report			
Creatine Phosphokinase Isoenzymes	See Special Report			
CARDIAC EVALUATION				
Cholesterol mg/dl	Age 20–30 130–230			
Triglyceride mg/dl	10–160			
T_4-RIA µg/dl	5.5-12.3			
Glucose mg/dl	65-115			
Uric Acid mg/dl	Male 2.1-7.8 Female 2.0-6.4			
HYPERTENSION				
Creatinine mg/dl	0.6–1.2			
Creatinine Clearance ml/min.	80–120			
Urinary Free Cortisol mcg/24 hr	75-410			
Metanephrines µgm/mg creatinine	Less than 1.82			
Chloride mEq/L	98–109			
CO_2 Content mM/L	24–30			
Sodium mEq/L	135–145			
Potassium mEq/L	4.0–4.8			
Glucose mg/dl	65-115			
ARTHRITIS				
E.S.R. %	40-54			
C-Reactive Protein	Negative			
Antinuclear Antibody Titer	Under 1:20			
Rheumatoid Factor Slide Latex Test				
Rheumatoid Factor Tube Latex Test				
Titer @ 56°C	1:20			
Titer @ 4°C	1:80			
Uric Acid mg/dl	Female 2.0-6.4 Male 2.1-7.8			
Total Proteins gm/dl	6.7-8.3			

TRANSITION PANEL				
Alkaline Phosphatase IU/L	Adult 4-13 Child 11-20			
Lactic Dehydrogenase IU/L	71-207			
Uric Acid mg/dl	Female 2.0-6.4 Male 2.1-7.8			
Gamma Glutamyl Traspeptidase IU/L	Female 3-55 Male 15-85			
Serum Calcium meq/L	4.5-5.4			
CBC $\bar{c}$ Differential	See Hematology Report Form			
HEPATIC				
SGO Transaminase IU/L	1-39			
Alkaline Phosphatase IU/L	Adult 4-13 Child 11-20			
Bilirubin Total mg/dl	0.1-1.2			
Bilirubin Direct mg/dl	0.0-0.2			
Prothrombin time Seconds	9.2-11.2			
Gamma Glutamyl Traspeptidase IU/L	Female 3-55 Male 15-85			
Partial Thromboplastic time sec.	15-38			
Total Proteins gm/dl	6.7-8.3			
Albumin gm/dl	3.7-4.9			
RENAL				
Urea Nitrogen (BUN) mg/dl	8–18			
Creatinine mg/dl	0.6–1.2			
Creatinine Clearance ml/min.	80-120			
Total Osmolality Serum mOsm/Kg	285–295			
Osmolality Urine mOsm/Kg	300-1000			
Osmolal Clearance ml/min				
Free Water Clearance ml/min				
Sodium mEq/L	135–145			
Potassium mEq/L	4.0–4.8			
Chloride mEq/L	98–109			
CO_2 Content mM/L	24–30			
Inorganic Phosphorus mg/dl	Adult 2.0–5.2 Child 4.0–7.0			
Calcium mEq/L	4.5-5.4			
Glucose mg/dl	65-115			
Uric Acid mg/dl	Female 2.0–6.4 Male 2.1–7.8			
Albumin gm/dl	Age 15–35 3.44–5.64 Age 35–55 3.22–5.10			
Total Proteins gm/dl	6.7–8.3			
Urinalysis	See Microscopy Report Sheet			
Urine Culture & Colony Count	See Bacteriology Report Sheet			

LEGEND

A. Laboratory Accident	N. Not done – see schedule
B. Confirmed	P. Patient not available
C. Clotted	Q. Quantity not sufficient
D. Moderate	R. Patient uncooperative
E. Recollect Specimen	T. Improperly labeled
F. Unit Notified	U. Test cancelled by physician
G. Specimen not received	W. Improper specimen
H. Hemolysis	X. Report to follow
J. Xanthochromic	Y. Obscured by dye
K. Marked (Over 100/hpf)	
L. Lactescence	
M. Minimum (Less than 0-1/hpf)	

______________________ M.D.
Clinical Pathologist

Fig. 3. Cumulative Report Form for Organ Panels (C).

ORGAN PANEL STUDIES

ORGAN PANEL STUDIES

STATE UNIVERSITY HOSPITAL SYRACUSE, N. Y.

CLINICAL PATHOLOGY	DEPT #	SEQ #	ROOM #
ORGAN PANEL STUDIES	31	8588	

SEND ONE REQUISITION TO INTIATE PANEL STUDY AND ANOTHER REQUISITION WITH EACH SPECIMEN

DIAGNOSIS	ORDERED BY M.D.	REQUISITIONED BY R.N.

COLLECTED BY	TIME AM PM

PATIENT CONTAGIOUS ☐ YES ☐ NO

ARTHRITIS	3486		MUSCLE (URINE RANDOM)	3676
CARDIAC INJURY	3494		PANCREATIC (2 HOUR URINE)	3700
CARDIAC EVALUATION	3528		PARATHYROID (24 HOUR URINE)	3734
CARDIO VASCULAR (HYPERTENSION, 24HOUR URINE)	3619		PULMONARY (ARTERIAL BLOOD)	3767
C.V.A. (URINE-RANDOM)	3551		RENAL (URINE COLONY CNT, 24 HOUR URINE)	3791
HEPATIC MANAGEMENT	3577		WATER & ACID BASE (2 HOUR URINE)	3858
INTESTINAL MALABSORPTION (3 DAY STOOL, D-XYLOSE 2 HOUR URINE)	3916			
METABOLIC	3643			

Fig. 4. An Example of a Service Request Form for Panels.

molecule. Such assays are demonstrably superior to those which are directed to the amino-terminal region of parathyrin for separating normal patients from those with primary hyperthyroidism *(4)*.

Prostatic Panel (Figure 1)

Increased serum acid phosphatase activity is a hallmark of carcinoma that has spread beyond the prostatic capsule. The effect of nonprostatic acid phosphatases is minimized by use of specific substrates. Serum acid phosphatase activity may increase in some cases after rectal examination *(5)*. Values for calcium and alkaline phosphatase are useful in evaluating the possibility of metastatic spread to bone. This screen also includes renal-function determinations because a high incidence of renal disease is associated with prostatism.

Pancreatic Panel (Figure 1)

Serum amylase assay is most valuable in the diagnosis of acute pancreatitis in the first 24 h. Urinary amylase and serum lipase are more sensitive than serum amylase two or more days after the onset of pancreatitis or in cases of relapsing pancreatitis. The ratio of amylase clearance to creatinine clearance is thought to be the most sensitive indicator of pancreatitis. This amylase/creatinine ratio can be calculated, with use of an untimed urine specimen for the two determinations *(5)*.

Thyroid Panel (Figure 1)

This complete thyroid profile features thyrotropin (thyroid stimulating hormone), the best diagnostic test for hypothyroidism, and triiodothyronine and (T_3) and thyroxine (T_4), which are useful laboratory tests for hyperthyroidism. Free thyroxine is included so that those physiologic or pathologic states associated with increased or decreased thyroid binding proteins can be identified; e.g., pregnancy or estrogen therapy *(7)*. In certain clinical situations where these diagnostic determinations may not be definitive, a test with thyroliberin, the thyrotropin-releasing hormone, is indicated: the serum thyrotropin response to intravenously injected thyroliberin is then determined *(7)*.

Hypopituitary Panel (Figure 2)

Pituitary tumors are responsible for 15 to 20% of all brain tumors and can be documented by pituitary hormone responsiveness to stimulation. The insulin tolerance test is the most sensitive procedure for documenting hypopituitarism. The response of cortisol and somatotropin (growth hormone) to the intravenous injection of insulin is used to assess the integrity of the pituitary. This provocative test can be performed in just over 2 h with very little risk to the patient; it has replaced the metyrapone test *(8)* for the documentation of pituitary insufficiency.

Microcytic Anemia Panel (Figure 2)

Iron-deficiency anemia, the most common form of anemia, can be identified by a low iron concentration in the serum and relatively high total iron-binding capacity. The inclusion of hemoglobin electrophoresis with serum iron and total iron-binding capacity should differentiate thalassemia (increased hemoglobin A_2 on electrophoresis) and anemia of chronic disease from iron deficiency (normal electrophoresis, low iron, but above-normal total iron-binding capacity).

Pulmonary Panel (Figure 2)

Functional disturbances of ventilation are reflected primarily by elevation of p_{CO_2} while defects in respiration reveal hypoxemia (depression of p_{O_2} and O_2 saturation). Chronic obstructive disease of the airways is often associated with ventilation and diffusion defects (elevated p_{CO_2} and decreased p_{O_2}). A complete blood count (CBC) is included to determine the severity of secondary polycythemia, which may be present.

Muscle Panel (Figure 2)

Heart and skeletal muscle contain relatively high activities of creatine kinase (EC 2.7.3.2). Damage of these organs causes a marked increase in the activity of this enzyme, and striking increases are observed in patients with myositis or Duchenne's muscular dystrophy. The MB isoenzyme of creatine kinase is increased in certain dystrophies *(10)*. In some patients with myositis, serum aldolase (fructose-biphosphate aldolase, EC 4.1.2.13) activity may be above normal when creatine kinase activity is normal. Radioimmunoassay for total T_4 is used to screen for hypo- or hyperthyroidism in patients with vague muscle complaints, weakness, and mild muscle wasting.

Malabsorption Panel (Figure 2)

Quantitative estimation of fecal fat is essential for determination of steatorrhea; the D-xylose absorption test reflects adequacy of carbohydrate absorption. Serum carotene is used as a screening test for fat-soluble vitamin absorption, and protein electrophoresis is valuable in identification of exudative enteropathy. Serum folate concentrations are decreased in many patients, although not all have megaloblastic anemia *(11)*.

Metabolic Panel (Figure 2)

Blood lipid (cholesterol and triglycerides) assays are essential for evaluation of patients with hyperlipidemia. Siperstein *(12)* has pointed out the difficulties of defining normal values for a 2-h postprandial blood glucose. Because of these difficulties, blood glucose measured during fasting is used as the most reliable aid for the diagnosis of diabetes mellitus. Radioimmunoassay of thyroxine is included to rule out secondary causes of lipid abnormalities. In addition, creatinine, calcium, and the enzymes alkaline phosphatase (EC 3.1.3.1) and lactate dehydrogenase (EC 1.1.1.27) make this panel a more comprehensive screening profile for the detection of other diseases in an apparently well individual. Hence it has many similarities in terms of composition and utility to our General Health Screen Panel (Figure 1).

Cardiac Injury Panel No. 1 (Figure 3)

Creatine kinase activity is increased in 95% of patients with acute myocardial infarction, and almost 100% of patients with myocardial infarction have increased lactate dehydrogenase activity in their serum. The former increases within 6 h after the onset of chest pain; serum lactate dehydrogenase increases by 48 h. If values for one or both of these enzymes are above normal, their isoenzymes are determined.

Cardiac Injury Panel No. 2 (Figure 3)

The MB isoenzyme of creatine kinase is almost specific for heart muscle and, coupled with the isoenzyme pattern of lactate dehydrogenase, in which fast-moving isoenzyme 1 exceeds isoenzyme 2, we have a powerful tool for diagnosing acute myocardial infarction *(13)*.

Cardiovascular-Hypertension Panel (Figure 3)

The best single determination for diagnosis of pheochromocytoma is urinary metanephrine. It is important to have a sensitive and specific test for this entity, which is a surgically correctable form of hypertension. Gitlow et al. *(17)*, in their experience with more than 16 000 patients, improved the diagnostic reliability of this determination. Measurement of urinary free cortisol is a sensitive evaluation of hypercortisolism. Blood glucose, measured during a fast, is the most reliable indicator of diabetes mellitus *(12)*. A creatinine clearance and serum electrolytes are included in this panel for the evaluation of renal function, which may be altered secondarily or primarily by hypertension. Renin profiling as a way of identifying volume and vasodilator constrictive components of hypertension leaves much to be desired *(18)*; current evidence suggests that drug treatment is best administered in an empiric fashion: diuretics first, an adrenergic blocking agent second, and vasodilator therapy third. For this reason, we do not advocate measurement of renin.

Cardiac Evaluation Panel (Figure 3)

Values for cholesterol and triglycerides may be used to screen for lipid disorders in individuals with a strong family history of atherosclerosis or those patients with evidence of coronary artery disease. High-density alpha-lipoproteins may be inversely related to the incidence of coronary artery disease. Glucose and thyroxine measurements are included to rule out secondary causes of hyperlipidemia.

Arthritis Panel (Figure 3)

The most common inflammatory diseases with joint symptoms—rheumatoid arthritis and systemic lupus erythematosis—are detected with rheumatoid factor and anti-DNA, respectively. The nonspecific tests such as sedimentation rate and C-reactive protein are included to differentiate osteoarthritis from inflammatory diseases with joint symptoms. Uric acid is measured for diagnosis of gout. Data on serum albumin and total proteins are useful for inflammatory joint diseases.

Transition Panel (Figure 3)

Metastic disease to the liver [lactate dehydrogenase, alkaline phosphatase, and γ-glutamyltransferase (EC 2.3.2.2)], to bone (alkaline phosphatase, calcium) or to the bone marrow (complete blood count and differential count) can be identified. This panel may also be used in the management of any patient with proven metastatic disease.

Hepatic Panel (Figure 3)

Aspartate aminotransferase (EC 2.6.1.1) is the most sensitive single enzyme assay for assessing hepatic cellular injury. Alkaline phosphatase is useful in determining the degree of hepatobiliary obstruction or interference with bile flow, as is γ-glutamyltransferase. Alcoholism and its related effects is the third leading cause of death in this country, so a special feature of serum γ-glutamyltransferase is its increase with only mild alcohol intake *(19)*. Partial thromboplastin time and prothrombin time are both quite useful as a liver-function test, especially for chronic liver disease, but also valuable as a prerequisite to liver biopsy, to assess hemostasis.

Renal Panel (Figure 3)

Serum urea (BUN) measurement is useful in following patients with end-stage renal disease undergoing dialysis because it reflects effectiveness of dialysis better than does serum creatinine. However, serum creatinine is a better single screening test for renal function, because it is virtually unaffected by protein ingestion or metabolism or by dehydration. Data on serum creatinine combined with the urinary creatinine can be used to calculate creatinine clearance—the most practical method for determining glomerular filtration rate. Values for serum albumin and total protein are included as an indication of severity of any proteinuria. Urine culture and urinalysis are used to detect infection, which may further compromise renal function.

Measurement of serum osmolality is very useful in following hypo- and hypernatremic states, which are usually associated with hypo- and hyperosmolality, respectively, as well as disturbances of water balance. In severe diabetes mellitus, measurement of serum osmolality is helpful in forewarning the physician against hyperosmolality, which may lead to intracellular dehydration, convulsions, and death. Urine osmolality is the most accurate method for assessing urine concentration, because renal and pituitary disturbances in function reflect changes in osmolality (number of solute particles), rather than specific gravity (mass of solute particles). Both serum and urine osmolality measurements generate the urine/plasma ratio (normal 4/1) which, with the urine volume, permit calculation of free water clearance. Knowledge of free water clearance aids in management of postoperative patients, patients with cardiac failure, cirrhosis, Addison's disease, and polyuric syndromes and those in situations such as dialysis.

In conclusion, organ panels represent a new form of health-care delivery that has achieved a secure status in our hospital, with indications that further expansion will enhance patient care by utilization of physician input and net gains in sensitivity, accuracy, precision, and (especially) efficiency and economy (Table 1). Furthermore, organ panels are responsive to medical problem solving and provide critical data for a Problem-Oriented Medical Record (POMR). Indeed, organ panels not only facili-

Table 1. **Advantages of Organ Panels**

Less cost
Increased efficiency
Problem oriented
Pedagogy (explain why for content)
Pathologist participation in ordering
More informed laboratory staff

tate, they enhance the relationship of the laboratory to the physician.

References

1. Korvin, C., Pearce, R., and Stanley, J., Admissions screening: Clinical benefits. *Ann. Intern. Med.* **83,** 197 (1975).
2. Purnell, D., Scholz, D., Smith, L., et al., Treatment of primary hyperparathyroidism. *Am. J. Med.* **56,** 800 (1974).
3. Benson, R., Riggs, B., Pickard, B., and Arnaud, C., Radioimmunoassay of parathyroid hormone in hypercalcemic patients with malignant disease. *Am. J. Med.* **56,** 821 (1974).
4. Roy, A., Brower, M., Hayden, J., Sodium thymolphthalein monophosphate. A new acid phosphatase substrate with greater specicity for prostatic enzyme in serum. *Clin. Chem.* **17,** 1093 (1971).
5. Salt, W., Schenker, S., Amylase—its clinical significance. A review of the literature. *Medicine* **55,** 269 (1976).
6. Vagenakis, A., and Braverman, L., Thyroid function tests—which one? (editorial) *Ann. Intern. Med.* **84,** 607 (1976).
7. Hershman, J., and Pittman, J., Control of thyrotropin secretion in man. *N. Engl. J. Med.* **285,** 997 (1971).
8. Kreiger, D., Glick, S., Silverberg, A., Kreiger, H., A comparative study of endocrine tests in hypothalamic disease. Circadian periodicity of plasma 11-OHCS levels, plasma 11-OHCS and growth hormone response to insulin hypoglycemia and metyrapone responsiveness. *J. Clin. Endocrinol. Metabl.* **28,** 1589 (1968).
9. Hilal, H., and McCurdy, P., A pitfall in the interpretation of serum iron values. *Ann. Intern. Med.* **66,** 983 (1967).
10. Goto, I., Creatine phosphokinase isoenzymes in neuromuscular disorders. *Arch. Neurol.* **31,** 116 (1974).
11. Baker, H., Frank, O., Sobotka, H., Mechanisms of folic acid deficiency in non-tropical sprue. *J. Am. Med. Assoc.* **187,** 119 (1964).
12. Siperstein, M., The glucose tolerance test: A pitfall in the diagnosis of diabetes mellitus. *Adv. Intern. Med.* **20,** 297 (1975).
13. Galen, R., Reiffel, J., Gambino, S., Diagnosis of acute myocardial infarction. *J. Am. Med. Assoc.* **232,** 145 (1975).
14. Miller, N., Forde, O., Thelle, D., The Tromsø Heart Study: High density lipoprotein and coronary heart disease; a prospective case-control study. *Lancet* **i,** 965 (1977).
15. Ahrens, E., Jr., The management of hyperlipidemia; whether, rather than how. *Ann. Intern. Med.* **85,** 87 (1976).
16. Mann, G., Diet-heart: End of an era. *N. Engl. J. Med.* **297,** 644 (1977).

17. Gitlow, S., Mendlowitz, M., Bertani, L., The biochemical techniques for detecting and establishing the presence of a pheochromocytoma—a review of ten years' experience. *Am. Jour. Cardiol.* **26,** 270 (1970).
18. Woods, J., Pittman, C., Pulliam, C., Werk, E., Waider, W. and Allen, C., Renin profiling in hypertension and its use in treatment with propanolol and chlorthalidone. *N. Engl. J. Med.* **294,** 1137 (1976).
19. Dragosics, B., Ferenci, P., Pesendorfer, F., and Wewalka, F.: Gamma-glutamyl transpeptidase (GGTP). Its relationship to other enzymes for diagnosis of liver disease. In *Progress in Liver Disease,* **V.** H. Popper and F. Schaffner, Eds. Grune and Stratton, York, 1976, pp 436–449.

Use of the Laboratory: Dollars and Sense

Edwin L. Overholt, M.D.

Department of Medical Education, Gundersen Clinic, Ltd.; and La Crosse Lutheran Hospital, 1836 South Ave., La Crosse, Wis. 54601

Constraining the role of the laboratory in clinical care emphasizes the physician's priorities. Over the years as a practitioner and teacher of clinical medicine, I have endeavored to challenge myself, house officers, and fellow staff to utilize the strengths of the laboratory "for cause, not because."

Laboratory procedures should assist the physician in obtaining a correct diagnosis and treatment, to monitor the concentration or effect of a therapeutic agent, and, where possible, to detect subclinical disease or risk factors. If the physician's highest priority is to serve his patient, our escalating inflation mandates that patient care must be tempered with a conscience of cost—and a more prudent use of our laboratory armamentarium must be practiced. The law of the marketplace—that consumer demand governs technological advance—is not operative in medicine. The consumer of this technology is not the patient. The physician who orders the procedures and thus creates the demand is the real consumer. The proliferation of automated analytical devices, although reducing the cost per test, has increased the opportunity for excessive and (or) unnecessary use of laboratory services. There are several reasons for our present dilemma of laboratory misuse:

1) Overutilization is generated by the physician's zeal, compulsiveness, or insecurity in his desire to identify deviations from good health.

2) During the past 30 years, we have all been students of our great teaching and research medical centers. Here, the inves-

tigator has become the predominant teacher and we have been stimulated by his development and use of diagnostic and therapeutic techniques to extend our general knowledge. Our "teaching institutions" have found it difficult to transfer this laboratory armory from the heady plane of new knowledge to the daily practice of medicine. New tests often receive immediate acceptance without critical evaluation. This attitude, developed during the early phase of the physician's training, too often tends to become a life-long pattern! If this judgement is too harsh, why is it that medical students, house staff, and practicing physicians are so uninformed about the limitations as well as the cost of laboratory tests?

3) Public expectations are far too high—complete documentation as a measure of quality as well as a defensive maneuver to blunt the malpractice crisis has blended into the life-style of the physician. His self-preservation is being challenged!

4) Third-party payment has tended to remove the physician and patient from direct conflict over cost.

5) A profit motive covertly arises.

The internist spends considerable time performing health examinations. The purpose of these periodic examinations is to detect conditions, risk factors, or disease in the preclinical or early clinical state that can be modified to the patient's benefit. The current major risk factors of cigarette smoking, obesity, hyperlipidemia, hypertension, and alcoholism must be emphasized in an effort to have our patients think in positive terms of health maintenance as contrasted with the complaint–response process between patient and physician *(1)*.

I support the recommendations *(2)* of the National Council of Preventive Medicine. Periodic evaluations should be performed at the teenage or early 20's, again at age 30, and every five years for the males until age 50. Thereafter, every two years should be sufficient. For the female, every two years after age 30 and yearly after age 50 should be appropriate. During this time-frame, many patients will have developed a problem that requires more frequent evaluation. In addition, many of my healthy patients desire yearly examinations and derive considerable satisfaction from being reassured that "all is well." In these patients, the "annual physical" may be the most effective therapeutic act I perform. Nevertheless, the physician in his teacher role should emphasize a more realistic approach in terms of the frequency of periodic examinations and the extent of complementary laboratory tests. In the "healthy" patient, a battery of tests from our laboratory arsenal is wasteful. This is not action "for cause, but because."

At the time of the *initial* periodic examination, most physicians as a minimum require a chest roentgenogram, random plasma glucose, cholesterol, complete blood count, and urinalysis. With automated analytic devices such as the SMA-12/60, calcium, urea, uric acid, total protein, bilirubin, aspartate aminotransferase, lactate dehydrogenase, albumin, inorganic phosphorus, and alkaline phosphatase are measured instead of just the glucose and cholesterol. This "package deal" is usually cheaper than the combined cost of the glucose and cholesterol. Also, an initial electrocardiogram is obtained if the patient is over age 40 to 50, for future reference should a cardiovascular problem eventually develop. Obtaining further electrocardiograms without supporting symptoms or signs is unwarranted. The 18-year Framingham study *(3)* disclosed that a normal electrocardiogram was recorded in 73% of the patients with ischemic heart disease before their myocardial infarction. The use of the ST segment response to exercise as a screening procedure in asymptomatic patients exemplifies the lack of appreciation of the predictive value of this test. In this population the predictive value of the positive exercise response (percentage of positive tests which actually predict important angiographically-demonstrable coronary artery disease) is low. Froelicher et al. *(4)* and Borer et al. *(5)* found that only 44% and 37%, respectively, of all positives in asymptomatic subjects are true positives. Of 451 men between the ages of 40 and 54, Froelicher et al. *(4)* observed 8.6% (39 men) with positive stress tests. However, only 3.8% had important angiographically-demonstrable coronary artery disease. The sensitivity was 58.8%, the specificity 93.6%. The predictive accuracy of the test was 25.6%. This is to be expected with any test applied to a study population when the prevalence of a disease is low. It is unwise to have the asymptomatic patient labor under the stress of such questionable information from a test with such a low predictive value.

For the early detection of lung cancer, the chest roentgenogram has not been an effective screening procedure, for it has not altered the outcome *(6–8)*. The five-year survival was 8% in patients who had received two roentgenograms yearly, in the hope of early detection. In the pilot study by the American Cancer Society and the Veterans Administration there were 438 false-positives, as compared to 97 true-positives. Thus, for every true-positive, more than four had to be evaluated to rule out carcinoma *(6)*. The prevalence of pulmonary tuberculosis is so low in the U.S. population that it now requires 100 000 chest roentgenograms in healthy populations to discover 13 new cases *(8)*. In the chronic bronchitic–emphysema patient, the evolution of pulmonary disease under treatment will be adequately substantiated if chest roentgenograms are obtained when new symptoms appear or there is poor response to treatment. Overall, neither lung cancer or chronic bronchitis has been greatly altered by our diagnostic and therapeutic attempts. The only effective approach is prevention by abstaining from cigarette smoking. Unfortunately, we have failed to alter the habit of cigarette smoking in most of our patients. Nevertheless, we must persist in the hope of discovering resectable carcinoma. In a sense, I "fine" those of my patients who have continued to smoke, a chest film at the time of their periodic examination. For the non-smoker without cardiopulmonary symptoms or signs, regular chest roentgenograms are unnecessary.

In the obese patient or when there is a family history of heart disease, cholesterol values can be measured every five years. We do not know whether our limited ability to modify cholesterol concentrations alters outcome, but our desire to slow the process of atherosclerosis should not be neutralized by this uncertainty. An above-normal cholesterol value should be followed by measuring triglycerides during a fast. At this juncture, lipoprotein phenotyping by electrophoresis is no longer advised by Frederickson *(9)*. The cholesterol and triglyceride concentrations can be used to detect, sort, and manage these disorders.

Except for the control of obesity, chemical diabetes (asymptomatic) requires no further therapy. There is little evidence that diabetic control alters the incidence and severity of vascular complications. The clinician may wish to use a 2-h postprandial plasma glucose value or the less-sensitive urine sugar test in addition to the fasting plasma glucose as a reinforcement technique for encouraging weight loss. This meets with only occasional success. Though the glucose tolerance test is more sensitive in detecting chemical diabetes, it does not have practical use as a screening test in the asymptomatic patient.

Cancer of the cervix is a slowly progressive disease, requiring three to 10 years to proceed from dysplasia to invasive carcinoma *(10)*. The yield of positive smears decreases rapidly in women who have had several previous negative Papanicolaou smears. Gray *(11)* concludes that after two annual negative smears, screening every three to five years should be sufficient. However, women of low social economic class or women who have frequent sexual contact are at higher risk and should have more frequent screening.

In summary, periodic health examinations, by replacing the more frequent annual physical, will save time for the physician and patient, and curtail cost. Further savings will accrue from limited use of laboratory services with each periodic examination. I have no difficulty in selling most of my patients on the use of limited laboratory studies and saving the cost of the repeated electrocardiogram, chest film, SMA-12/60 battery, and complete blood count. This amounts to at least $65 in savings. Indeed, they are most appreciative, and the doctor–patient relationship is enhanced. Occasionally, to reassure a patient I will have some of these tests done. In these instances, it proves useful because of its placebo effect. Thus, after initial periodic examination in the well patient, all that should be needed is a hemoglobin, a Papanicolaou cervical smear in women, and stool checks for occult blood if the person is over 40 years of age.

We accept as routine the search for microcytic anemia and the administration of iron with hemoglobin concentrations of less than 100 g/liter. Though Elwood *(12)* has found it impossible to correlate the symptoms of palpitations, dizziness, fatigue, or headache until the hemoglobin values is $<$70 to 80 g/liter, we will not abandon this traditional approach. Returning the hemoglobin to its normal range is appropriate, if for no other reason than assuring a maximum oxygen carrying capacity should acute bleeding subsequently occur. Moreover, in elderly people, especially middle-aged men with iron-deficiency anemia, as many as half had demonstrable pathology on complete barium gastrointestinal series *(13)*. The outpatient with stable disease—i.e., previous myocardial infarction, unchanged angina, previous congestive failure or controlled hypertension, diabetes, or rheumatoid arthritis—represents another challenge. Although those patients will receive more frequent complete examinations, they should require only a few well-directed laboratory studies. Worsening of the disease will prompt more detailed laboratory evaluation. This overall streamlining does not detract from the quality of care. Rather, it reflects our limited ability to maintain health in our patients.

Turning to the hospitalized patient. we have seen routine admission studies expanded by virtue of advances in technology. Routine complete blood counts, VDRL, urinalysis, chest films, electrocardiogram (if the subject is over 40 years of age), and a chemistry profile of 12 to 20 or more tests have become commonplace. It is appealing to obtain several test values at low cost from one sample of blood but the process of "screening for disease" has many shortcomings. After a thorough history and physical examination, most of our patients have a recognized problem and should require more relevent laboratory procedures for the confirmation of disease or its progress with treatment. Laboratory studies should be a logical extension of the history and physical examination; that is, problem-oriented rather than process-oriented! Routine procedures applying to all patients cannot be defended! Each patient's problem should direct the type of laboratory procedures required. This should allow selection of appropriate laboratory tests under circumstances where the prevalence of a specific disease is much greater. Remember, the predictive value of such selected tests will be much greater because positive tests will more likely be true positives. The hospital or the physician should not be expected to perform extensive laboratory screening programs on all hospitalized patients. We must remember that many of our patients are relatively well, being admitted for surgical procedures such as hernia repair, appendectomy, cholecystectomy, severe bone fractures, hemorrhoidectomies, etc.; on the medical services we see thrombophlebitis, peptic ulcer, pneumonia, heart failure, diverticulitis, etc. Few of these patients require "screening for disease." Indeed, most have had this done previously.

In our hospital, the routine assessment of the serum Na^+, K^+, Cl^- or CO_2 is discouraged. A SMA-12/60 battery, which assays calcium, inorganic phosphorus, glucose, urea, uric acid, cholesterol, total protein, albumin, total bilirubin, alkaline phosphatase, lactate dehydrogenase, and aspartate aminotransferase is obtained in 50% of the admissions. Here, economics almost forces the clinician to obtain these multiple test values, because obtaining only two or three of the above tests costs the patient more than the SMA-12/60—this in spite of the fact that a single test can be done with equal ease! Much of this information is not needed to assess the patient's problem. Moreover, false-positive results or borderline results mislead the clinician and can provoke additional laboratory testing. Before repeating or requiring additional tests, the clinician should review the history and physical examination. This should allow some of these unsolicited laboratory data to be disregarded.

After our initial laboratory requests, whether they be individual or as a SMA-12/60, a major challenge is to curtail their unnecessary repetition. They often are repeated more frequently than the illness warrants, or after the disease process has stabilized or cleared. In the teaching hospital, tests tend to be ordered more frequently, as a measure of caution in an environment abounding with laboratory facilities and experts. Such habitual ordering can lead to a stereotyped approach. For emphasis, here are a few challenges that I regularly encounter:

• Beyond a hemoglobin determination, what does a complete blood count offer in the majority of our patients who do not have infection?

• Why should the patient who has no cardiopulmonary symptoms or signs or even a known but clinically stable cardiopulmonary problem have chest roentgenograms or electrocardiograms?

• What help does a clinician get from an extensive general chemistry profile on a patient with a specific disease process—for example, a myocardial infarction, congestive failure, or stroke?

• Why get two or three chest roentgenograms done for a patient with pneumonia whose clinical course is satisfactory during his hospital stay?

• Why get serial, as often as two to three times a week, measurements of bilirubin, aspartate aminotransferase, and lactate dehydrogenase on patients with hepatitis whose clinical course is uncomplicated? Why not just look at his conjunctiva for the extent of icterus and as a rough guide the improvement of his clinical course as evidenced by his appetite?

• Why get creatine kinase and lactate dehydrogenase values or isoenzyme fractionation on a patient who has typical chest pain and acute injury pattern evidenced on electrocardiogram?

• Why should a patient with obvious respiratory distress from chronic airway obstruction have a pre- and post-bronchodilation spirogram when, at best, most of the needed information can be obtained from a simple spirogram?

• Why repeat the gastrointestinal series after the patient with duodenal ulcer has become asymptomatic?

• Occasionally, extensive testing may be useful in a diagnostic problem, but in most patients, how does it improve our diagnostic ability or improve our patient's care?

At the Gundersen Clinic–La Crosse Lutheran Hospital Medical Complex, there are no mandatory admission laboratory studies. Both the director of the clinical laboratory (an internist–hematologist) as well as the director of the chemistry section (a Ph.D.) encourage the prudent use of the laboratory. Previous outpatient laboratory and roentgenographic studies are reviewed and there is continuing emphasis on non-repetition unless there is concern over recent changes in the patient's problem. All our house staff and staff are issued easily carried cards which list the cost of commonly used laboratory and x-ray studies. A cost conference is held every other month analyzing all expenses from several randomly selected medical records. At the monthly medical and surgical autopsy conferences, the total cost of each patient's care is reported. The director of medical education holds teaching rounds on two different units four days out of the week and in this process reviews the use of laboratory studies. In addition, all staff members are encouraged to assess the relevancy of diagnostic studies and consultations on their daily teaching rounds. Finally, we are most fortunate in having many interdisciplinary teaching conferences, which stimulate challenging discussion of laboratory utilization. This broad emphasis developed over the past decade is a continuation of this group's approach to quality control.

Table 1 compares 1973 inpatient data to 1976. The total number of hospitalized patients increased by 14.7%, but the average hospital stay decreased by 13%. During the same period, emergency ward visits increased by 37%. The laboratory tests (excluding roentgenographic studies) decreased slightly (−1.6%) despite an increase of 7172 emergency ward visits and 2221 hospital patient discharges. It should be remembered that the SMA-12/60 is counted as 12 separate tests and the complete blood count as four separate tests.

Table 2 shows the outpatient laboratory volume for the same period. Again, there was no change in volume that could not be attributed to an increase in patient load. During this period, 18 young physicians were added to the staff. It might be anticipated that they would utilize the laboratory more intensively. It's my opinion that this is offset by our continued encouragement to utilize laboratory facilities prudently.

Table 3 reflects roentgenographic examinations, and again discloses that there has not been an increase in the total number

Table 1. **Data for 1973–1976 on Inpatients**[a]

Date	*EW patients*	*Hosp. tests (includes all EW testing)*	*Patient days*	*Patient discharges*	*Average patient stay*
1973	19 284	417 772	115 776	15 084	7.7 days
1974	21 654	425 360	121 814	16 355	7.5 days
1975	23 594	420 945	117 674	16 677	7.0 days
1976	26 456	411 094	116 271	17 305	6.7 days
	(+37%)	(−1.6%)	(+.4%)	(+14.7%)	(−13%)

[a] SMA 12/60 counted as 12 separate tests; CBC counted as 4 separate tests. All laboratory services are tallied: blood bank, immunology, hematology, urinalysis, microbiology, surgical pathology, chemistry. EW, emergency ward.

Table 2. **Total Clinic Laboratory Test Volume and Patient Visits**[a] **(tests assigned by origin of request, not location of determination)**

Date	*No. of MD's*	*Clinic tests*	*% increase*	*Clinic patient visits*	*% increase*
1973	80	246 881		234,848	
1974	88	277 703	12%	264,368	12%
1975	93	309 162	25%	276,588	18%
1976	98	317 594	29%	295,248	27%
	(% increase: 23%)				

[a] SMA 12/60 counted as 12 separate tests; CBC counted as 4 separate tests. All laboratory services are tallied.

Table 3. **X-ray Examinations**

Date	*Hosp. exams*	*Hosp. discharges*	*Clinic exams*	*Clinic patient visits*
1973	28 505	15 084	29 372	234 848
1974	29 402	16 355	34 113	264 368
1975	29 564	16 677	36 815	276 588
1976[a]	29 820	17 305	40 283	295 248
	(+4%)	(+14.7%)	(+37%)	(+27%)

[a] Total patient days in hospital 1976 vs. 1973 was +0.4%

of such procedures in hospitalized patients, in contrast to the clinic, where there was an increase of 37% during a four-year period in which the patient load increased 27%. This modest increase in roentgenographic procedures probably reflects the continuing emphasis on outpatient evaluation rather than hospitalizations.

Table 4 shows the use of aspartate aminotransferase (AST) determinations for different months. In March 1977 the coronary care protocol excluded the AST. After reviewing the standing orders in the coronary-care protocol, it was concluded that total lactate dehydrogenase (LDH) and creatine kinase (CK) and (or) their isoenzymes should be sufficient to assess myocardial damage. A prompt reduction in AST volume was noted in April and May. All admission standing orders and (or) routine protocols involving laboratory tests should regularly be reevaluated. We surveyed the use of the SMA-12/60 the past month. Forty-eight percent of the hospitalized patients had one series, the rest did not. More importantly, only 0.3% of those who had an SMA-12/60 had a repeat study.

During the past decade, the La Crosse Lutheran Hospital, staffed by physicians from the Gundersen Clinic, has emerged as a major teaching hospital. It is well known that the zeal and inexperience of house staff and medical students leads to overutilization of the laboratory. Moreover, the staffs of teaching hospitals are less inclined to pursue a more disciplined approach to the use of diagnostic facilities. This paradox stems from the traditional concept that this will allow a better learning experience as well as the discovery of new concepts. Despite the above pressures, utilization of the laboratory has not increased at La Crosse Lutheran Hospital or the Gundersen Clinic. Indeed,

Table 4. **Aspartate Aminotransferase Test Volume/Month (Selected Months)**

	February	*April*	*May*
1976	220	298	229
1977	199	66	76

(The admission protocol dropped this test March 1977)

more patients are being served in the Clinic and Hospital with *less* utilization of the laboratory. We attribute this to the *daily* applied pressure of the approach previously alluded to. Many of us have changed from the ritualistic approach of "laboratory screening" to more careful use of the laboratory. We do not think this has detracted from the quality of patient care.

There has been limited research in the use of the laboratory by clinicians. Granier and Liptzin *(14)*, at Strong Memorial Hospital, Rochester, N.Y., evaluated the use of laboratory services for 855 patients from the medical section (including the intensive care unit) over a three-month period. Seven tests, including radiology, were performed per patient per hospital day during the first week. They concluded there was considerable unnecessary duplication of tests in most patients. They compared the rate of selected hospital charges from 1966 to 1970 relative to 1965. Laboratory charges rose considerably faster than the cost of other patient-care functions and there was a 95% increase in chemistry test volume. After this evaluation, a systematic attempt was made to inform all physicians and medical students of the cost of laboratory tests and the areas of apparent overuse. During the ensuing nine months, the number of laboratory tests remained constant, apparently blunting the previous upward trend.

More recently, Dixon and Laszlo *(15)* estimated the effectiveness of utilization of the clinical chemistry laboratory by the medical house staff in a teaching hospital. Although 30% of the patients had initially normal chemical values, an average of two additional profiles were obtained when there was no further significant clinical change in the patient's status. Chemistry 12/60 test profiles were often ordered when only one or two values were of interest. Moreover, tests were commonly requested daily on patients having a protracted illness and such data were of no benefit to patient care. After this evaluation, interns were arbitrarily limited to an average of eight chemistry and (or) hematologic tests per patient per day. Overall, it was possible to reduce the number of laboratory tests per patient by about 25%.

In a 575-bed general hospital, Korvin et al. *(16)* assessed the clinical benefits from 20 routine clinical and hematologic tests. Only 10% of the abnormal values led to 87 new diagnoses in 79 patients from the 1000 patients who were assessed during a six-month period. Of this latter group, 30 might have benefited had the abnormal findings been followed, 39 had findings or diagnosis of no lasting significance, and in 14 asymptomatic (chemical) diabetes was discovered.

Eisenberg and Goldfarb *(17)* evaluated the usefulness of admission prothrombin times in 301 patients on a Veterans Administration service. They previously had surveyed four medical services in Veteran Administration hospitals in different parts of the country and found 78% of the patients had admission prothrombin times measured; 44 (14.6%) of the 301 patients had prolonged prothrombin times. All but three of the 41 patients with prolonged prothrombin times had pertinent history or physical findings, or both, such that it would be anticipated to be abnormal. Pertinent data included alcohol abuse with known liver disease, increased ease of bleeding, and (or) physical findings of liver disease or abnormal bleeding tendency. Eisenberg *(18)* then reviewed all patients on two different medical services from two medical schools who operated parallel medical services at the same hospital. One unit was subjected to a six-week teaching program emphasizing the clinical utility of the prothrombin time; the other did not receive the educational program. There was no further formal educational effort for the next 12 months. The medical service that did not receive the educational program sustained the previous high volume of prothrombin time requests. The other medical unit reduced prothrombin time requests on admission from 79% to a nadir of 55% six months later. However, the volume of prothrombin time determinations gradually rose over the next 12 months to the approximate level that existed before the educational program. Thus, a continuing-education program is needed if modification of clinical behavior is to be sustained.

Skipper et al. *(19)* queried medical students as to their knowledge concerning the cost of laboratory studies. Only 27% of the first-year students and 34% of the senior students came within 25% of the actual cost. My own experience with young physicians indicates that these figures are appropriate and that many of the staff physicians suffer from this same inadequacy.

Freeborn et al. *(20)* analyzed the use of the clinical laboratory for outpatients in the Kaiser–Portland Health Plan. Five percent of their patient population from 1967 through 1970 were evaluated by computer analysis. Outpatient laboratory procedures increased by 18% per patient. The ratio of laboratory procedures per patient visit showed a steady increase. Physicians in the department of medicine had the highest ratio of laboratory tests per patient visit and accounted for over 60% of the total test requests. An increasing number of laboratory tests and office visits were in the preventive medicine category. This occurred without any automated multi-test device! There was considerable variation in the way individual physicians used the laboratory, ranging from very intensive to very low use patterns. Moreover, there was a consistent relationship between a physician's use of laboratory over this time frame—a low user remained a low user. Overall, physicians who went to school a generation ago when the laboratory services were less developed used the laboratory less. There was a good correlation between the chiefs of services' utilization of the laboratory and the use by their colleagues in their departments. If the chief was a high utilizer, so were most of the other physicians.

Schroeder et al. *(21)* studied the variations in total laboratory use per physician as disclosed by auditing the outpatient records of 33 faculty internists. The mean annual laboratory charges generated by each physician were compiled and distributed among the 33 physicians. Each physician could identify his own code and compare his use and cost of the laboratory with that of his colleagues. This study was repeated during the immediate three months after the cost audit was released.

The initial audit disclosed a 17-fold difference between physicians in annual laboratory fees per patient. There was a strong correlation between the frequency of visits per patient and laboratory charges. Of interest was the observation that a patient with hypertension seen by "high test volume" physicians did not have a higher percentage of complications to explain the increased frequency of visits and increased laboratory studies.

After the audit release, the overall laboratory charges for the group of physicians decreased approximately 30%, along with a 5% decrease in visits. Changes in behavior affecting patient laboratory charges was most marked in the full-time staff physician. To what extent these alterations in behavior were sustained is unknown.

How can our profession change its ways? At the very least, our medical schools and teaching institutions must teach awareness of the relative cost of laboratory tests along with their effectiveness and limitations. This will not occur unless our major teaching institutions "bite the bullet." Faculty, medical students, and house staff must believe and practice these concepts. They must move from lip service to the arena of their daily practice. Clinical investigation protocols must continue, but the difference between patient care and investigation should be emphasized.

All postgraduate courses should integrate prudent use of the laboratory, along with their emphasis of updating information on diagnosis and management of diseases. The ritual of laboratory screening for disease must be curtailed. Our limited ability to maintain health must be appreciated. In so doing, we will recognize the real value of laboratory services. For the ill patient, we must learn to obtain the laboratory tests directed by the patient's problem. Additional tests should be requested as dictated by the clinical course. Single tests or small combinations of tests should not be relatively overpriced. Advances in automated testing permits less expensive, logical, small groups of test combinations rather than all-inclusive blocks of blood values. Flexibility in ordering test combinations rather than artificial completeness should be available at competitive prices. This will allow the clinician to obtain needed laboratory studies and meet the challenge of cost effectiveness. This, then, is doing things "for cause, not because."

References

1. Breslow, L., and Somers, A. R., The lifetime health-monitoring program. A practical approach to preventive medicine. *N. Engl. J. Med.* **296,** 601–608 (1977).
2. *Preventive Medicine USA*, Prodist, New York, 1976.

3. Margolis, J. R., Clinical features of unrecognized myocardial infarction—silent and symptomatic. *Am. J. Cardiol.* **32,** 1–7 (1973).
4. Froelicher, V. F., Thomas, M., Pillow, C., and Lancester, M. C., An epidemiological study of asymptomatic men screened with exercise testing for coronary heart disease. *Am. J. Cardiol.* **34,** 770 (1974).
5. Borer, J. S., Brensike, J. F., Redwood, D. R., et al., Limitations of the electrocardiographic response to exercise in predicting coronary artery disease. *N. Engl. J. Med.* **293,** 367 (1975).
6. Lilienfeld, A., An evaluation of radiologic and cytologic screening for the early detection of lung cancer. A cooperative pilot study of the American Cancer Society and the Veterans Administration, *Cancer Res.* **26,** 2083–2121 (1966).
7. Boucot, K. R., and Weiss, W., Is curable lung cancer detected by semiannual screening? *J. Am. Med. Assoc.* **224,** 1361–1365.
8. Tenret, J., Is a systematic annual chest film useful? *Conn. Med.* **39,** 604–607 (1975).
9. Fredrickson, D. S., It's time to be practical. *Circulation* **51,** 209–211 (1975).
10. Petersen, O., Spontaneous course of cervical precancerous conditions. *Am. J. Obstet. Gynecol.* **72,** 1063–1071, 1956.
11. Gray, L. A., The frequency of taking cervical smears. *Obstet. Gynecol. Survey* **24,** 909–913 (1969).
12. Elwood, P. C., Some epidemiological problems of iron deficiency anemia. *Proc. Nutr. Soc.* **27,** 14–23 (1968).
13. Andrews, J., Aspects of malnutrition in the elderly. *Proc. Nutr. Soc.* **27,** 196–201 (1968).
14. Griner, P. F., and Liptzin, B., Use of the laboratory in a teaching hospital. Implications for patient care, education, and hospital costs. *Ann. Intern. Med.* **75,** 157–163 (1971).
15. Dixon, R. H., and Laszlo, J., Utilization of clinical chemistry services by medical house staff. *Arch. Intern. Med.* **134,** 1064–67 (1974).
16. Korvin, C. C., Pearce, R. H., and Stanley, J., Admissions screening: Clinical benefits. *Ann. Intern. Med.* **83,** 197–203 (1975).
17. Eisenberg, J. M., and Goldfarb, S., Clinical usefulness of measuring prothrombin time as a routine admission test. *Clin. Chem.* **22,** 1644–1647 (1976).
18. Eisenberg, J. M., An educational program to modify laboratory use by house staff. *J. Med. Educ.* **52,** 578–581 (1977).
19. Skipper, J. K., Smith, G., Mulligan, J. L., and Garg, M. L., Medical students' unfamiliarity with the cost of diagnostic tests. *J. Med. Educ.* **50,** 683–684 (1975).
20. Freeborn, D. K., Baer, D., Greenlick, M. R., and Bailey, J. W.,

Determinants of medical care utilization: Physicians' use of laboratory services. *Am. J. Public Health* **62,** 846–853 (1972).

21. Schroeder, S. A., Kenders, K., Cooper, J. K., and Piemme, T. E., Use of laboratory tests and pharmaceuticals. Variation among physicians and effect of cost audit on subsequent use. *J. Am. Med. Assoc.* **225,** 969–973 (1973).

The Application of the Laboratory to Hypertensive Patients

Harold D. Itskovitz, M.D.

Medical College of Wisconsin, 8700 West Wisconsin Ave., Milwaukee, Wis. 53226

The clinical evaluation of hypertensive patients begins with a process of selection: which patients are to be selected for which tests? Although the clinician has been provided with a base of knowledge and diagnostic tools to pinpoint many of the pathophysiologic and biochemical abnormalities responsible for hypertension, when applied to a large population of hypertensive subjects, more than 95% of patients probably will not be helped by them. These 95% of hypertensive individuals have been diagnosed as having primary or essential hypertension. However, our inability to identify a specific cause of hypertension does not interfere with the ability to treat hypertensive patients, because high blood pressure can usually be controlled with nonspecific hypotensive drugs.

When one considers that 20 to 50% of our adult population has high blood pressure *(1)*, that the complete diagnostic workup of a hypertensive individual may cost $1000, and that in 95% of cases this expensive workup will be of little help for managing the patient, then it is apparent that there is a potential waste of billions of dollars in the nonselective application of the diagnostic laboratory to hypertensive patient in general *(2–4)*. In the context of economic realities the physician who evaluates hypertensive patients must compromise his traditional desire to practice complete scientific medicine and perform a full diagnostic workup. Concepts such as cost effectiveness and cost/benefit ratios enter into the medical decision-making process.

To this end, the use of screening tests is an essential first step in the spectrum of the hypertensive workup to help select the appropriate diagnostic procedures for any given hypertensive patient. Two additional components of the spectrum of a hypertensive evaluation are listed in Table 1. These include: *(a)* the diagnostic workup to determine a possible specific etiology for high blood pressure, particularly as this may lead to the discovery of curable hypertension, and *(b)* an assessment of the course of the disease and its complications. If physicians are to treat hypertensive patients for a lifetime they must be provided with the information to indicate that the therapeutic approach or drugs that have been selected are beneficial to the patient.

Table 2 lists the screening tests used to recognize various secondary forms of hypertension along with the relative frequency of each type. To be effective, a screening test should be

Table 1. **Evaluation of Hypertension—Spectrum of Studies**

1. Screening procedures—guide to diagnostic workup
2. Diagnostic studies for secondary hypertension leading to a possible curable form of hypertension
3. Assessment of complications and course of hypertensive condition as a guide to prognosis and therapy

Table 2. **Screening for Secondary Hypertension**

Type of hypertension	*Relative frequency %*	*Screening procedures*	
		Primary screen	*Secondary screen*
Renal:			
Parenchymal disease	<3	History and physical exam., urinalysis, serum creatinine	IVP
Arterial lesion (amenable to surgical cure)	<1		
Adrenal:			
Primary aldosteronism	<1	Serum K^+ (<3.5 mmol/l)	
Cushing's (syndrome or disease)	Infrequent	History and physical exam.	
Adrenogenital syndrome	Rare	History and physical exam.	
Pheochromocytoma	<0.05	History and physical exam. (see text)	24-h urinary catecholamines, metanephrines, VMA
Neurologic	Infrequent	History and physical exam.	
Coarctation of aorta	Infrequent	History and physical exam. (arm BP ≥ leg BP)	
Primary (essential)	>95	Neg. evaluation	

simple, inexpensive, safe, relatively noninvasive, and of sufficient sensitivity to detect most individuals with the potential problem. In most types of hypertension, a good history and physical examination represent the most useful screening procedures available. Other useful screening tests include urinalysis and measurement of serum creatinine concentration for detection of renal parenchymal disease and measurement of serum potassium concentration for detection of patients with possible aldosterone-secreting tumors of the adrenal gland. In the latter case, hypokalemia (serum potassium concentration < 3.5 mmol/liter) has been found in more than 90% of patients with hypertension secondary to an aldosteronoma. Although measurements of catecholamines, metanephrines, or vanillyl mandelic acid (VMA) in urine have been used as screening tests for pheochromocytoma, these measurements are not now recommended for every patient with hypertension in view of the infrequency of pheochromocytoma relative to the expense of the procedures and the high number of false-positive reports observed, which require further expensive diagnostic studies. However, such measurements of catecholamines and metabolites in urine have an important place as a second echelon of screening patients who have, or are expected to have, a higher frequency of pheochromocytoma (history of paroxysmal attacks, familial occurrence of pheochromocytoma, presence of neurofibromatosis, paradoxical hypertensive response to hypotensive drugs, postural hypotension). Similarly, the intravenous urogram (IVP) is useful as a screening test for renal artery stenosis, but only as a second-echelon procedure because of its expense relative to the incidence of curable renal hypertension. The IVP is reserved for hypertensive patients who have a history of renal disease or trauma; the young hypertensive; those with severe, accelerated, or worsening hypertension; and patients with abnormal kidneys or in whom an abdominal bruit is heard on physical examination.

When the physician determines that it would be useful to study a patient for a possible secondary form of hypertension after the screening procedure, or in a case of severe hypertension which is not responsive to therapy, then additional laboratory and radiographic studies may be scheduled. The nature of these studies varies, depending on the type of secondary hypertension that the physician believes to be present (renal, adrenal, neurologic, etc.). Generally, three types of studies are used to assure the presence or absence of a secondary form of hypertension that may be amenable to cure: *(a)* anatomic tests, to determine the type and location of the lesion; *(b)* functional tests, to measure whether the lesion interferes with organ function or is causing a physiologic aberration; and *(c)* hormone or pressor studies, to determine if the lesion is producing a substance that may be responsible for an elevated blood pressure (Table 3). To be predictive of secondary hypertension with a high potential for cure, results for all three categories of tests should be positive. For example, angiographic demonstration of a renal arterial lesion does not predict that the demonstrated lesion is in fact responsible for a patient's hypertension, because many such lesions have no functional significance and can even be found among a high proportion of normotensive patients at postmortem examination. In this case, surgical correction of the renal arterial lesion or nephrectomy may not be curative of hypertension if the blood pressure is elevated for another reason. Thus it has been reported that patients with renal lesions who are undergoing surgery for expected cure of hypertension experience less than a 50% rate of success when the decision to operate is made solely on an anatomic basis *(5)*.

When the renal lesion is associated with a renal functional abnormality, as detected by differential renal function studies, the surgical cure rate of renal hypertension approaches 70% *(6, 7)*. Further, when the renal anatomic lesion is associated with increased renin production by the abnormal kidney, an 80 to 90% surgical cure rate for hypertension has been reported *(8)*. Conversely, the presence of high plasma renin activity without an anatomic test to localize the source of high renin does not indicate that a curable form of hypertension may be present, because the high plasma renin value may be secondary to bilateral renal parenchymal disease.

The anatomic, functional, and hormonal tests available for the diagnosis of various specific forms of hypertension are also listed in Table 3. Proper interpretation of these tests depends not only on their accuracy and specificity for a given form of hypertension but also on knowledge of the physiologic state of the patient at the time the tests are made. Thus plasma renin activity may be

Table 3. **Evaluation for Secondary Hypertension**

Type of hypertension	*Anatomic studies*	*Functional studies*	*Hormone or pressor studies*
Renal disease:			
Parenchymal (tumor, cyst, pyelonephritis)	IVP or radioactive scan. Renal biopsy Renal angiogram	IVP, creatinine clearance (↓)	Peripheral and differential renal venous renin measurement (unilateral vs bilateral ↑)
Arterial disease: (arteriosclerosis, fibromuscular hyperplasia, etc.)	IVP or radioactive scan. Renal angiogram	IVP and/or radioactive renogram Rarely differential renal function study	Peripheral and differential renal venous renin measurement (unilateral ↑)
Adrenal:			
Primary aldosteronism (aldosterone tumor, hyperplasia)	Nephrotomogram Arteriogram or adrenal scan Adrenal venography	Plasma vol. (↑) Hypokalemic alkalosis Plasma pH (↑) Urinary $Na^+ + K^+$ (↑)	Plasma or urine aldosterone or metabolites (↑) Response to DOCA or saline load (little effect) Plasma renin (↓); response to low salt diet or diuretics (little effect) Plasma and urinary cortisol or metabolites (normal)
Cushing's (syndrome or disease) (tumor, hyperplasia)	X-rays, skull and abdominal Radioactive or computerized scans of skull, brain, and abdomen Arteriogram or scan of adrenal	Polycythemia Hypokalemic alkalosis Elevated blood sugar Cushingoid appearance	Plasma and/or urinary cortisol and metabolites (↑) Plasma adrenocorticotropin (↑ or ↓)
Adrenogenital syndrome (tumor, hyperplasia)		Abnormal secondary sex characteristics and development	Plasma or urinary mineralocorticoids (↑) Plasma and urinary cortisol (↓) Plasma renin (↓) Plasma adrenocorticotropin (↑)
Pheochromocytoma (tumor)	Nephrotomogram Arteriogram or scan of adrenal	Plasma volume (↓) Hypotensive response to IV phentolamine Pressor response to IV histamine or glucagon	Plasma catecholamines (↑) 24-h urinary catecholamines, metanephrine, VMA (↑)
Coarctation of aorta	Arteriogram	Arm BP $\geq$ leg BP	
Neurologic disease: (tumor, hematoma)	X-rays, radioactive or computerized scans of skull and brain	Abnormal cerebrospinal fluid pressure and chemistries Functional deficit	May have ↑ plasma or urinary catecholamines and metabolites

high, normal, or low in patients who are normotensive or hypertensive, depending on their posture, activity, the presence or absence of drugs which alter renin release, and the patient's recent salt intake and state of hydration. It is imperative that each situation be controlled and that studies be obtained under standardized conditions to assure that the results are interpretable. Standardized conditions are necessary, not only for measurements of renin and other hormones, but also for measurements of physiologic function, which must be assessed in patients with possible secondary hypertension.

Several comments are in order concerning the procedures listed in Table 3. Generally, where there is a choice, those studies that are the least invasive should be scheduled. Thus in the case of a potential renal arterial lesion, the IVP or radioactive renogram are preferred to the differential renal function study for assessment of individual kidney function, because the differential renal function study requires ureteral catheterization. The differential renal function study is reserved for patients in whom the simpler tests do not provide the necessary quantitative information. In the case of hormonal or pressor assays, especially if these involve comparative studies from different locations, it may be necessary to stimulate a greater production of the chemical pressor agent in order to accentuate differences in secretion between the site of the lesion and other sites under evaluation. For this reason, differential renal-vein renin studies are usually obtained after salt depletion or diuretic therapy. Also, since pressor enzymes or hormones may be secreted at one site and circulate in the plasma to another site, high concentrations may be found in the peripheral blood and the venous effluent of normal organs. Thus relative differences in venous concentrations of renin derived from each kidney are more useful than absolute measurements and comparisons are also necessary with concentrations measured in peripheral or central venous plasma. Generally, renal-venous renin ratios > 1.5 (diseased kidney : normal kidney) are considered significant. The more important elements of the diagnostic workup as a prerequisite to surgery are the hormonal or pressor assays and the anatomic studies to localize a potential lesion.

The functional studies are the least important element in the diagnostic triad because these may be abnormal secondary to a variety of nonspecific causes. However there are situations, especially when the exact mechanism of hypertension is not understood, when the demonstration of an anatomic lesion alone may suffice as a criterion for surgery (coarctation of the aorta). In patients with a possible pheochromocytoma the situation is reversed. This condition is of sufficient danger if left alone that exploratory surgery may be indicated to find a suspected tumor on the basis of an abnormal hormonal assay alone, whether or not a tumor can be demonstrated by radiographic techniques.

The third component in the evaluation spectrum of hypertensive patients is the assessment of the patient's course and status of vital organs (Table 4). Because high blood pressure injures the patient by causing accelerated arteriosclerosis, the laboratory is utilized to evaluate those organs most likely to be damaged by the arteriosclerotic process. Thus an electrocardiogram and chest roentgenograms are obtained to determine cardiac status, and a urinalysis and serum creatinine concentration to evaluate renal function. Concentrations of potassium, uric acid, glucose, triglycerides, and cholesterol are also measured, to determine whether additional risk factors may be present that play a role in the arteriosclerotic process and to monitor potential biochemical aberrations that may be produced by the drugs used in treating hypertension. These studies are obtained during the initial evaluation of the hypertensive patient and at periodic intervals thereafter, with a frequency that depends on the individual situation. In general, it has been recommended that such studies be scheduled at least once per year.

In summary, physicians who are responsible for the evaluation and care of hypertensive patients have available to them many sophisticated and useful diagnostic tools and therapeutic interventions. In most cases the diagnostic procedures are of

Table 4. **Assessment of Complications and Course of Hypertension**

1. Complete blood count	
2. Serum creatinine concentration Urinalysis (early a.m.)	Determine renal status
3. Serum K^+ Blood sugar Serum uric acid Serum cholesterol Serum triglycerides	Assess arteriosclerotic risk factors and potential complications of antihypertensive therapy.
4. Electrocardiogram Chest x-ray	Determine cardiac status

little help to the management of patients, and their indiscriminate use is not cost-effective. A physician must know when to apply these and how to use them correctly in the individual patient. Also, he must use the laboratory appropriately as a guide in the proper management of patients over the long term. An overall approach that may be recommended on the basis of this discussion is summarized in Table 5. This illustrates a step-wise approach to the study of hypertensive patients, encompassing the three components of the evaluation process. Thus the initial workup (Step I) includes the elements necessary for screening and the long-term followup of most hypertensive individuals. Additional screening procedures which are not cost effective for Step I are listed in Step II, to be utilized in patients with a higher expected incidence of secondary hypertension. Finally, Step III, which includes those specific studies necessary for a complete diagnostic evaluation, is reserved for patients in whom there is a high suspicion of a specific cause of hypertension that may be amenable to surgical cure.

Table 5. **Stepwise Approach to Evaluation of the Hypertensive Patient**

Step I	
Indicated for all hypertensive patients:	1. History and physical exam.
Includes screening tests for secondary hypertension and information required for long term followup.	2. Complete blood count 3. Blood components Serum K^+, creatinine, uric acid, cholesterol, triglycerides Blood sugar 4. Urinalysis 5. Electrocardiogram 6. Chest x-ray (optional)
Step II	
Additional screening tests indicated for patients considered for renal hypertension or pheochromocytoma.	Intravenous urogram or radioactive renogram 24-h urinary catecholamines, metanephrine, VMA
Step III	
Especially indicated for patients considered for surgical cure of secondary hypertension on basis of preliminary evaluation.	Specific anatomic, functional, and hormonal studies relative to suspected type of secondary hypertension (Table 3).

References

1. Itskovitz, H. D., Kochar, M. S., Anderson, A. J., and Rimm, A. H., Patterns of blood pressure in Milwaukee. *J. Am. Med. Assoc.* **238,** 864–868 (1977).
2. Weinstein, M. C., and Stason, W. A., *Hypertension, a Policy Perspective,* Harvard University Press, Cambridge, Mass., 1976.
3. McNeil, B. J., Varady, P. D., Burrows, B. A., and Adelstein, S. J., Cost effectiveness in the diagnosis and treatment of hypertensive renovascular disease. *N. Engl. J. Med.* **293,** 216–221 (1975).
4. Report of the Joint National Committee on Detection, Evaluation, and Treatment of High Blood Pressure. *J. Am. Med. Assoc.* **237,** 255–261 (1977).
5. Combined staff conferences. Recent advances in hypertension, *Am. J. Med.* **39,** 616–645 (1965).
6. Dustan, H. P., Page, I. H., Poutasse, E. F., and Wilson, L., An evaluation of treatment of hypertension associated with occlusive renal arterial disease. *Circulation* **27,** 1018–1027 (1963).
7. de Mendoca, P. P., and Young, J. D., Jr., Renovascular status after renal surgery for hypertension. *J. Am. Med. Assoc.* **201,** 592–599 (1967).
8. Kirkendall, W. M., Fitz, A. E., and Lawrence, M. S., Renal hypertension. Diagnosis and surgical treatment. *N. Engl. J. Med.* **276,** 479–485 (1967).

Unraveling the Laboratory Model of a Syndrome: The Osmolality Model

Harry F. Weisberg, M.D.

Division of Biochemistry, The Ida Soref-David and Ruth Coleman Department of Pathology and Laboratory Medicine, Mount Sinai Hospital Center, P. O. Box 342, Milwaukee, Wis. 53201

A question recently submitted to the *Journal of the American Medical Association* by an anonymous physician stated, "In our hospital, serum and urine osmolality determinations, though quite helpful in managing patients with fluid and electrolyte problems, renal disease, and other conditions, are requested infrequently, as a rule, even for patients whose urine specific gravity is affected by the presence of solutes such as protein and glucose. What is your opinion concerning the value of requiring osmolality determinations as a routine in patients such as those mentioned?" *(1)*.

More important than Loeb's reply *(1)* that osmolality should not be a *routine* test is the fact that serum (plasma) and urine osmolality measurement is not readily available, even though some clinical laboratories were doing it some 30 years ago. The Chemistry Section of the College of American Pathologists' Survey Program has included serum osmolality since 1973 *(2)*. Of the more than 10 000 hospital and commercial laboratories in the United States, only 2238 participated in 1973, with an increase of about 12% by 1975 to reach 2618 participating laboratories. Of these, the percentage of laboratories submitting osmolality data were 30 and 44.6%, respectively, for 1973 and 1975—an increase of about 50%.

General Background

Let us review the physicochemical background of osmolality. Some pertinent references to the subject are by Holmes *(3–7)*, Weisberg *(8–9)*, and the recent monograph published by the American Physiological Society *(10)*.

The three phases of water—gas, liquid, and solid—exist in an equilibrium. Within any *single* phase the system is *bivariant*—both temperature and pressure can vary, within limits, without affecting the number of phases present. If *two* phases are present, the system is *univariant*; that is, at any given temperature or pressure, the other is fixed along the respective "curves." The fusion or melting or freezing "curve" separates the solid and liquid phases; the sublimation curve separates the solid and gas phases; and the vaporization or dew-point curve separates the liquid and gas phases. At normal atmospheric pressure (1 atm, ≈ 101 kPa) the boiling point of water is 100 °C and the melting (or freezing) poing is 0 °C (by definition, *one* locus along the sublimation curve, at the "double point").

If all *three* phases are present the system is *invariant*, being fixed at only one set of values, the "triple point." In 1948 it was proposed that the customary centigrade scale be called the Celsius scale after its developer (and to conform with the customary usage in Europe). When the Celsius scale was redefined in 1960 so that the freezing point was at the triple point (0.0075 °C), conforming with the physical scale, it was more reason to abandon the term centigrade since 100 divisions were no longer present. When exposed to "air" at 760 mmHg[1] water freezes at 0 °C, but when measured in an evacuated vessel under its own vapor pressure (4.58 mmHg) water freezes at the triple point.

The physicochemical ("colligative") properties of a solution, such as osmotic pressure, vapor pressure, boiling point, and freezing point of the solvent, depend upon the *number* of particles in solution at a certain temperature, the nature of the particles not being relevant *(10a, 10b)*. Because 1 mole of a substance contains 6.023×10^{23} molecules, equimolecular concentrations of the substances present in solution in the molecular (undissociated) state exert the same osmotic pressure, etc. When 1 mole of nonelectrolyte is dissolved in 1 kg of water, the osmotic

[1] 1 mmHg ≈ 133 Pa

pressure of the solution is *increased* by 22.4 atm (17 000 mmHg), the vapor pressure is *decreased* by 0.3 mmHg (the vapor pressure of pure water is 17.5 mmHg at 20 °C), the boiling point is *increased* by 0.52 °C, and the freezing point is *decreased* by 1.86 °C.

Osmosis (derived from the Greek *osmos,* "a push") is defined as the process of fluid movement through a semipermeable membrane from the outside into the compartment in which the nonpermeable solute is present. Osmosis is an important mechanism in the regulation of the blood volume and in the transfer of fluids into the vascular system at the venous end of the capillaries. *Osmotic pressure* is defined as that pressure which must be exerted on a solution on one side of a semipermeable membrane to oppose the action of osmosis due to the presence of the nonpermeable solute *(11)*, or as that pressure which must be exerted on a solution to make the vapor pressure of the solution equal to that of the pure solvent at the same temperature *(12)*. The van't Hoff approximation formula for osmotic pressure is

$$\pi \cong RTC\alpha \tag{1}$$

in which π is the *effective osmotic pressure*, $C\alpha$ is the *molar* concentration of solute α in a two-component solution, R is the general gas constant, and T is the Kelvin temperature. For an *ideal* solution near 0 °C a depression of the freezing point of 0.00186 °C has the same significance as an osmotic pressure of about 0.0224 atm, which is the pressure exerted at 37 °C by a column of water about 23.3 cm in height *(12)*.

One gram formula weight of an un-ionized substance—e.g., 180 g of glucose—dissolved in 1 liter of water will exert a certain osmotic pressure under standard conditions of temperature and pressure (STP). The term *osmole* refers to the *weight* of any element or compound that will exert the same osmotic pressure; one gram formula weight of any un-ionized substance equals one osmole. An aqueous solution containing 180 g of glucose per liter of solution contains 1 osmol and has an osmolarity of one. For more precise work in physical chemistry, *molal* solutions are used (e.g., 180 g of glucose or 1 osmol dissolved in 1 kg of solvent, e.g., water); such a solution has an *osmolality* of 1. More important for glucose is the fact that 180 g of glucose dissolved in 1 kg of water (volume at 20 °C, about 1001.8 ml) displaces water so that the final volume is about 1128 ml. The molarity of a solution (moles per liter of solution) will vary with the temperature because liquids expand and contract with changes in temperature; therefore, a specific temperature is specified, usually 20 °C. The molality of a solution (moles per kilogram of solvent) and the mole fraction do not vary with temperature.

The word "os-millimole" was introduced by Gamble *(13)* but the definition and (or) name has been altered by various authors. The adjective "molar" is descriptive of values pertaining to the "mole" concentrations and evolved into osmolar and milliosmolar. According to Wolf *(14–16)*, the term "osmolarity" is old, ill-defined, no longer in good usage, and should be discarded. Osmolality, however, is considered to be the most convenient index of osmotic ("particle") concentration and includes all of the colligative properties *(16)*. In equation 2

$$\underset{\text{(osmol/kg)}}{\text{Osmolality}} = \phi \text{nc} \qquad (2)$$

ϕ is the osmotic coefficient, n is the number of particles (nonelectrolytes) or ions (electrolytes) formed upon solution (dissociation) of the solute, and c is the number of moles per kilogram of solvent. Since osmolality is proportional to osmotic pressure and the freezing point depression (Δ °C) of solutions below that of water, it can be described in terms of the "molal freezing-point depression" or molal depression constant *(17)* of water ("1.86"), the basis of "cryoscopic" methods. Osmolality is numerically equal to the molality of an ideal solution (nonelectrolyte) having a similar freezing-point depression; this is expressed in equation 3.

$$\text{Osmolality} = \frac{\Delta}{1.86} \qquad (3)$$

Another measure of osmotic pressure is *osmosity* (equation 4), the numerical value of the *molarity of a solution of sodium chloride* having the same osmotic pressure *(15)*.

$$\text{Osmosity} = \frac{\text{osmolality}}{1.86} \qquad (4)$$

A milliosmole (mosmol) is one-thousandth of an osmole. The proper abbreviations for osmole and milliosmole are osmol and mosmol, respectively; older abbreviations were mosM and mosm for millosmolar and milliosmolal, respectively, whereas mOsm was the non-committed type of abbreviation used by some authors. For monovalent ions such as sodium, 1 mmol (or 1 mEq) is equal to 1 mosmol; for bivalent ions, 1 mmol (or 2 mEq) is equal to 1 mosmol. One gram formula weight of an ionized substance—e.g., 58.4 g of sodium chloride dissolved in 1 kg of water—will have twice as many particles (if completely ionized) as does an osmolal solution of glucose; therefore, the concentration for the saline is 2 osmol and the theoretical freezing point should be depressed to −3.72 °C. To calculate the concentration of protein as milliosmoles per liter, the concentration in grams per liter is divided by 80; in contrast, to calculate the protein as milliequivalents per liter, the concentration in grams per liter is multiplied by 0.243. The total milliosmolar concentration of normal plasma electrolytes is about 290 mosmol per liter of plasma (Table 1); on the basis of per liter of plasma *water*, the theoretical concentration is about 310 mosmol.

Values for freezing point depression by cryoscopy refer to *kilogram of water*, because the non-water components of serum or plasma do not freeze. About 310 mosmol/liter of plasma water (osmolarity) or 310 mosmol/kilogram of plasma water (osmolality) is the theoretical value for the "electrolytes." At the low solute concentrations and the relatively constant temperature of the body fluids, the difference between molar and molal concentrations is negligible.

Raoult's law (1884) states that the lowering of vapor pressure (in dilute solutions) is proportional to the mole-fraction of solute present in the mixture. The mole-fraction is more easily calculated from the molality than from molarity; for aqueous solutions, as the solute concentration approaches zero, at 4 °C the molality and molarity approach each other. At room temperature the difference is only 0.18% or an "error" (milliosmolar per liter of solution minus milliosmolal per kilogram of solvent) of about 0.5 mosmol for plasma determinations and about 2 mosmol for normally concentrated urines; at 37 °C the differences would be about 1 mosmol for plasma and 4 mosmol for urine. Osmolal concentration is not equivalent to urine relative density (specific gravity) *(19–21)* but, in normal circumstances (no glycosuria and no proteinuria), a rough approximation is 40 mosmol for one "unit" of relative density; thus relative density values of 1.010, 1.020, and 1.030 are roughly equivalent to 400,

Table 1. **Representative Normal Serum Electrolytes**[a]

Electrolyte	*mg/liter of plasma*	*mEq/liter of plasma*	*mmol or mosmol/ liter of plasma*
Cations			
Sodium	3260	142	142
Potassium	200	5	5
Calcium	100	5	2.5
Magnesium	24	2	1
Total	3584	154	150.5
Anions			
Bicarbonate	605[b]	27	27
Chloride	3657	103	103
Phosphate	34	2	1
Sulfate	16	1	0.5
Organic acids	175	5	5
Proteinate	65000	16	2
Total	69487	154	138.5
Total electrolytes	73071	308	289

[a] Modified from Weisberg *(18)*
[b] Milliliters of CO_2 per liter

800, and 1200 mosmol, respectively. Relative density and osmolality have been reported to have the following relationship *(22)*.

$$\text{osmol/kg} = [42.5 \pm 5] \cdot (\text{rel. density} - 1.000) \tag{4}$$

If a substance behaves as an ideal electrolyte (completely dissociated), the osmolality will depend upon the concentration of the substance in the solution times the number of ions or particles into which the substance dissociates. This is not so in the body, for more elaborate considerations must be made for the dissociation, which depends not only on the concentration but on the *activity* of the substance, which, in turn, may depend upon the pH, temperature, pressure, and even the concentration of the salt *(23)*. Other errors may be introduced by the presence of body enzymes, which break larger molecules into smaller fragments and so increase the osmolality without modifying the solute content; or ion-complexing substances may be present, decreasing the osmolality. Simple substances such as sodium chloride have been described by the Debye-Hückel theory to have an activity less than their concentration. The *activity coefficient* for sodium chloride is 0.80 at 0.1 formal (molar) concentration, 0.90 at 0.01 formal, and 0.96 at 0.001 formal; only at more dilute solutions does the coefficient approach one. The presence of a "common-ion" between two salts in a solution diminishes the dissociation. Thus, *osmolality is not necessarily a linear function of the amount of solutes (23)*. The ionic activity of electrolytes (sodium, potassium, and chloride) in plasma is affected by the displacement of water by plasma proteins *(24–26)* and by lipids *(27)*. In addition to the volume-displacement effect, chloride reportedly is bound by albumin *(24)*. Activity coefficients in plasma (approximately 0.15 formal) are 0.75 to 0.80 for sodium, potassium, and chloride, and 0.50 for calcium *(24–25)*.

The osmotic pressure exerted by the plasma nonelectrolytes can be neglected, except in cases of diabetes mellitus with excess glucose or of uremia with excess nonprotein nitrogenous compounds. A blood-sugar value of 1 g/liter equals 1/180 osmol, or 5.5 mosmol/liter; a blood sugar concentration of 5 g/liter equals 27.5 mosmol per liter. Urea has a gram formula weight of 60, of which only 28 is attributable to nitrogen. A serum urea nitrogen concentration is 0.11 g/liter; thus, 0.11/28 osmol equals 4 mosmol/liter. If the serum urea nitrogen rises to 1.5 g/liter the concentration would be 53 mosmol/liter.

The theoretical concentration of the "electrolytes" in terms of milliosmoles is 289 per liter of plasma, which becomes 308.8 when expressed per liter (or kilogram) of plasma water (Table 2). Expressed as per liter of plasma water, the total *electrolyte* concentration is about 330 mEq, in contrast to 309 mosmol/kg (liter) of plasma water. The decreased total of the latter is due to lesser osmolar concentrations of multivalent ions (calcium, magnesium, phosphate, sulfate, some organic acids, and especially proteinate). The Gibbs–Donnan equilibrium holds for the passage of "electrolytes" into the interstitial space. While the exact values for the intracellular osmole concentrations may vary, the total will be equal to that of the interstitial compartment, but the intracellular and interstitial total (317.2 mosmol/kg of water) is less than the intravascular value (318.7 mosmol/kg of plasma

Table 2. **Water Compartment Osmolalities**[a]

	IVF		ISF	ICF
	mosmol/l plasma	*mosmol/kg plasma H_2O*	*mosmol/kg H_2O*	*mosmol/kg H_2O*
Total "electrolytes"	289	308.8	307.3	230.5
Glucose	5	5.4	5.4	—
Urea nitrogen	4	4.3	4.3	4.3
Creatinine	0.2	0.2	0.2	9
Other compounds	—	—	—	73.4
Total "non-electrolytes"	9.2	9.9	9.9	86.7
Total milliosmoles	298.2	318.7	317.2	317.2
Corrected mosmol (0.93)	—	286.8	285.5	285.5
Osmotic pressure (mmHg) (mosmol × 19.3)	—	5535	5510	5510

[a] IVF, intravascular fluid compartment; ISF, interstitial fluid compartment; ICF, intracellular fluid compartment

water). The respective concentrations are 285.5 and 286.8 when corrected for the "activity" of the electrolytes and other forces acting on the molecules and ions (Table 2). The difference of 1.3 mosmol in the corrected osmolality values is essentially due to the plasma proteins and accounts for the difference of 25 mmHg (the plasma colloid "oncotic" pressure). The "osmolality" is converted by equation 5 into millimeters of mercury to express the "osmotic pressure" *(28)*.

$$\underset{\text{(mmHg)}}{\text{Osmotic pressure at 37 °C}} = 19.3 \times (\text{mosmol/kg}) \qquad (5)$$

If erythrocytes were placed in distilled water it would require 778 kPa (7.7 atm) of water pressure to prevent water from entering the cells, because the cells are osmotically active *(11)*. When the erythrocytes are in plasma, however, there are enough nondiffusible plasma proteins present to balance the osmotic activity of the blood-cell contents.

Methodology

Osmotic pressures can be obtained indirectly from *concentration estimates* by use of conductivity meters or specific-ion electrodes; these techniques cannot be used to monitor osmotic pressures of solutions with more than one salt and (or) significant amounts of nonelectrolytes *(29)*.

"Membrane" osmometers function on the basis of osmosis and are used primarily to determine number-average molecular weights in the range of 20 000 to 1 000 000, especially in polymer chemistry. Membrane osmometry has been used to measure the colloid osmotic pressure of blood *(30)*; this technique has been extended to be on-line and to also measure total osmotic pressure, albeit with a possible resolution value of 3 to 15 mosmol *(31)*.

Methods based on measuring the temperature effect (cryoscopy or ebullioscopy osmometers for freezing point or boiling point, respectively) are not as sensitive as membrane osmometers; present instruments make use of thermistors capable of detecting a change of ± 0.001 °C *(32–34)*. Physiological solutions are relatively concentrated (plasma is about 0.15 molar concentration) and extrapolation to infinite dilution is often difficult or impossible to carry out with precision; but the measured results are correlated to much more complex (clinical) phenomena *(34–37)*. An osmometer using ultramicro quantities has been described and utilized for body fluids *(38)*. An automated osmometer incorporating a tray containing 30 samples was described by Forman and Changus *(38a)*.

Johnson and Hoch *(39)* recalculated the quantities of sodium chloride needed to make up accurate osmolal standards for cryoscopy. For standards of 100, 300, 500, 1000, and 1400 mosmol/kg, the respective amounts of sodium chloride per kilogram of water at 20 °C are 3.094, 9.476, 15.93, 32.12, and 44.98 g. As an additional convenience, I have calculated the amount of sodium chloride needed to make up identical standards on the basis of per liter of solution at 20 °C; the respective quantities are 3.083, 9.429, 15.824, 31.758, and 44.26 g.

The vapor pressure method of osmometry was described by Hill *(40)* and Baldes *(41)*. For calibration standards, saline, sucrose, or mannitol can be used. Measurements can be made at any convenient temperature; for example, the plasma osmolality can be determined at 37 °C. The calibration constant of the instrument is independent of the nature of the solute *(42)*. A differential vapor pressure apparatus has been designed for research in analytical chemistry *(43)*.

A different type of vapor pressure (dew point) osmometer has been introduced (the Wescor osmometer). It measures water potential (which includes *osmotic* and *matric* water effects) as a measure of relative vapor pressure and, therefore, osmolality *(44–45)*.

Practically all of the osmometers used in clinical laboratories utilize the cryoscopic principle. Since 1973, some inroads have been made by the Wescor apparatus; however, this instrument has a greater coefficient of variation than do the big three—those manufactured by Advanced Instruments, Fiske, and Precision (Osmette). In no case was this significant; in 1975 the CV varied from 1.8 to 2.8%. Utilizing the "all-methods" mean, the Fiske osmometer had a positive bias of up to 2 mosmol/kg on 12 serum specimens, while the Advanced and Osmette osmometers had both positive and negative differences, to a lesser degree; the Wescor, however, had a bias ranging from +2.5 to −4.5 mosmol/kg *(2)*. The pros and cons of the various instruments must be debated with the different manufacturers *(45a)*. However, it must be pointed out that the Wescor osmometer does *not* detect volatiles, e.g., alcohol *(9, 46–48)*.

Both serum and plasma have been used for determining the "osmotic pressure." Table 2 illustrates that the total milliosmoles (or the osmotic pressure) is practically the same for plasma (based on water content), interstitial fluid, and intracellular fluid. The earliest reference to *whole-blood* osmometry by the freezing point technique is that of Gettler and Baker *(49)*, who reported a value of about 307 mosmol. Table 3 shows osmolality data obtained with a DigiMatic osmometer, Model 3DII (Advanced Instruments, Inc., Needham Heights, Mass. 02194) on hospitalized patients, primarily those undergoing cardiovascular (by-pass) surgery with multiple specimens on many of the subjects. Duplicate specimens of whole blood and of plasma (sodium heparin anticoagulant) were obtained. Because repeat determinations were done after thawing (and hemolysis of the cells), similar repeat determinations were done on the plasma specimens. The hematocrit values are low because of dilution of the blood while the patient was on the pump; in addition, blood from central blood vessels has a hematocrit approximately 91% of that of peripheral blood. The data in Table 3 show that the theoretical "equality" of osmotic pressure of the body fluid compartments does in fact exist. The test can therefore be performed

Table 3. **Plasma vs. Whole Blood Osmolality (Advanced Instruments, Inc., Osmometer)**

	n	*Mean*	*SEM*	*r*	
Plasma					
1	209	296	0.996	0.983	0.975
2	209	296	0.996		
1R[a]	171	297	0.994	0.984	
2R	166	297	0.993		
Whole blood					
1	211	299	0.998	0.973	
2	211	299	0.998		
1R	211	296	0.998	0.971	
2R	208	296	0.998		
Hematocrit					
1	51	31.3	0.994	0.999	
2	51	31.4	0.994		

[a] R, repeat determination (after thawing)

on whole blood as well as on plasma or serum; the advantage of a lesser volume is balanced by the disadvantage that the whole blood cannot be used for other tests (as could be done with serum or plasma).

According to Dormandy *(52)* the distinctions between osmolality, osmolarity, and osmotic pressure are not as important as those between the different types of osmolalities. *True* osmolality is a cardinal property of the solvent; it represents the effect of solutes on the vapor pressure of any solvent at any temperature. *Measured* osmolality is usually in terms of freezing-point depression because it is the most convenient, and does reflect changes in true osmolality. The *recorded* osmolality by cryoscopy determines changes in terms of millidegrees rather than milliosmoles, which would reflect the true osmolality only if the biological specimens were pure water solutions of sodium chloride. The *calculated* osmolality assumes ideal conditions in the plasma or urine, such as complete ionization, and does not take into account the unknown solutes that are present in abnormal conditions *(52)*.

Loeb *(1)* stated that plasma osmolality need not be routinely measured, because the calculated osmolality is sufficient in most cases. The first reported formula that I have found is by Warburg *(53)*, which states

$$\text{Calculated osmol} = 2\ \text{Na} + 2\ \text{K} + \text{glucose} + \text{urea} \qquad (6)$$

in which all concentrations are expressed as moles per liter. Similar formulas were previously tabulated by me *(8–9)*. The most commonly used formula is by Holmes *(3)*,

$$\text{Calculated osmol} = 1.86\ \text{Na} + \frac{\text{BS}}{18} + \frac{\text{UN}}{2.8} \qquad (7)$$

in which sodium is expressed as milliequivalents (millimoles) per liter and the sugar (BS) and urea nitrogen (UN) concentrations are expressed in milligrams per deciliter. About 22 formulas have been published *(8, 9, 55)* and, for "normal" concentrations, the normal value ranges from 273 to 304 mosmol/kg. The formula that I use *(8)* is

$$\text{Calculated osmol} = 2\ \text{Na} + \frac{\text{BS}}{20} + \frac{\text{UN}}{3} \qquad (8)$$

which incorporates a correction for the water content of the plasma. Other formulas *(55–57)* have a similar correction by dividing the calculated osmolality by the water content of the plasma. Formula 8 has met with favor, with comments such as ". . .ease of mental calculation. . ." *(55)* and ". . .fortuitously simple equation to arrive at an extremely accurate prediction of the plasma osmolarity. . ." *(58)*. The calculated osmolality is usually lower than the determined osmolality, but the difference will vary with the particular formula used.

Roscoe *(59)* found calculated values for urine osmolarity to be higher, especially in alkaline urines, than the determined values. He reports the possibility that the solute "activities" are less in alkaline urines than in more acid ones.

Dormandy *(52)* states that, "The attempt to bring true osmolality into line with calculated osmolality has been part of an egregious campaign to establish cryoscopy as a quick and easy alternative to estimating sodium, urea, glucose, and other solute concentrations. Admittedly freezing point determination is quick, but it is no quicker than modern methods which measure these concentrations more accurately." He goes on, with pointed humor, "True also that the technic is easy, but the handling of most laboratory apparatus nowadays calls for the split-level intelligence of a chimpanzee (for when the apparatus works) and a highly trained engineer (for when it breaks down)."

Other Formulas

The urinary/plasma (U/P) ratio of a substance is a measure of how much of the filtered water has been removed in forming the final urine; for example, a high U/P ratio for creatinine (e.g., 100) is seen in a concentrated urine, whereas a low U/P ratio (e.g., 5) is present in a dilute urine. The ratio of U_{osmol}/P_{osmol} ranges from 0.2 to 4.7; in normal circumstances, after overnight dehydration, the U/P osmol ratio usually exceeds 3. In conditions of poor concentrating ability the U/P osmol ratio is decreased (but still greater than 1) in such conditions as secondary amyloidosis of the kidney, cystic disease of the renal medulla, cystinosis, hypercalcemia, inappropriate secretion of antidiuretic hormone, potassium deficiency, and sickle cell anemia. In cases of intrinsic renal disease the U/P osmol ratio is about 1. The ratio is decreased below 1 in a normal response to a water load diuresis, in severe pyelonephritis, and in the various types of diabetes insipidus.

As analyzed by Homer Smith, the volume of urine excreted per minute consists of the water required to excrete the urinary solutes as a solution isotonic with the plasma (osmolal clearance) and the "excess" of water (free water clearance) per se.

$$\dot{V} = C_{osmol} + C_{H_2O} \tag{9}$$

The formula of the osmolal clearance (C_{osmol} or C_{osm}) is that of any urinary "clearance" formula.

$$C_{osmol} = \left[\frac{U_{osmol}}{P_{osmol}}\right]\dot{V} \tag{10}$$

The free water clearance (C_{H_2O}) is that amount of water excreted in the urine in excess of that needed to have the urine isosmotic with the plasma (U/P = 1). Thus from equation 9,

$$C_{H_2O} = \dot{V} - C_{osmol} = \dot{V}\left[1 - \frac{U_{osmol}}{P_{osmol}}\right] \tag{11}$$

The free water clearance is a quantitative value of the diluting ability of the kidney (milliliters of nonosmotically obligated water excreted per minute). Under ordinary circumstances the free water clearance has a negative value; in concentrated urines the value becomes more negative, whereas in dilute urines (diuresis) the free water clearance becomes positive.

When the urine is concentrated, the osmolal clearance (ml/min) is greater than the urine volume (ml/min), imparting a negative value to the free water clearance (see the above formula 9). Such a "negative free water clearance" can be regarded as the "medullary water reabsorption" during concentration or as the rate of "tubular water reabsorption" ($T^{C}_{H_2O}$) in excess of that which would result in urine isosmotic with plasma. The "negative free water clearance" is a quantitative value of the concentrating ability of the kidney (milliliters of non-osmotically obligated water *reabsorbed* per minute).

$$-C_{H_2O} = T^{C}_{H_2O} = C_{osmol} - \dot{V} = \dot{V}\left[\frac{U_{osmol}}{P_{osmol}} - 1\right] \qquad (12)$$

Water reabsorption is primarily under the control of vasopressin, the antidiuretic hormone (ADH), which under normal conditions is regulated by the plasma osmolality. Under basal conditions (water *ad libitum*), the normal concentration of plasma arginine vasopressin is 2.7 ng/liter (±1.4 SEM) at an osmolality of 287 mosmol/kg (±2 SEM) *(60–61)*. After fluid deprivation the plasma osmolality increases to 292 mosmol/kg with arginine vasopressin concentrations of 5.4 ng/liter, whereas under conditions of rapid water loading the respective values are 282 mosmol/kg and 1.1 ng/liter *(61)*.

Applications

Holmes *(3)* reported that normal ranges of solute excretion in urine vary with dietary intake. Healthy men had higher values than did women for concentration of solutes per kilogram of urine and for total daily excretion (Table 4). Obviously, the number of subjects in each group is not very large. The women may have a lesser solute output, related to the tendency for women to eat less than men. This difference is more obvious on comparing "healthy" and "hospitalized" subjects. A "patient"

Table 4. **Solute Excretion in Urine of Healthy Men and Women, in Comparison to Hospitalized Patients**[a]

		Solute excretion	
	n	*mosmol/kg*	*mosmol/d*
Healthy males	9	838 (392–1090)	1088 (767–1628)
Healthy females	16	745 (301–1093)	654 (433–1146)
"Convalescent" patients	12	549 (273–896)	596 (340–900)
"Bed" patients	6	562 (288–884)	547 (261–900)

[a] Values expressed as average (and ranges). Modified from Holmes *(3)*.

in the hospital is more likely to have impaired concentration of the urine, in addition to consuming less food.

Figure 1 illustrates the relationship between urinary solutes and urine volume as related to the dietary intake. An average adult diet will yield 1.2 osmol of solutes in the course of 24 hours; the 1200 mosmol will represent about 60 g of solids. The glomerular filtrate issuing from the proximal renal tubules has a solute concentration of about 0.3 osmol/liter or a relative density (specific gravity) of about 1.008. Figure 1 shows that 4 liters of urine water are required to excrete the total 1.2 osmol at such low concentrations. The usual normal concentration of solutes in urine is 0.8 to 1.3 osmol/liter, corresponding to a relative density of 1.020 to 1.032. These values are similar to those of Holmes *(3)* for healthy men. With normal renal function, the urine can be concentrated so that the same amount of solutes will be excreted in a smaller volume. For each additional unit of relative density there is more conservation of water, but there is also a greater workload presented to the kidneys. The usual increase of relative density from 1.008 to 1.020 will reduce the volume of urine required to excrete the solids from 4000 ml to 1500 ml, a saving of 2500 ml of urine water. However, a similar increase of relative density of 12 units from 1.020 to 1.032 will only achieve a saving of 600 ml of urine water (from 1500 ml to 900 ml) because of the greater energy expenditure required to concentrate urine at the higher relative density. The normal maximal concentration achieved by the kidneys is a relative density of 1.035, or a solute concentration of 1.4 osmol/liter. If it were possible to achieve a concentration of 1.8 osmol/liter (rela-

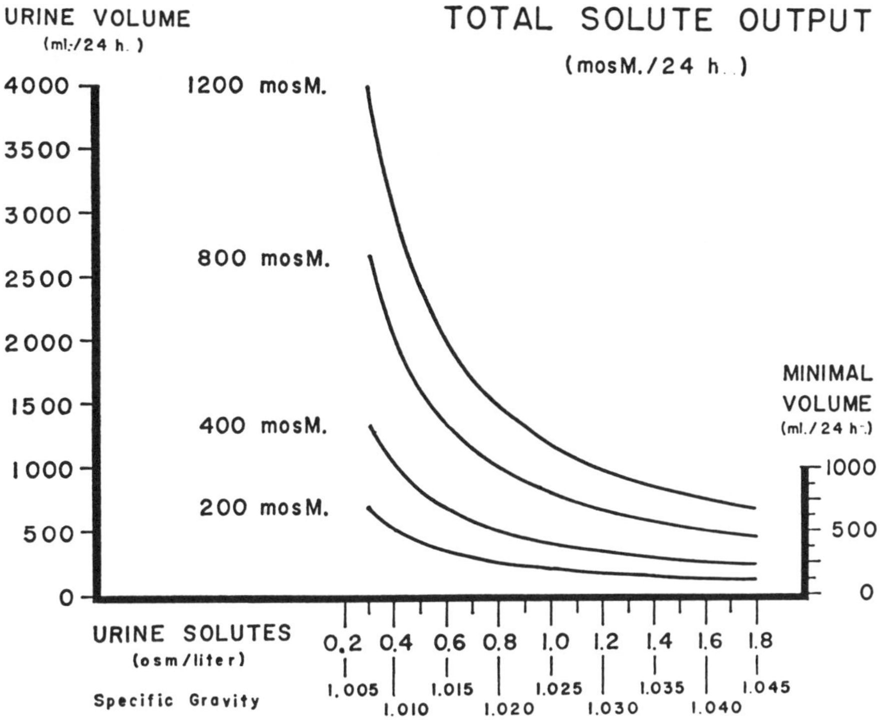

Fig. 1. Urinary solutes and urine volume as related to the dietary intake.

tive density of 1.045) the additional saving of water would be only 375 ml for the extra "concentration."

In the fasting patient, 0.8 osmol of solutes is produced in the course of a day by the breakdown of body tissues. This is contained in 2500 ml of glomerular filtrate and, with normal kidney function, can be concentrated in the distal renal tubules into a minimal volume of about 600 ml with a relative density of 1.030. The ingestion or infusion of 100 g of glucose will prevent some tissue breakdown, resulting in less water (about 350 ml) being needed to excrete the smaller amount of solutes produced. The individual on a high-carbohydrate, low-protein, low-salt diet produces only 0.2 osmol of solutes per 24 hours. This is contained in about 700 ml of water as glomerular filtrate, but can be excreted, with normal renal concentration, in about 150 ml of urine with a relative density of 1.030 (Figure 1).

It should be obvious that an evaluation of renal function (medical or surgical) should include not only the volume (as a sign of oliguria) but the concentration, expressed in terms of osmolality or relative density. In a similar fashion, reporting the concentration of sugar excreted by an uncontrolled diabetic (either "plus" or in grams per liter) is not sufficient; in addition the *volume* of the particular specimen should be recorded.

With normal kidney function, after a 14-h restriction of fluids, one should achieve a concentration of more than 800 mosmol/kg *(19)*. A value of 600 to 800 mosmol/kg indicates fair renal function, 400 to 600 is indicative of impaired renal function, and a value < 400 mosmol/kg is seen with severe renal impairment.

Various values have been reported for normal serum (plasma) osmolality *(8)*. Our normal range is 285 ± 5 (SD) mosmol/kg. Even under conditions of surgery, the values for whole blood fall within this range (Table 3).

Applications of serum and urine osmometry have been made under various situations. Table 5 lists some examples for plasma, urine, and other fluids. It was mentioned earlier that the Wescor osmometer does not detect volatiles such as ethanol. This is so because ethanol (the solute) will act as part of the solvent vapor; actually an increase in the vapor pressure caused by ethanol will seem to be a decreased osmolality *(46, 48, 80–81)*. A more important aspect of osmolality regarding ethanol is illustrated in Figure 2. The ethanol determinations were performed in response to physicians' requests for diagnostic value (none were on a legal basis regarding driving); 71 determinations were done in a six-month period. Values less than 0.1 g/liter were considered negative; 54 had ethanol detected in the blood and 17 were negative. The increase in plasma osmolality as a result of ethanol being present is expected *(56, 58, 74, 80–81)*. However, it is important to note that time and money can be saved by measuring plasma osmolality before the ethanol determination. If the osmolality is normal, the odds are that no ethanol is present *(9)*; ethanol could be present in the face of a normal osmolality if the subject had a low osmolality—due to chronic malnutrition or illness, drugs, severe hyponatremia, or water intoxication—before ingesting the alcoholic beverage. The practice of the osmolality screen has a greater advantage, however, as evident from the data in Figure 2, which shows a negative ethanol with 445 mosmol/kg and the

Table 5. **Some Applications of Osmometry**[a]

A. Plasma

1. Evaluate hypo/hypernatremia (hypo/hyperosmolality)
2. Antidiuretic hormone function
3. Toxicology (e.g., screen for alcohol in emergency room)
4. Substitute for multiple electrolyte determinations
5. Determine solute retention in oliguria/solute loss in polyuria
6. Δ-osmol
   a. Dialysis decisions (with relative density)
   b. Shock/trauma
   c. Coma
   d. Appropriate values regarding alcohol concentration
7. Follow therapy
   a. Intravenous
      (1) CHO
      (2) mannitol
      (3) saline
      (4) TPN
   b. Tube feedings
8. Osmolal and free water clearances
9. Determine solute absorption from burned body surface
10. Evaluate metabolic disturbances
    a. liver disease
    b. hyperosmolar non-kinetic coma

B. Urine

1. Evaluate solute excretion in renal disease
2. DD diabetes insipidus and psychogenic water drinking
3. Concentration/dilution test
4. Metabolites of small vs. large relative molecular mass (with relative density)
5. Osmolal and free water clearances
6. Evaluate solute replacement in polyruia
7. SIADH
8. Hypokalemia
9. Routine (instead of relative density) in patients after x-ray contrast media studies

C. Other

1. Hemodialysis
   a. check bath concentrations
   b. vs plasma: need to remove water
   c. disequilibrium syndrome
2. Check concentration of solutions
   a. anesthesia
   b. baby formulas
   c. "cardiac arrest" solution
   d. reconstitution of lyophilized material (e.g., control serum)
3. Check chemical purity of solution
4. Monitor chromatography eluates
5. Amniotic fluid in relation to fetal maturity
6. Milk adulteration
7. Postmortem (vitreous humor)

Table 5. (Continued)

8. Check "wash" after thawing/washing frozen erythrocytes
9. Total carbon dioxide content
10. Screening for cystic fibrosis (instead of sweat electrolytes)

[a] Based on references *3, 35–37, 62–79b*..

very low ethanol (0.25 g/liter) with 378 mosmol/kg. The regression equation for our data is $y = 0.00286x - 0.823$ ($r = 0.88$), which is very similar to that reported by Champion et al. *(81)*.

Table 6 lists various causes of hypoosmolality ("hyponatremia"); Table 7 lists various conditions with defects of urine

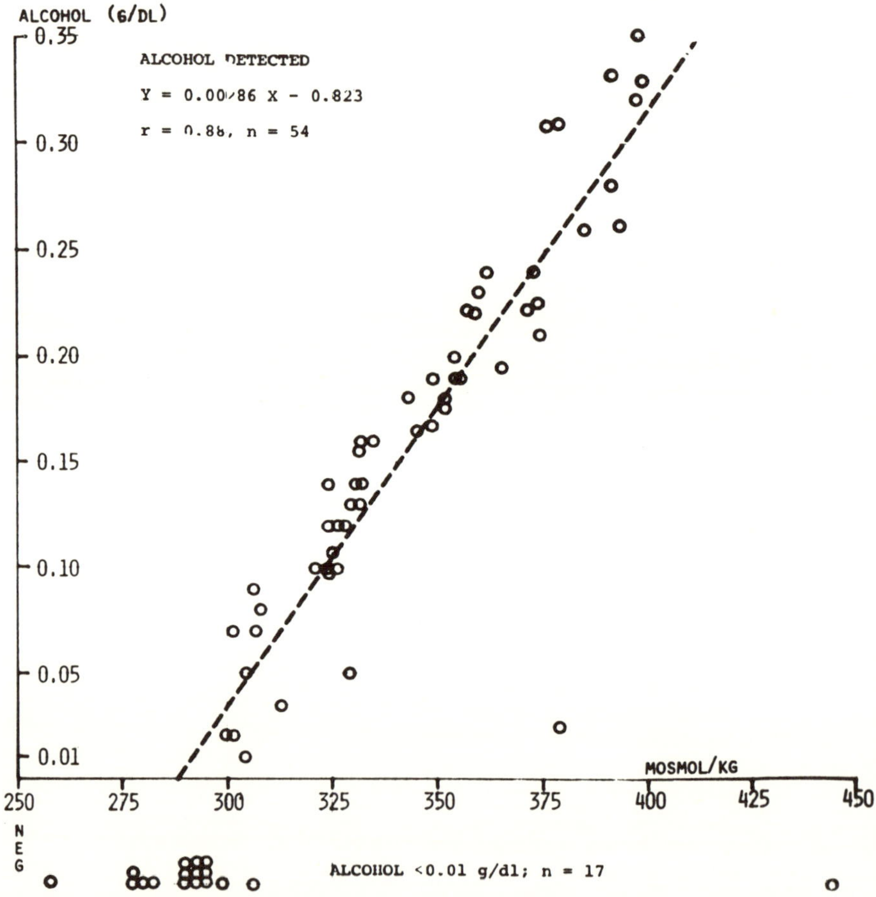

Fig. 2. Response of plasma osmolality to increasing concentrations of alcohol.

Alcohol determined with Ethyl Alcohol Stat-Pak (Calbiochem, La Jolla, CA, 92037) and osmolality determined with DigiMatic osmometer, model 3DII (Advanced Instruments, Needham Heights, MA, 02194).

concentration; and Table 8 lists various causes of hyperosmolality.

The difference between the determined osmolality and the calculated plasma osmolality has been called Δ-osmol, Δ-osmolality, and osmolal discriminate (mOsm-D) *(71)*. As mentioned previously, the amount of difference will vary with the formula used; usually the determined is greater than the calculated osmolality. However, the calculated may be larger because of errors in the actual determination of sodium, glucose, and urea nitrogen, or in the calculations. One potential source of error is the possible reaction between glucose oxidase (EC 1.1.3.4) and oxazepam (7-chloro-1,3-dihydro-3-hydroxy-5-phenyl-2H-

Table 6. **Causes of Hypoosmolality ("Hyponatremia")**[a]

- I. *Spurious or Pseudo or "Delusional" Hyponatremia*
  - A. Osmolality ↑
    - 1. Plasma diluted by i.v. fluid
  - B. Osmolality normal
    - 1. Lab error
      - a. Dilutor ("timed" aliquot)
      - b. Hyperviscosity of plasma
      - c. Nonlinear response at high Na
    - 2. Plasma water displacement
      - a. Hyperlipemia
        - i. Hepatic disease
        - ii. Idiopathic
        - iii. Nephrotic syndrome
        - iv. Uncontrolled diabetes mellitus
      - b. Hyperproteinemia
        - i. Macroglobulinemia
        - ii. Multiple myeloma
  - C. Osmolality ↑ (N)
    - 1. Diabetes mellitus
    - 2. Iatrogenic
      - a. Hypertonic CHO
      - b. Mannitol
    - 3. Renal insufficiency
- II. *True Hyponatremia*
  - A. Overhydration ("dilutional" hyponatremia)
    - 1. Without edema
      - a. SIADH (nonosmotic stimulation)
        - i. ADH production by tumor
          - (a) Carcinoma
            - (i) Duodenum
            - (ii) Lung
            - (iii) Pancreas
            - (iv) Prostate

Table 6. (Continued)

- (b) Hodgkin's disease
- (c) Leukemia
- (d) Lymphosarcoma
- (e) Thymoma

ii. Pulmonary ("salt wasting") Disease
- (a) Abscess, lung
- (b) Cavitation (e.g., *Aspergillus*)
- (c) Pneumonia
- (d) Tuberculosis
- (e) Ventilation R_x

iii. CNS Disorders
- (a) Brain tumor or abscess
- (b) Bulbar polio
- (c) Central pontine myelinosis
- (d) Cerebral atrophy
- (e) Cerebral vascular thrombosis
- (f) Encephalitis
- (g) Idiopathic
- (h) Infectious polyneuritis (peripheral)
- (i) Landry–Guillain–Barré syndrome
- (j) Meningitis (Tbc or purulent)
- (k) Psychosis, acute
- (l) Porphyria, intermittent, acute (with or without hypochloremia)
- (m) Schizophrenia, acute
- (n) Subarachnoid hemorrhage
- (o) Subdural hematoma
- (p) Systemic lupus erythematosus
- (q) Trauma

iv. Drugs
- (a) Amitryptyline
- (b) Carbamazepine
- (c) Chlorpropamide
- (d) Chlorthalidone
- (e) Clofibrate
- (f) Cyclophosphamide
- (g) Fluphenazine
- (h) Oxytocin
- (i) Thiazides
- (j) Thiothixene
- (k) Vasopressin
- (l) Vincristine

b. Other non-osmotic stimuli

i. Anxiety, fear, pain, and trauma

ii. Autonomic nervous system (β-adrenergic) stimulation

iii. Idiopathic
- (a) Infections
- (b) Metabolic edema of females
- (c) Postoperative (especially mitral commissurotomy)
- (d) Transient hyponatremia
    - (i) Hyperventilation

Table 6. (Continued)

(ii) Obscure CNS disease
(iii) Obscure hypovolemia
c. Endocrine deficiency
i. Adrenal
ii. Pituitary
iii. Thyroid
d. Excess water
i. "Acute" (exogenous) intoxication
(a) Enteral
(i) "Psychogenic" (e.g., beer potomania, "contests," schizophrenia)
(ii) "Miner's cramps"
(iii) Water enema (in megacolon)
(b) Parenteral
(i) Excess parenteral CHO/W (without electrolytes)
(ii) Transplacental in newborn
(iii) Transurethral irrigation
ii. "Chronic" (endogenous)
(a) Iatrogenic
(i) Anesthesia per se
(ii) Barbiturates
(iii) Excess therapy with ADH
(iv) Excess water + oxytocin in obstetrics
(v) Meperidine
(vi) Morphine
(vii) Stress (e.g., surgery)
(b) Water shift (ECF hyperosmolality)
(i) Hyponatremia (glucose "remains" in ECF)
(ii) Non-electrolytes (e.g., diabetes mellitus, iatrogenic, renal insufficiency)
(iii) Potassium depletion
(iv) "Sick-cell" syndrome
2. With edema ("distributional" or "depositional" hyponatremia)
a. Normal GFR and renal hemodynamics
i. Idiopathic nephrotic syndrome
ii. Local accumulations
(a) Bowel obstruction
(b) Burns
(c) Lymphedema
(d) Peritonitis
(e) Trauma
iii. Oxytocin R_x
iv. Sodium shift into erythrocytes
(a) Liver diseases
b. Abnormal GFR and renal hemodynamics
i. CHF
ii. Excess aldosterone
iii. Hepatic cirrhosis
iv. Nephrotic syndrome
v. Renal failure

Table 6. (Continued)

B. Euhydration
  1. Diuretic R_x
  2. Osmostat reset
    a. Chronic debilitating conditions
      i. Anorexia nervosa
      ii. Carcinoma
      iii. Tuberculosis
    b. Cirrhosis
    c. Malnutrition
    d. Potassium depletion, severe
    e. Toxemia of pregnancy

C. Dehydration or "depletional" hyponatremia (Especially with inadequate and hypotonic fluid therapy)
  1. GI Tract
    a. Diarrhea
    b. Fistula, external (biliary, intestinal, or pancreatic)
    c. Iatrogenic
      i. Cation-exchange resins
      ii. Enemas (esp. megacolon and chronic constipation)
      iii. Inadequate intake (e.g., low salt diet)
      iv. Intubation (biliary or intestinal)
      v. Lavage
      vi. Laxatives
      vii. Paracenteses
    d. Malabsorption syndrome
    e. Malnutrition
    f. Rectal mucorrhea
    g. Steatorrhea
    h. Vomiting, chronic
  2. Hemorrhage (severe)
  3. Lungs
    a. Bronchorrhea in pulmonary adenomatosis
  4. Renal
    a. Osmotic diuresis
      i. Endogenous
        (a) Azotemia
        (b) Glycosuria
      ii. Exogenous (iatrogenic)
        (a) Hypertonic carbohydrate solution
        (b) Glycosuria
    b. Subarachnoid ureteral anastomosis (for hydrocephalus)
    c. Tubular reabsorption of Na decreased
      i. Adrenal insufficiency (glucocorticoids $\downarrow$)
      ii. Azotemia, prerenal
      iii. Diuresis
        (a) Drugs, various (see *Clin. Chem* **21**(5): April 1975)
        (b) Diuretic therapy per se
        (c) In R_x of "intoxications"
        (d) Lithium (therapy of manic psychoses) toxicity
      iv. Renal ("salt-losing") disease

Table 6. (Continued)

(a) Acute tubular necrosis (recovery phase)
(b) Fanconi syndrome
(c) Glomerulonephritis, chronic
(d) Hypercalcemia (?)
(e) Hypokalemia (?)
(f) Medullary cystic disease
(g) Nephritis, interstitial
(h) Nephrotic nephritis (diuresis phase)
(i) Obstructive uropathy
(j) Polycystic kidneys
(k) Pyelonephritis, chronic
(l) Renal tubular acidosis (proximal)

5. Skin
   a. Burns
   b. Extensive skin lesions
   c. Sweating
      i. Abnormal
         (a) Adrenal insufficiency
         (b) Cystic fibrosis (mucoviscidosis)
      ii. Normal (reaction to heat stress)

[a] Modified from Weisberg *(82)*, Goldberg *(83)*, Fanestil *(84)*, and Berl et al. **(85)**.

Nonstandard abbreviations used:

ADH—antidiuretic hormone	IV—intravenous
CHO—carbohydrate	N—normal
CHF—congestive heart failure	Na—sodium
CNS—central nervous system	R_x—treatment
ECF—extracellular fluid	SIADH—syndrome of inappropriate secretion of ADH
GFR—glomerular filtration rate	W—water

1,4-benzodiazepin-2-one) *(88)*. Rubin et al. *(89)* reported patients with various disease states in whom the determined osmolality exceeded the calculated value by 40 to 125 mosmol/kg; 98% died within two weeks of the hyperosmolality being found. Boyd and Mansberger and their associates *(65, 67)* reported large differences in patients who were admitted to the hospital in hemorrhagic shock; the difference returned to normal in those who survived, but it persisted at about 30 mosmol/kg in those who died. The osmoles may be endogenous (idiogenic) as mentioned, or may be exogenous due to alcohol, drugs, etc. The Δ-osmol may be increased in the presence of a "normal" determined osmolality if water displacement has occurred due to the increased concentration of protein or triglycerides in the plasma.

There is a definite relationship between plasma antidiuretic hormone and plasma osmolality. Below a certain threshold value for plasma osmolality (280 mosmol/kg), antidiuretic hormone is

Table 7. **Defects of Urine Concentration**[a]

- I. *Loop of Henle (Countercurrent Multiplier)*
  - A. Diuretics
    - 1. "Loop"
    - 2. Mercurial
  - B. Protein deficiency
    - 1. Malnutrition
- II. *Collecting Tubule (ADH)*
  - A. Diabetes insipidus
  - B. "Nephrogenic" D. I.
    - 1. Hereditary
    - 2. Hypercalcemia
    - 3. Amyloidosis
    - 4. Therapy
      - a. Colchicine
      - b. Demeclocycline
      - c. Lithium
      - d. Xanthine
  - C. Water diuresis
- III. *Vasa Recta (Countercurrent Exchanger)*
  - A. Sickle cell anemia
  - B. Hyperviscosity syndrome
  - C. Arterial hypertension
- IV. *"Mixed" (Osmotic Diuresis)*
  - A. Diabetes mellitus
  - B. Iatrogenic
    - 1. Glycerol
    - 2. High-protein tube feeding
    - 3. Mannitol
    - 4. Sodium sulfate
  - C. Lactic acidosis, chronic
  - D. Fanconi syndrome
    - 1. Primary
    - 2. Secondary
      - a. Cadmium
      - b. Cyclamates
      - c. Outdated tetracycline
      - d. Wilson's disease
- V. *Not Classifiable*
  - A. Impaired concentration predominant
    - 1. Potassium depletion
    - 2. Interstitial nephritis
    - 3. Papillary necrosis
    - 4. Acute urinary tract infection
    - 5. Iatrogenic
      - a. Phenacetin
      - b. Amphotericin
      - c. Propoxyphene
  - B. General decreased renal function (impaired concentration possible)
    - 1. Amyloidosis
    - 2. Multiple myeloma

Table 7. (Continued)

3. Sjogren's syndrome
4. Polycystic kidneys
5. Medullary cystic disease
6. Obstructive uropathy (post-obstructive diuresis)
7. Pyelonephritis
8. Radiation nephritis
9. Transplanted kidney
10. Sarcoidosis
11. Periarteritis nodosa (?)

[a] Modified from Jamison and Maffly *(86)* and Brunner and Thiel *(87)*.

Table 8. **Causes of Hyperosmolality [Modified from Weisberg *(82)*]**

I. *Hypernatremia*
 A. Dehydration (Hemoconcentration)
  1. Excess loss (vs. Na loss)
   a. GI tract
    (1) Diarrhea
    (2) Vomiting
   b. Iatrogenic
    (1) Dialysis
    (2) Diuretics
    (3) Drugs, various (See *Clin. Chem.* **21,** No. 5, April 1975)
    (4) Hyperosmolar infusions
   c. Renal
    (1) Chronic renal insufficiency
    (2) "Diabetes insipidus;"
     (a) neurogenic
     (b) partial (e.g., hypokalemia, hypercalcemia, drug therapy)
     (c) nephrogenic
    (3) Nephrotic syndrome
    (4) Dialysis
     (a) malfunction of proportioning system
     (b) loss of water
    (5) Osmotic diuresis
     (a) diabetes mellitus
     (b) urea (e.g., post-obstructive uropathy, diuretic phase of acute renal failure, etc.)
   d. Respiratory
    (1) Dyspnea
    (2) Fever
    (3) Hyperventilation
    (4) Oxygen therapy (dry)
    (5) Tracheostomy
   e. Skin
    (1) Environmental heat loss
    (2) Open therapy of burns
  2. Inadequate intake
   a. Adipsia

Table 8. (Continued)

b. Inability to swallow
c. Nausea and (or) vomiting
d. Physical disability
    (1) Fracture
    (2) Infancy
    (3) Old age
    (4) Stroke
e. Stupor and (or) coma
f. Surgery of head & neck regions

B. "Encephalogenic"
1. After prefontal leukotomy
2. Carcinoma, primary
3. Cerebral lacerations
4. Cerebral vascular accident
5. Essential hypernatremia (altered osmoreceptors)
6. Head injury
7. Meningitis
8. Metastases to brain
9. Pituitary tumors
10. Poliomyelitis, bulbar, severe
11. Subarachnoid hemorrhage
12. Subdural hematoma

C. Excess of sodium
1. "Alkaline" salts
2. Na salts of drugs (See *Clin. Chem.* **21,** No. 5, April 1975)
3. "Salt poisoning" (infants)
4. Dietary "supplements"
5. Excess "saline" (i.v.)
    a. Therapeutic abortion (NaCl)
    b. Post-cardiac arrest ($NaHCO_3$)
    c. Lactic acidosis therapy ($NaHCO_3$)
6. Excess mineralocorticosteroids
    a. Aldosterone
        (1) Conn's syndrome
        (2) Familial pseudoaldosteronism
        (3) Iatrogenic
    b. Cortisol
        (1) Cushing's syndrome
        (2) Iatrogenic
    c. Desoxycorticosterone
7. "Contamination"
    a. Detergents
    b. "Wrong" anticoagulants

II. Non-sodium causes

A. Alcohol
B. Glucose
C. Mannitol
D. Urea nitrogen
E. Idiogenic osmoles
    1. Shock
    2. Trauma

suppressed to low or undetectable concentrations, whereas above this value the hormone concentration increases in proportion to the increase in osmolality. A water load of 20 ml/kg body weight will usually lower plasma osmolality to an average of 282 mosmol/kg, starting a maximum diuresis in 1 to 1.5 h, whereas an infusion of hypertonic saline will cause release of arginine vasopressin when a value of about 287 mosmol/kg is achieved. Thus the change in plasma osmolality from full diuresis to start of antidiuresis is about 3% *(90)*.

Table 9 lists various factors that affect the release of antidiuretic hormone. It had been considered that plasma volume (as monitored by low-pressure baroreceptors in the left atrium and pulmonary veins) was a more important factor and would take precedence over the effect of the osmoreceptors. As mentioned above, once the osmolality threshold is passed, the concentration of arginine vasopressin increases linearly with the rise in osmolality. However, Robertson *(61)* reports his studies showing that the rise in arginine vasopressin has a curvilinear relation to the fall in blood volume; little or no increase in it was detected until the blood volume (of rats) was decreased by 8 to 10%. In man, the "hypovolemic threshold" is about 10 to 20% *(61)*. The antidiuretic hormone system is important in maintaining the *tonicity* of body fluids; blood volume is defended only when the hypovolemia is of sufficient degree to be life-threatening, at which time the tonicity will be compromised to conserve volume.

The high-pressure baroreceptors of the carotid sinus and aortic arch—blood-pressure monitors—have a threshold for stimulation of arginine vasopressin when the blood pressure drops by 5%. Robertson *(61)* demonstrated that the osmoreceptor system and the hemodynamic control system (blood volume and blood pressure) interact. Animal studies have shown that the hemodynamic system does not abolish the effect of a concurrent osmotic stimulus. What does occur, however, is that the osmoreceptor threshold value is decreased by about 5 mosmol/kg in the presence of hypovolemia or hypotension. As the hypovolemia or hypotension becomes more severe, much more arginine vasopressin is present in the plasma, exceeding maximum antidiuresis. At this point the primary function of arginine vasopressin may be to increase vascular tone *(61)*. Table

9 lists many additional factors affecting vasopressin secretion and release. Even the psychological stress of venipuncture in healthy subjects may evoke a vasovagal reaction characterized by a transient fall in blood pressure with a marked increase in release of arginine vasopressin *(61)*.

Vasopressin release is related to plasma osmolality; both should be evaluated for a proper diagnosis regarding posterior pituitary function. Injection of hypertonic saline causes an increase in plasma osmolality and a concomitant rise in plasma arginine vasopressin concentration. An injection of hypertonic mannitol or sucrose to cause the same rate of rise of osmolality will result in increases of arginine vasopressin similar to those for hypertonic saline, even though the sodium concentration increases with saline injection and falls with injection of mannitol or sucrose *(61)*. The intracarotid injection of hypertonic urea causes an increase in osmolality but has no effect on arginine vasopressin; this correlates with the fact that hypertonic urea will not inhibit a water diuresis in conscious dogs *(93)*. In normally hydrated humans, infusion of hypertonic glucose will have the identical effects as mannitol on osmolality, volume, and sodium concentration but will cause a paradoxical *fall* in the arginine vasopressin concentration. Thus it is the tonicity and (or) hydration volume of cells in the brain rather than in peripheral tissues that controls water balance *(61)*. It is not yet known whether solutes other than glucose can evoke the same paradoxical effect. It must be pointed out that the diuresis seen in uncontrolled diabetes mellitus (or other hyperglycemic states) can be due to increased excretion of carbohydrate but also to the suppression of arginine vasopressin secretion *(61)*.

Katz *(94)*, writing on hyponatremia, suggested that the clinician use the rule that for each 100-mg increase in glucose/100 ml of serum, one should expect a decrease of 1.6 mosmol in serum sodium per liter. This was in contrast to what had been used for many years, namely, an expected decrease of 2.8 mosmol/liter. His reasoning was based on several assumptions: no free glucose present in the intracellular fluid, sodium present only in the extracellular fluid, the solute concentration in the intracellular fluid unchanged, and osmotic equilibrium was achieved only by water leaving the intracellular fluid, without any reference to possible changes in intake or excretion. In cases of hyponatremia

Table 9. **Factors Affecting Secretion and (or) Release of Antidiuretic Hormone (ADH)**

Increased	*Decreased*
A. *Osmoreceptors*	
1. ECF Osmolality $\uparrow$ (> "287") (ICF $\downarrow$ in SHT nuclei)	1. ECF osmolality $\uparrow$ (> "282") (ICF $\uparrow$ in SHT nuclei)
a. Dehydration	a. Overhydration
b. Hyperosmolar infusions	b. Infusions containing only CHO
B. *Non-osmotic*	
1. "Volume" baroreceptors deflated	1. "Volume" baroreceptors distended
a. Low pressure baroreceptors of left atrium and pulmonary veins	
(blood volume decreased)	(blood volume increased)
(1) hypovolemia (hemorrhage, dehydration)	(1) hypervolemia
(2) upright position	(2) reclining position
(3) positive-pressure breathing	(3) negative-pressure breathing
(4) mitral commissurotomy	(4) acute cold exposure (peripheral vasoconstriction and shift of blood centrally)
b. Carotid sinus and aortic arch (high pressure baroreceptors)	
(blood pressure decreased)	(blood pressure increased)
(1) ECF (or TBV) $\downarrow$	(1) ECF (or TBV) $\uparrow$
(2) Cardiac output $\downarrow$	(2) Cardiac output $\uparrow$
2. Drugs	2. Drugs
a. β-adrenergics	a. α-adrenergics
b. Anticonvulsants	b. Alcohol
c. Barbiturates	c. Atropine
d. Carbamazepine	d. Chlorpromazine
e. Clofibrate	e. Diphenylhydantoin
f. Chlorpropamide	f. Epinephrine (?)
g. Cyclophosphamide	g. Reserpine
h. Demerol	
i. Isoproterenol	
j. Morphine	
k. Nicotine	
l. Vincristine	
3. Emotional stress ("psychoses") and/or "alarm" reaction	3. Hypnotic suggestion and (or) psychologic conditioning
4. Hypothalamic blood temperature $\uparrow$	4. Hypothalamic blood temperature $\downarrow$
5. p_{O_2} $\downarrow$ (stimulation of carotid body chemoreceptors)	5. CO_2 inhalation
6. Acute pain	
a. Surgical procedures	
b. Traction on viscera	
7. Anesthesia, especially ether	
8. Nausea and (or) vomiting	
9. Section of afferent limb of neurohumoral reflex arc	
a. Vagus nerve (for left atrium & pulmonary veins)	

Table 9. (Continued)

b. Glossopharyngeal nerves (for carotid sinus)

Nonstandard abbreviations used:
CHO—carbohydrate
CO_2—carbon dioxide
ECF—extracellular fluid
ICF—intracellular fluid
p_{O_2}—partial pressure of oxygen
SHT—supraoptic hypophyseal tract
TBV—total blood volume

Modified from Weisberg *(18)*, Kleeman and Fichman *(91)*, and Schrier and Berl *(92)*.

it is important to know the plasma osmolality (whether determined or calculated) rather than the expected decrease of sodium caused by the hyperglycemia. Serum sodium values may be normal or high in the presence of severe hyperglycemia *(95)*.

Arieff et al. *(96)* discuss the effects of hyperosmolality on the brain. Thus, in experimental animals, restlessness and increased irritability are seen at values of 330 to 350 mosmol/kg, ataxia and tremulousness at 375 to 400, tonic spasm and focal or grand mal seizures at 400 to 420, and death when values exceed 420 mosmol/kg. Similarly, the sensorium of patients is depressed when the blood sugar exceeds 6.0 g/liter (without ketosis); when the average osmolality values were 330 mosmol/kg the patients were alert, at 360 the patients were obtunded, at 385 the patients were stuporous, and at an average of 410 the patients were in coma.

The brain solutes are increased in the presence of hyperosmolality but the types of reactions vary with the cause and duration of the increased osmolality *(96)*. After 1 h of hypernatremia (>170 mosmol/liter) the brain solutes (control at 300 mosmol/kg) became about 380, the cause being increased concentrations of the electrolytes (sodium, potassium, and chloride). After 4 h of hypernatremia the brain solutes increased further, to about 400 mosmol/kg, with 40% of the total increment being due to "idiogenic" osmoles. After a week of hypernatremia the brain solutes concentration was 390 mosmol/kg, and 52% of the change was due to idiogenic osmoles.

In animals with plasma osmolality of about 350 mosmol/kg induced by glucose or mannitol, some of the carbohydrate is found in the brain. However, after 4 h of hyperglycemia more than half of the increased brain osmolality (to about 370 mosmol/kg) is due to idiogenic osmoles, with lesser contributions from the glucose or electrolytes. With mannitol infusion for 4 h the major increase of the brain osmolality is due to electrolytes and mannitol *(96)*. Thus the brain's production of idiogenic osmoles is a response to the hyperglycemia rather than to the hyperosmolality per se. Similarly, for animals in acute renal failure for three to four days the increase in brain osmolality is due to urea whereas in chronic renal failure (two to three weeks) about half of the increase in brain solutes is due to idiogenic osmoles.

Arieff et al. *(96)* conclude their studies by stating that hyperosmolal states induced by the infusion of mannitol or glycerol (exogenous materials) result in loss of brain tissue water and the secondary increase of electrolytes in the brain. The brain is unable to regulate its intracellular fluid volume and no idiogenic osmoles are generated. If the plasma osmolality is increased by endogenous substances (e.g., glucose, urea, or sodium) the brain can maintain its intracellular fluid volume, after an initial period of disequilibrium. About half of the increased brain solute is due to idiogenic osmoles; the increased intracellular solutes enables the intracellular volume to be intact in the face of the elevated plasma osmolality *(96)*.

Moses and Miller *(90, 97)* have reported a simple system for differentiating hypotonic polyuria, after excluding conditions such as drug toxicity, hypercalcemia, and hyperkalemia (see Table 7). It would be best to determine the actual arginine vasopressin concentration in the plasma, but for clinical purposes the scheme presented in Figure 3 is satisfactory. A random plasma specimen is evaluated for osmolality or sodium concentration, which will distinguish between a normal water overload and diabetes insipidus. The most important aspect of the procedure is the standardized dehydration test (Figure 3). Fluids are restricted overnight, but if the subject has severe renal disease or is unable to go without fluids all night the "dehydration" is started in the morning. Hourly urine specimens are collected in the morning until three consecutive specimens reach a plateau—

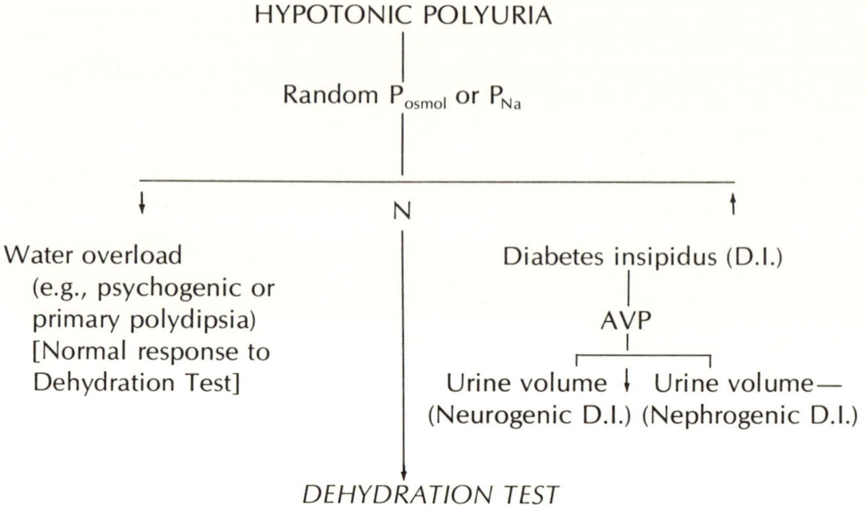

DEHYDRATION TEST

(a) Restrict fluids until 3 consecutive hourly urines reach plateau of osmolality (differ by less than 30 mosmol/kg)

(b) Determine P_{osmol}

(c) Administer 5 u AVP (arginine vasopressin) subcutaneously

(d) Collect urine (2 periods of 30 min) and determine U_{osmol}

Figure 3. Flow-chart to distinguish various types of hypotonic polyuria.

Based on references *90* and *97*. Also see Table 10.

urine osmolality does not differ by more than 30 mosmol/kg. At this time a blood sample is drawn for plasma osmolality and then 5 units of arginine vasopressin are administered subcutaneously. Two urine specimens (30 min each) are collected and osmolality is again determined. Some of the data reported by Miller et al. *(97)* is abstracted in Table 10; ranges are given to show the spread of the data. (For other details the reader is referred to the original article.) The important diagnostic features are that a normal individual will not increase the urine osmolality after arginine vasopressin, but the value may actually decrease. Patients who had primary (psychogenic) polydipsia showed a slight increase in urine osmolality after arginine vasopressin. Severe neurogenic diabetes insipidus patients had an increase of more than 50%, whereas the nephrogenic diabetes insipidus subjects had a lesser increase. Those individuals with "partial" diabetes insipidus had increases greater than normal but much less than those for severe diabetes insipidus.

Table 10. **Results of Dehydration Test to Distinguish Various Types of Hypotonic Polyuria**[a]

Diagnosis	*n*	U_{osmol} *plateau*	P_{osmol}	U_{osmol} *after AVP*	U_{osmol} *% change*
"Normal"					
Normal	10	866–1446	281–294	596–1451	−36 to +0.4
Primary polydipsia	7	467–943	279–294	684–1070	+1 to +13
Diabetes insipidus					
Partial	11	327–729	286–303	397–807	+9 to +67
Severe	18	70–282	291–332	190–963	+51 to +792
Nephrogenic (congenital)	2	96–151	302–320	139–210	+3 to +17

[a] Data abstracted from Miller et al *(97)*, showing ranges. Also see Figure 3.

Baek and his colleagues *(98–101)* have utilized the free water clearance (formula 11 above) to predict and follow the course of acute renal failure in postoperative patients. Figure 4 summarizes their schema. If the free water clearance is zero or positive (abnormal), plasma osmolality is measured, to distinguish between a physiological response to a water overload (decreased plasma osmolality) and a potential acute renal failure (normal or increased plasma osmolality). They advocate the intravenous injection of 80 to 400 mg of furosemide; if the free water clearance then becomes negative the patient has normally functioning kidneys. If the C_{H_2O} does not change or becomes more positive after the furosemide it proves the presence of acute renal failure. They also claim *(98–101)* that the intravenous administration of 50 mg of antidiuretic hormone per kilogram of body weight will be useful in showing that the renal tubules are not ischemic (if the C_{H_2O} becomes negative). The free water clearance and maximum tubular reabsorption of water (formula 12, above) were used to show that an initial diagnosis of "renal failure" (high blood urea nitrogen and acidosis) in infants was really attributable to high dietary protein and electrolytes *(102)*.

Table 11 differentiates the two major types of oliguria. The major problem with such distinctions between laboratory procedures is the "gray zone." As discussed above, the surgeons have been using the measurement of free water clearance. However, nephrologists *(103)* would favor the ratio of sodium clearance to creatinine clearance,

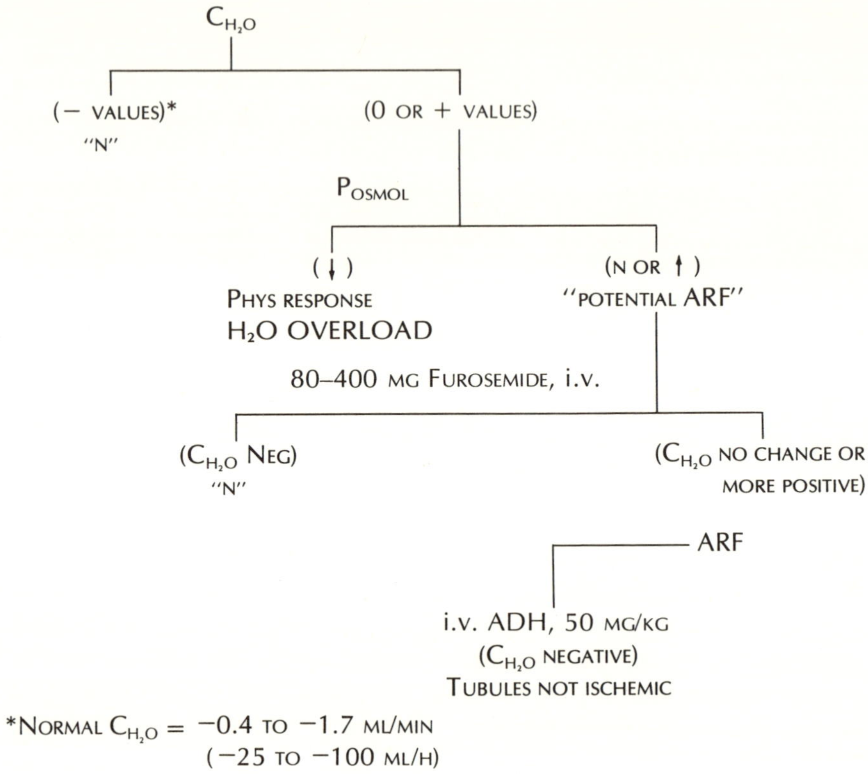

Fig. 4. Flow chart for use of free water clearance (C_{H_2O}) in diagnosis of acute renal failure.

Furosemide and antidiuretic hormone are also utilized in the diagnosis. Based on references *98–101*. N, normal; ARF, acute renal failure

$$\frac{\dfrac{U_{Na}}{P_{Na}} = \dot{V}}{\dfrac{U_{cr}}{P_{cr}} = \dot{V}} \tag{13}$$

which in reality is the ratio of the sodium and creatinine U/P ratios, because the term $\dot{V}$ cancels out—thus the need of a timed urine specimen is negated.

$$\frac{U/P_{Na}}{U/P_{cr}} \tag{14}$$

Table 11. **Differential Diagnosis of Oliguria (Less Than 20 ml/h per 70 kg Individual)**[a]

Hypoperfusion (*hypovolemic kidney*)		*Acute vasomotor nephropathy (VMN)* (*"conduit" syndrome: diffuse nephron disease*)
Pre-natal azotemia (PRA)		Acute tubular necrosis (ATN)
Shock		Acute renal failure (ARF)
Congestive heart failure		or insufficiency (ARI)
Acute glomerulonephritis (AGN)		Chronic renal insufficiency (end-stage)
Nephrotic syndrome (esp. with hypertension)		Hemoglobinuric nephropathy ("transfusion reaction")
Hepato-renal syndrome		Crush syndrome ("shock kidney")
Acute urinary tract obstruction (UTO)		Chronic UTO
"Spot" urine:		
Sp. gr.	>1,020	<1012
Na (mmol/liter)	<15 (or 20)	>40
U/P ratios:		
Creatinine	>40	<15 (or 20)
Urea	>20	<5
Osmolal	>1.2 (or 1.5)	<1.1
C_{H_2O}	"N" (− values)	0 or + values
FE_{Na} and (or) renal failure index (RFI)	<1%	> 1% (or 2 or 3)

[a] If no diuretic therapy, no azotemia, and no urinary tract obstruction is present. Modified from Weisberg *(8, 18, 82)*.

The final expression for the FE_{Na} is multiplied by 100 to obtain an answer in terms of percentage.

$$FE_{Na} = \frac{U_{Na} \times P_{cr}}{P_{Na} \times U_{cr}} \times 100 \qquad (15)$$

Others use the renal failure index (RFI), which is essentially the FE_{Na} except that the plasma sodium and the factor of 100 are omitted from equation 15.

Summary

The osmolality determination has been available for many years, but even with the advent of more sophisticated, compact instruments it is not utilized often enough. Some of the indications where it has been used are listed in Table 5; additional applications will be added in time. "Routine" measurement of urine osmolality, to replace measurement of the relative density of urine, is not indicated except when it is known (or suspected) that the patient has had radiopaque materials, is a diabetic with glycosuria, or shows proteinuria (renal disease).

Some institutions incorporate plasma (serum) osmolality as part of their "electrolyte panel" (some offer measurements of pH and blood gases as part of the same panel). If this is the course they choose, it would be more advantageous to go further and calculate the plasma osmolality with formula 8 (glucose and urea nitrogen concentrations are needed). This would be of use to the physician since the Δ-osmol will be increased in conditions such as alcoholism, drug intoxications, and in the presence of idiogenic osmoles such as appear in shock or trauma. I believe that plasma osmolality is indicated as the first screening procedure in the emergency room if the patient is stuporous or comatose—in such circumstances an increased plasma osmolality may be a very important clue to possible drug overdosage or acute alcoholism.

A hospitalized patient who has a high plasma osmolality should (and would) have the plasma sodium determined. Signs of vascular overload, with hypernatremia, are indicative of excess sodium added to the body, e.g., sodium bicarbonate in the treatment of lactic acid acidosis or cardiac arrest, excess sodium chloride entering the blood stream during an induced abortion, incorrect sodium concentration in a dialyzer bath, "excess" therapy (in presence of poor urinary output) with the usual sodium salt of barbiturates or antibiotics, etc. A high plasma osmolality with a normal sodium concentration may be due to hyperglycemia, azotemia, or alcoholism. If the determined and calculated osmolalities "agree," the trouble is probably sugar or urea alterations; if not, it may be due to alcoholism or other exogenous compounds such as mannitol. In a similar fashion, a low sodium concentration and low osmolality would be consistent with an isotonic loss of salt from the body.

The more important uses of osmolality determinations are evident in the "exact" data for urinary concentration-dilution tests, calculation of osmolal clearance, free water clearance (or tubular water reabsorption), differential diagnosis of various degrees of diabetes insipidus, etc. At the present time osmolality is of greater benefit in supplying an answer to a specific question raised by the physician or clinical chemist rather than in use as a screen for unknown disease processes.

References

1. Questions and Answers. (Loeb, J. N.), Serum and urine osmolality determination. *J. Am. Med. Assoc.* **238,** 1187 (1977).
2. Juel, R., Serum osmolality; a CAP survey analysis. *Am. J. Clin. Pathol.* **68,** 165 (1977).
3. Holmes, J. H., Measurement of osmolality in serum, urine and other biologic fluids by the freezing point determination. In *Workshop on Urinalysis and Renal Function Studies: Pre-Workshop Manual,* CCE Council on Clinical Chemistry, Am. Soc. Clin. Path., Chicago, Ill., 1962, pp 1–21.
4. Holmes, J. H., Diagnostic value of 24-hour urinary excretion patterns. *ibid.,* pp 108–139.
5. Holmes, J. H., Measurement of solute concentrations in body fluids by the freezing point technique. In *Workshop on Urinalysis and Renal Function Studies: Technique Manual,* CCE Council on Clinical Chemistry, Am. Soc. Clin. Path., Chicago, Ill., 1962, pp 1–7.
6. Holmes, J. H., Diagnostic use of urinary excretory patterns. *ibid.,* pp 47–61.
7. Holmes, J. H., Critique of cases #1 to #15. *ibid.,* pp 62–68.
8. Weisberg, H. F., Osmolality, *CCE Council on Clinical Chemistry, Check Sample CC–71,* Am. Soc. Clin. Path., 1972, pp 1–49.
9. Weisberg, H. F., Osmolality—calculated "delta," and more formulas. *Clin. Chem.* **21,** 1182 (1975). Letter to the Editor.
10. Andreoli, T. E., Grantham, J. J., and Rector, F. C., Jr., Ed. *Disturbances in Body Fluid Osmolality,* Am. Physiol. Soc., Bethesda, Md., 1977, 349 pp.

10a. Hammel, H. T., Colligative properties of a solution. *Science* **192,** 748 (1976).

10b. Andrews, F. C., Colligative properties of simple solutions. *Science* **194,** 567 (1976).

11. Randall, J. E., *Elements of Biophysics,* Year Book Publishers, Chicago, Ill., 1958.
12. Clark, W. M., *Topics in Physical Chemistry,* 2nd ed., Williams and Wilkins, Baltimore, Md. 1952.
13. Gamble, J. L., McKhann, C. J., Butler, A. M., and Tuthill, E., An economy of water in renal function referable to urea. *Am. J. Physiol.* **109,** 139 (1934).

14. Wolf, A. V., *The Urinary Function of the Kidney*, Grune & Stratton, Inc., New York, N. Y., 1950.

15. Wolf, A. V., *Thirst: Physiology of the Urge to Drink and Problems of Water Lack*, Charles C Thomas, Springfield, Ill., 1958.

16. Wolf, A. V., Urinary concentrative properties. *Am. J. Med.* **32,** 329 (1962).

17. Abele, J. E., The physical background to freezing point osmometry and its medical-biological applications. *Am. J. Med. Electronics* **2,** 32 (1963).

18. Weisberg, H. F., *Water, Electrolytes, and Acid-Base Balance*, 2nd ed., Williams & Wilkins, Baltimore, Md., 1962.

19. Jacobson, M. H., Levy, S. E., Kaufman, R. M., Gallinek, W. E., and Donnelly, O. W., Urine osmolality. *Arch. Intern. Med.* **110,** 83 (1962).

20. Frank, M. N., Dreifus, L. S., Rarick, F., and Bellet, S., Urinary osmolar concentration in the hydropenic state as a measure of renal tubular function; a test for early renal impairment. *Am. J. Med. Sci.* **233,** 121 (1957).

21. Galambos, J. T., Herndon, E. G., Jr., and Reynolds, G. H., Specific-gravity determination—fact or fancy? *N. Engl. J. Med.* **270,** 506 (1964).

22. Kaitz, A. L. and London, A. M., Osmolar urinary concentrating ability and pyelonephritis in hospitalized patients. *Am. J. Med. Sci.* **248,** 7 (1964).

23. Karhausen, L. R., Osmolarity. *Nature* **194,** 1234 (1962).

24. Rock, R. C., Ionic activity of electrolytes in serum. *Bull. Pathol.* **9,** 109 (1968).

25. Dahms, H., Rock, R., and Seligson, D., Ionic activities of sodium, potassium, and chloride in human serum. *Clin. Chem.* **14,** 859 (1968).

26. Dahms, H., Automated potentiometric determination of inorganic blood constituents (Na^+, K^+, H^+, Cl^-). *Clin. Chem.* **13,** 437 (1967).

27. Albrink, M. J., Hald, P. M., Man, E. B., and Peters, J. P., The displacement of serum water by the lipids of hyperlipemic serum; a new method for the rapid determination of serum water. *J. Clin. Invest.* **34,** 1483 (1955).

28. Guyton, A. C., *Textbook of Medical Physiology*, 3rd ed., W. B. Saunders, Philadelphia, Pa., 1966.

29. Freedman, F. B., and Corbin, T. P., Theoretical aspects of on-line osmometry. *J. Assoc. Adv. Med. Instr.* **5,** 350 (1971).

30. Prather, J. W., Gaar, K. A., Jr., and Guyton, A. C., Direct continuous recording of colloid osmotic pressure of whole blood. *J. Appl. Physiol.* **24,** 602 (1968).

31. Chinard, F. P., Derivation of an expression for the rate of formation of glomerular fluid (GFR); applicability of certain physical and physico-chemical concepts. *Am. J. Physiol.* **171,** 578 (1952).

32. Bowman, R. L., Trantham, H. V., and Caulfield, P. A., An instrument and method for rapid, dependable determination of freezing point depression. *J. Lab. Clin. Med.* **43,** 310 (1954).

33. Crawford, J. D., and Nicosia, A. P., A method for the determination of freezing point of small quantities of biological fluids. *J. Lab. Clin. Med.* **40,** 907 (1952).

34. Lott, P. F., and Millich, F., Instrumentation for osmometry. *J. Chem. Educ.* **43,** A295 (1966).

35. Lazar, M. Z., Osmolality for the clinician. *Hartford Hosp. Bull.* **19,** 16 (1964).

36. Lobdell, D. H., A practical guide to clinical osmometry. *St. Vincent's Hosp. Med. Bull.* (Bridgeport, Conn.) **8,** 7 (1966).

37. Warhol, R. M., Eichenholz, A., and Mulhausen, R. O., Osmolality. *Arch. Intern. Med.* **116,** 743 (1965).

38. Prager, D. J., and Bowman, R. L., Freezing point depression; New method for measuring ultramicro quantities of fluids. *Science* **142,** 237 (1963).

38a. Forman, D. T., and Changus, G. C., An automated osmometer. *Clin. Chem.* **14,** 38 (1968).

39. Johnson, R. B., Jr., and Hoch, H., Osmolality of serum and urine. *Standard Methods Clin. Chem.* **5,** 159 (1965).

40. Hill, A. V., A thermal method of measuring the vapour pressure of an aqueous solution. *Proc. Roy. Soc. London, Series A,* **127,** 9 (1930).

41. Baldes, E. J., A micromethod of measuring osmotic pressure. *J. Sci. Instru.* **11,** 223 (1934).

42. Adicoff, A. and Murbach, W. J., Some aspects of thermoelectric vapor pressure osmometry. *Anal. Chem.* **39,** 302 (1967).

43. Farm, R. J. and Bruckenstein, S., A sensitive and versatile differential vapor pressure apparatus. *Anal. Chem.* **40,** 1651 (1968).

44. Hunter, D. T. and Campbell, E. C., Biologic fluid osmolar studies using a thermocouple psychrometer. Presented at *Symposium on Thermocouple Psychrometers,* Utah State Univ., March 17–19, 1971.

45. Campbell, E. C., Campbell, G. S., and Barlow, W. K., A dewpoint hygrometer for water potential measurement. *Agric. Meteorol.* **12,** 113 (1973).

45a. Wolf, A. V. and Prentiss, P. G., Osmotic concentrations; Comparison of vapor pressure and freezing point methods. *J. Lab. Clin. Med.* **72,** 688 (1968).

46. Dorwart, W. V., and Chalmers, L., Response to reference 9, *Clin. Chem.* **21,** 1185 (1975). Letter to the Editor.

47. Rocco, R. M., Volatiles and osmometry. *Clin. Chem.* **22,** 399 (1976). Letter to the Editor.
48. Barlow, W. K., Volatiles and osmometry (cont.). *Clin. Chem.* **22,** 1230 (1976). Letter to the Editor.
49. Gettler, A. O., and Baker, W., Chemical and physical analysis of blood in thirty normal cases. *J. Biol. Chem.* **25,** 211 (1916) (quoted in reference *50*).
50. Waymouth, C., Osmolality of mammalian blood and of media for culture of mammalian cells. *In Vitro* **6,** 109 (1970).
51. Weisberg, H. F., Doberneck, R., and Grzys, M., unpublished data.
52. Dormandy, T. L., Osmometry. *Lancet* **i,** 267 (1967).
53. Warburg, E. J., Carbonic acid compounds and hydrogen ion activities in blood and salt solutions; a contribution to the theory of the equation of L. J. Henderson & K. A. Hasselbalch. *Biochem. J.* **16,** 153 (1922) (quoted in reference *54*).
54. Siggaard-Andersen, O., *The Acid-Base Status of the Blood,* 4th ed., Munksgaard, Copenhagen, 1974, 229 pg.
55. Dorwart, W. V., and Chalmers, L., Comparison of methods for calculating serum osmolality from chemical concentrations, and the prognostic value of such calculations. *Clin. Chem.* **21,** 190 (1975).
56. Glasser, L., Sternglanz, P. D., Combie, J., and Robinson, A., Serum osmolality and its applicability to drug overdose. *Am. J. Clin. Pathol.* **60,** 695 (1973).
57. Edelman, I. S., Leibman, J., O'Meara, M. P., and Birkenfeld, L. W., Interrelations between serum sodium concentration, serum osmolarity and total exchangeable sodium, total exchangeable potassium and total body water. *J. Clin. Invest.* **37,** 1236 (1958).
58. Loeb, J. N., The hyperosmolar state. *N. Engl. J. Med.* **290,** 1184 (1974).
59. Roscoe, M. H., A comparison of calculated and estimated osmolarities of urine. *J. Clin. Pathol.* **13,** 514 (1960).
60. Robertson, G. L., Athar, S., and Shelton, R. L., Osmotic control of vasopressin function. In *Disturbances in Body Fluid Osmolality,* T. E. Andreoli, J. J. Grantham, and F. C. Rector, Jr., Eds., Am. Physiol. Soc. Bethesda, Md., 1977, pp 125–148.
61. Robertson, G. L., The regulation of vasopressin function in health and disease. *Recent Prog. Hormone Res.* **33,** 333 (1977).
62. Stevens, S. C., Neumayer, F., and Gutch, C. F., Serum osmolality as a routine test. *Nebraska Med. J.* **45,** 447 (1960).
63. Sawinski, V. J., Goldberg, A. F., and Goldberg, N. B., Osmolality of spinal anesthetic agents. *Anesthesiology* **27,** 86 (1966).
64. Haroway, A. W., Jr., and Becker, E. L., Clinical application of cryoscopy. *J. Am. Med. Assoc.* **205,** 94 (1968).

65. Boyd, D. R., and Mansberger, A. R., Jr., Serum water and osmolal changes in hemorrhagic shock. *Am. Surgeon* **34,** 744, (1968).

66. Cowley, R. A., Attar, S., La Brosse, E., et al., Some significant biochemical parameters found in 300 shock patients. *J. Trauma* **9,** 926 (1969).

67. Mansberger, A. R., Boyd, D. R., Cowley, R. A., and Buxton, R. W., Refractometry and osmometry in clinical surgery. *Ann. Surg.* **169,** 672 (1969).

68. Miles, P. A., and Pearson, J. W., Amniotic fluid osmolality in assessing fetal maturity. *Obstet. Gynecol.* **34,** 701 (1969).

69. Jetter, W. W., Clinical osmometry. *P. Med.* **72,** 75 (1969).

70. Desmond, J., Serum osmolality and plasma electrolytes in patients who develop dilutional hyponatremia during transurethral resection. *Can. J. Surg.* **13,** 116 (1970).

71. Boyd, D. R., Addis, H. M., Chilimindris, C., Lowe, R. J., Folk, F. A., and Baker, R. J., Utilization of osmometry in critically ill surgical patients. *Arch. Surg.* **102,** 363 (1971).

72. Redetzki, H. M., Hughes, J. R., and Redetzki, J. E., Differences between serum and plasma osmolalities and their relationship to lactic acid values. *Proc. Soc. Exp. Biol. Med.* **139,** 315 (1971).

73. Lohaus, A. W., and Skelton, J. B., Cryoscopic osmolality determinations during pregnancy. *Obstet. Gynecol.* **38,** 63 (1971).

74. Redetzki, H. M., Koerner, T. A., Hughes, J. R., and Smith, A. G., Osmometry in the evaluation of alcohol intoxication. *Clin. Toxicol.* **5,** 343 (1972).

75. Mattar, J. A., Weil, M. H., Shubin, H., and Stein, L., Cardiac arrest in the critically ill. III. Hyperosmolal states following cardiac arrest. *Am. J. Med.* **56,** 162 (1974).

76. Abrams, C. A. L., Phillips, L. L., Berkowitz, C., Blackett, P. R., and Priebe, C. J., Jr. Hazards of overconcentrated milk formula. *J. Am. Med. Assoc.* **232,** 1136 (1975).

77. Smithline, N., and Gardner, K. D., Jr., Gaps—anionic and osmolal. *J. Am. Med. Assoc.* **236,** 1594 (1976).

78. Murty, B. S. R., Kapoor, J. N., and De.Luca, P. P., Compliance with USP osmolarity labeling requirements. *Am. J. Hosp. Pharm.* **33,** 546 (1976).

79. Rodrigo, F., Shideman, J., Mc Huey, R., et al, Osmolality changes during hemodialysis. *Ann. Intern. Med.* **86,** 554 (1977).

79a. Glick, J. H., Jr., Osmometric estimation of vial-to vial variation in contents of lyophilized sera. *Clin. Chem.* **23,** 781 (1977). Letter to the Editor.

79b. Paxson, C. L., Jr., Adcock, E. W., III, and Morriss, F. H., Jr., Osmolalities of infant formulas. *Am. J. Dis. Child.* **131,** 139 (1977).

79c. Oh, M. S., Whang, E. S. M., and Carroll, H. J., An osmometric method for measurement of total CO_2. *Kidney Int.* **8,** 56 (1975).

79d. Webster, L., and Lochlin, H., Cystic fibrosis screening by sweat analysis. *Med. J. Aust.* **1,** 923 (1977).
80. Robinson, A. G., and Loeb, J. N., Ethanol ingestion—commonest cause of elevated plasma osmolality? *N. Engl. J. Med.* **284,** 1253 (1971).
81. Champion, H. R., Caplan, Y. H., Baker, S. P., et al., Alcohol intoxication and serum osmolality. *Lancet* **i,** 1402 (1975).
82. Weisberg, H. F., *Interpretation of Electrolyte and Acid-Base Imbalance,* Private mimeographed printing, 3rd ed., 1971.
83. Goldberg, M., Water control and the dysnatremias. In *The Sea Within Us*, N. S. Bricker, Ed., Searle & Co., San Juan, P. R., 1975, pp 14–25.
84. Fanestil, D. D., Hyposmolar syndromes. In *Disturbances in Body Fluid Osmolality*, T. E. Andreoli, J. J. Grantham, and F. C. Rector, Jr., Eds. Am. Physiol. Soc., Bethesda, Md., 1977, pp 267–284.
85. Berl, T., Anderson, R. J., McDonald, K. M., and Schrier, R. W., Clinical disorders of water metabolism. *Kidney Int.* **10,** 117 (1976).
86. Jamison, R. L., and Maffly, R. H, The urinary concentrating mechanism. *N. Engl. J. Med.* **295,** 1059 (1976).
87. Brunner, F. P., and Thiel, G., Diagnostik und Pathophysiologie der renalen Konzentrationsdefekte. *Schweiz. Med. Wochenschr.* **106,** 1229 (1976).
88. Zileli, M. S., Telatar, F., Deniz, S., et al., Pseudohyperosmolar non-ketoacidotic coma due to oxazepam intoxication. *Clin. Toxicol.* **5,** 337 (1972).
89. Rubin, A. L., Braverman, W. S., Dexter, R. L., Vanamee, P., and Roberts, K., The relationship between plasma osmolality and concentration in disease states. *Clin. Res. Proc.* **4,** 129 (1956).
90. Moses, A. M., Diabetes insipidus and ADH regulation. *Hosp. Practice* **12,** 37 (1977).
91. Kleeman, C. R., and Fichman, M. P., The clinical physiology of water metabolism. *N. Engl. J. Med.* **277,** 1300 (1967).
92. Schrier, R. W., and Berl, T., Nonosmolar factors affecting renal water excretion. *N. Engl. J. Med.* **292,** 81 and 141 (1975).
93. Verney, E. B., The antidiuretic hormone and the factors which determine its release. *Proc. Roy. Soc. London, Series B,* **135,** 25 (1947).
94. Katz, M. A., Hypoglycemia-induced hyponatremia—calculation of expected serum sodium depression. *N. Engl. J. Med.* **289,** 843 (1973).
95. Lavender, S., and McGill, R. J., Nonketotic hyperosmolar coma and frusemide therapy. *Diabetes* **23,** 247 (1974).
96. Arieff, A. I., Guisado, R., and Lazarowitz, V. C., Pathophysiol-

ogy of hyperosmolar states. In *Disturbances in Body Fluid Osmolality*, T. E. Andreoli, J. J. Grantham, and F. C. Rector, Jr., Eds. Am. Physiol. Soc., Bethesda, Md., et al., 1977, pp 227–250.

97. Miller, M., Dalakos, T., Moses, A. M., et al., Recognition of partial defects in antidiuretic hormone secretion. *Ann. Intern. Med.* **73,** 721 (1970).

98. Baek, S. M., Brown, R. S., and Shoemaker, W. C., Early prediction of acute renal failure and recovery. I. Sequential measurements of free water clearance. *Ann. Surg.* **177,** 253 (1973).

99. Baek, S. M., Brown, R. S., and Shoemaker, W. C., Early prediction of acute renal failure and recovery: II. Renal function response to furosemide. *Ann. Surg.* **178,** 605 (1973).

100. Baek, S. M., Makabali, G. G., Brown, R. S., and Shoemaker, W. C., Free-water clearance patterns as predictors and therapeutic guides in acute renal failure. *Surgery* **77,** 632 (1975).

101. Brown, R. S., Leahy, M. S., and Farber, E. R., Free water clearance. Use in early prediction of acute renal failure and recovery. *CCE Council on Clinical Chemistry Check Sample ACC-18*, Am. Soc. Clin. Path., 1976, pp 1–6.

102. Zoppi, G., and Zamboni, G., Mechanism of diet-induced uraemia and acidosis in infants. *Eur. J. Pediat.* **125,** 197 (1977).

103. Espinel, C. H., The FE_{Na} test. Use in the differential diagnosis of acute renal failure. *J. Am. Med. Assoc.* **236,** 579 (1976).

PANEL DISCUSSION

Dr. Hicks: How do you reconcile within your institution the reduction in test ordering, which is obviously desirable from one viewpoint, with the institution's need for income to cover its costs?

Dr. Overhold: The hospital administrator has to recognize what his total income is and whether he likes it or not, or I or my colleagues like it or not, he has to increase the room rent if he cannot pay his bills. My point is that obviously, if we make an effort, use clinical judgment, and look at the process of information we are getting from our laboratory in terms of management of our patients, we will in turn decrease, or at least not increase, the otherwise escalating use of the laboratory. New tests will come along, others should be dropped, and it will be a problem for the administrator to meet his budget and if he can't, he will have to increase the room rent. At least the laboratory should have a reasonable profit, but it has to be put in perspective with respect to the total mission, painful as that may be.

Dr. Hicks: Because it comes from New Zealand, I think I will just read out a comment that may be pertinent.

At Auckland Hospital in the University of Auckland School of Medicine in New Zealand, we have instituted a cost-awareness program and within the first three months the department of medicine wards cut their test-ordering 40%. It should be noted that the laboratories in New Zealand do not have to support the other hospital activities financially.

Dr. Henry, is it possible to reconcile the practice of organ profiling with Dr. Overholt's approach to ordering of useful laboratory studies only, and has the cost benefit of organ profiling been studied in your laboratory?

Dr. Henry: I think it is possible to reconcile these two approaches, if you develop some levels of prior review and screening. In other words, I think the classic example is with cardiac injury. If you do not have measurements of total LDH or CK, then you don't bother with the isoenzymes.

We have not generated hard data on the cost/benefit ratio. We did consider our revenue in relation to the organ panels where there was some concern expressed by the hospital administrator that there would be a drop in revenue related to these decreased charges.

We are a state institution, and we were told that we couldn't charge less for individual measurements than the community hospital, for fear of being accused of receiving a state subsidy; therefore, individual laboratory measurements and examinations are all priced equal to or higher than whatever the community rate is, but with the panels we were able to decrease the cost, and interestingly enough, after an initial slight dip, the revenue picked up within a quarter, so the impact on the revenue to the hospital was virtually nil.

We did a very preliminary and minimal study on a medical service where there are two 30-bed units adjacent to one another. Organ panels were used in one but not in the other, and there appeared to be no difference in the total bill going out to the patients; but there appeared to be slight but significant decrease in hospital stay.

Dr. Hicks: Dr. Itskovitz, do you believe the usual level of sodium intake is of importance in the genesis of essential hypertension? If so, does measurement of sodium excretion have a place in the evaluation, or monitoring of, essential hypertension. Another question is whether there is any value in measuring sodium and potassium in unselected urines.

Dr. Itskovitz: Many epidemiologic studies have indicated that those who have a very high salt intake are more likely to have hypertension, so I do believe that the level of salt intake does relate to hypertension. I do not measure urinary sodium. The only time I would measure it would be to indicate why a particular patient may not be responding to therapy. If the patient is placed on therapy and his blood pressure still remains high, then one question may be whether the patient is taking in too much salt. In that situation I might measure urinary sodium, otherwise not. I do not consider measurement of urinary sodium and potassium to be a good screen for anything.

Dr. Hicks: Would the panelists comment on the concept that overutilization of laboratory tests is less of a problem than misutilization; i.e., the wrong procedures are requested because of cost and simplicity of performance, but the right tests are not ordered because they are considered complicated and expensive?

Dr. Weisberg: As a result of a new affiliation with the medical school, we have quite a few full-timers locally and a new housestaff who are much more vigorous, and I think that unfortunately it is the other way around. They want a test because it was just seen in a newspaper or in a recent journal, or because it is there. The best example is one test that I had called about this past week. The doctor wanted a certain drug level measured and I asked him where he had heard of it, and he said the only place he knew that it has been done was a research place in Denver.

Dr. Henry: I think this is a very complicated area, because when you talk about utilization, you can have either excessive utilization or under-utilization, or inappropriate utilization, and I think we have to focus on the effective utilization of the laboratory, through education. We must share with our colleagues the most cost-effective and efficient approach to the diagnostic or therapeutic management of the laboratory. Although we are talking about diagnosis primarily here today, we see the major effort in our institution being directed towards management. That is, the laboratory work being generated in our setting is applied more to patient-care management than in diagnosis. When you get into this area, I think the opportunities for less-than-effective utilization are significantly different than they are in the diagnostic application of a laboratory, particularly in the area of so-called standing orders. In tertiary care centers, with institutional and staffing requirements, and fast turnover, there is a tendency to develop standard operating procedures in depth: certain selected measurements and examinations are done perhaps more frequently and over more extended intervals than might be necessary than in other centers. These situations can best be handled by more frequent review, and understanding of their effectiveness and cost-effectiveness in particular.

Dr. Hicks: Do your pediatricians order organ profiles? Do they order the General Health Screen Panel?

Dr. Henry: No. In our institution the pediatricians have been conspicuous by their lack of interest in profiles or panels of *any* sort; interest has been generated mostly from our internists, and even more directly from the surgeons in all the surgical specialties. We do not have in our institution an obstetrical and neonatal service, but for the standard pediatric service, the answer is no.

Dr. Weisberg: How old are those patients? What is the age of the children in the pediatric service? Because as we heard yesterday about the screening of the newborn, most youngsters are going to be fairly well anyhow. You wouldn't have to be screening them unless symptoms for liver disease or something like that are present.

Dr. Henry: Well, ours is really a tertiary-care pediatric facility, and they are the usual pediatric range from a few months on up.

Dr. Hicks: One can imagine interest in some sort of organ profile, perhaps in response to chemotherapy or something of that type. It is something that we have considered but haven't started.

Dr. Weisberg: Let me speak on something close to my heart. For the past three months I have been collecting every urine that I pass; collecting it, not just discarding it, and the reason is as follows. Creatinine clearance, especially in the case of the cardiovascular surgical unit, must be determined before the patient goes to surgery, but mind you, this is just for the record, because we get the urine as the patient is being wheeled down to the surgical suite. We have 24-h urines, 12-h urines, and 4-h urines, and, for the renal people, the 1-h creatinine clearance. The big problem is that when you order 24-h clearance the urine specimen comes back very nicely from the nurse, 6:00 a.m. to 6:00 a.m. Now, I can't urinate at 6:00 a.m. every day, and neither can you. I will make a bet on it. And so I asked them to give me the exact time of the first urine collection and the time of the last urine; invariably the last urine is collected at the exact time as the first urine, twenty-four hours. Remember in the creatinine clearance calculation the volume per minute is on top, so this tends to magnify that number and we are getting an erroneous value.

There comes a question that was raised about using the creatinine value with, for instance, your AutoAnalyzer. At a level of 1 to 2 mg, the standard deviation that we have is 0.05. Therefore you can have a 0.1 mg difference, plus or minus, for your 95% confidence limits, and in a normal individual that one-tenth of a milligram will make a difference of about 10 to 15 ml per 1.73 m^2 per minute in the creatinine clearance.

If you have a severely ill patient whose creatinine clearance is reduced, that little variation is not going to make much difference. But for normal individuals, with a creatinine concentration below 2 mg, the clearance can vary anywhere from 50 to 125 ml/minute per 1.73 m^2. The upper limit of normal depends upon the individual. Determinations of creatinine clearance based upon serum creatinine alone, in which corrections for age and sex are sometimes made, should not be used until the creatinine clearance has been done as a baseline, and then one can use the serum creatinine value, realizing the variations.

Dr. Hicks: Dr. Weisberg, I have seen very few cases of urinary osmolality of 1000, even in severe renal failure. What is the validity of measuring urinary osmolality?

Dr. Weisberg: The urine osmolality is not measured as such. No place that I know of does it routinely to replace specific gravity, especially when it is measured with a TS meter.

Some people, however, measure plasma osmolality routinely whenever they have electrolytes measured. I don't do that, and they are not doing it for money. It is usually as a requested test, but you can do it as a screen. You need urine osmolality when you do U/P ratios or osmolar clearance. We use specific gravity as our routine analysis on urinalysis, and the only time we measure urine osmolality is for the untimed urine where the nephrologist wants it that way, or where we do the calculated clearances.

Dr. Hicks: Dr. Henry, I am going to try and combine several questions. If two or more panels are ordered with overlapping tests, how do you handle that? Does the same test get done twice? Secondly, why in one panel—for example, the general health screen—do you consider you need the urea and the creatinine, whereas on another metabolic screen you only use one of these?

Dr. Henry: If individual patients have more than one panel requested, and there is the possibility of duplication of the same determinations, of course a measurement is only performed once, even though it may be a part of more than one panel, but the results would be recorded in both of the two different panels, so there is no duplication in terms of technical effort and time. Overlapping panels usually are quickly identified in advance and therefore no unneeded effort is made.

To answer the question about the use of both urea and creatinine in the general health assessment versus the use of creatinine alone in the metabolic panel, we think that the creatinine datum is superior because it doesn't reflect protein intake or hydration.

It has been more in response to our clinical colleagues, particularly those in our ambulatory medical care unit, who have asked for the BUN, that we included it in the general health assessment panel this past spring. We may abandon the urea nitrogen determination when they have more confidence in the creatinine.

Dr. Hicks: Dr. Henry, have third-party payers accepted totally the concept of organ panels? We tried to set up an immunological panel for children with immunological problems and then Group Health Insurance would not accept it.

Dr. Henry: Yes. It depends. Now on our inpatients, for the Blues [Blue Cross and Blue Shield] we have a per diem and there is no separation other than in the charging. The patient pays so much per day regardless of whether one laboratory test is done or 20. A correction is made at the end of each chart in which the institution rebills the Blues for any deficit in terms of the total charges that have transpired for that area and there has been no problem.

As for outpatients, they have also been identified in terms of the Blues reimbursement schedule and we haven't had any feedback to the contrary.

I have heard that in other parts of the country there is a separation in terms of the reimbursement for automated versus non-automated, and I suspect that if we were pressed in this area, that we could really justify this, because there has been a significant reduction in terms of the individual test cost as opposed to their assembly in a panel.

Dr. Hicks: Dr. Itskovitz, in view of recent evidence indicating primary hyperparathyroidism as a cause of curable hypertension, should we be doing routine serum calciums on all patients presenting with hypertension?

Dr. Itskovitz: I would say not. If you want to do routine serum calciums to rule out hyperparathyroidism, that is one thing. As for this being a cause of hypertension, I am not sure that it is, and it is just adding another factor to our expensive workup of all these relatively normal people. So I would only do a serum calcium routinely if I were interested in hyperparathyroidism, as a routine part of the hypertensive workup.

Dr. Hicks: Do you allow the medical staff to request more than one organ panel at a time? It appears that you have an excellent opportunity to develop a problem-oriented approach to laboratory use.

Dr. Henry: Yes. We do get feedback between one of our attending physicians and the residents in clinical pathology whenever anything appears to be askew; that is, too many organ panels, or apparent misunderstanding of their selectivity or application. There is consultation, discussion, and review to clarify and confirm this. The efficacy of the panels is really in the ability to validate this portion of the data base, because if you have agreed upon a selected number and variety of measurements and examinations, it is relatively easy to go back and correlate them with the particular diseases or organ dysfunction that you are going after and reject or accept them.

More importantly, we had to update our hepatic panels. Someone alluded earlier to the sensitivity of the prothrombin time, and indeed this appears to be the most sensitive test in chronic liver disease as opposed to the other measurements in examination. So there is no question in my mind that it is in the validation of our data base in the laboratory in problem orientation, problem-oriented medical records, or the system, where the real strength and the future of this will reside.

Dr. Hicks: Dr. Henry, I have three general questions. One is how you handle stat [urgent] tests, whether you have a separate stat laboratory; the other is how you deal with the fact that physicians may feel that some of their discriminatory powers have been taken away; and the last is what percentage of your work load do you guess is part of an organ profile?

Dr. Henry: We do not have a stat lab. per se. We did at one time and we have published about the role and the place of the stat lab., but because of our concurrent computerization and automation, we have been able to blend these in with geographic proximity to those areas of the laboratory where the demands for immediate turnaround or stat service have been documented.

We only have 375 beds in our hospital, but the number of

critical-care beds is far in excess of the usual proportion. Because of this we have emphasized rapid response for all tests. We don't even hassle over the stats versus the regular requests, and indeed we try to identify the source of the request, rather than the nature of the request, to provide the appropriate immediacy of response.

Where we have had a problem with stats is in the clinics, but even there we have been able to identify and anticipate their needs, so that patients are coming by to see us before they go to the clinic, and we have anticipated what would be otherwise stat requests. Thus, literally most of the essential data are available to the physician before the patient leaves from his clinic visit.

Whether or not we have eliminated or reduced the discriminatory ability of the physician in selecting laboratory data, I really can't say, because the input I would get would be strictly subjective and hearsay. I certainly had some people tell me that this in effect is what we have done and have accused me directly of doing this. My response is: you don't have to use it; you do have the regular assembly of individual measurements and examinations if you so prefer; if you want to change any one of these panels, by all means let me have the documentation, the references and other information, and we are more than willing to update, revise, modify, amplify, or reduce any panel where the information would suggest it should be done.

The percentage of the work load that is stat, interestingly enough, has stayed fairly constant; it is somewhere between 25 and 30% of our total work load.

Dr. Weisberg: It is interesting that Dr. Henry is doing what we used to do years ago, and I am sure there are several physicians in the audience who know that when we worked in a clinic we would order the test we wanted the patient to have done before he came back the next time. Nowadays things have changed, not only in the clinic but also in the so-called primary care units associated with the hospital. The doctor doesn't make up his mind which laboratory test to order until he sees the patient; then he wants that answer before the patient leaves, within the next half hour. Dr. Henry, what is your turnaround time for those "expensive care" units and so on?

Dr. Henry: Oh, it varies anywhere from, say, blood gases in 6 to 8 minutes to the electrolytes all within 15 or 20 minutes. Our big problem, as you probably well know, is getting the specimen to the laboratory, and more importantly, getting the specimen clotted and the serum separated, more than it is in doing the analysis.

Dr. Overholt: Let me comment about that. The clinician can anticipate and should know the laboratory studies he wants done on the patient, or the physician should have the patient come in early and get his laboratory studies. Results would be available, at least most of them, by the time he sees the patient, or if they aren't available, it is because they are being run later in the day. It is very convenient for me, at least, to tell my patient that I will communicate with him either by letter or by telephone, depending on the distance. They accept that very well.

Dr. Itskovitz: Of course a lot depends on whether it is an acute problem or a chronic problem.

Dr. Overholt: But the clinician can't anticipate that either. The laboratory is part of the anticipatory team in terms of delivering the appropriate care to the patient. If it is an emergency problem or an acute problem, that is a different issue.

Dr. Hicks: We have a category in our laboratory which is probably not in most laboratories, partly because of our location, and partly because a lot of the patients are from low socioeconomic groups and can't keep coming back to the hospital. We have something that is called the "patient waiting" category. We have stickers saying that the patient is waiting, and the technologists appreciate what it is like to sit around waiting. The priority of these tests comes next in line after those from the intensive-care unit, and it has created quite a good feeling between us and the outpatient department.

There are several questions to Dr. Overholt, maybe about six or seven which relate to whether any misdiagnoses have resulted from using your approach, whether there has been an increase in malpractice suits, whether there has been any effect on the cost of malpractice insurance in your hospital; maybe you will comment on the whole question of malpractice when so many tests are not ordered.

Dr. Overholt: It goes back to the key combination or thought process of clinical judgment, and in my part of the country one never gets sued for decent clinical judgment. Documentation of a reasonable number of studies obviously is appropriate and I like Dr. Itskovitz's approach. He started off on a very complex subject, showed you all the variables one has to consider, and then milked it down to a very palatable approach. Finally, he wrapped it up by saying if you really want to be frugal, you only need about three tests. I don't know what is wrong with that.

Now, I have to admit that I am in a very fortunate part of the country. In the first place, the Midwest is unique. In Minnesota and Wisconsin there are large clinics in which we function as a team, and our patients accept us under this kind of circumstances.

Secondly, I am in a great rural area, and most of my patients are farmers, they work 12 to 16 hours a day, they are not overly demanding, they are quite intelligent, and all they expect is a fair shake, and if I give them that fair shake through appropriate communication they are aware of my desire to help them. They are quite appreciative of the fact that they are not getting a variety of studies that I don't think will be contributory to their welfare, their patient care.

While I am on the subject, I don't have any real hangups with the profiling concept in this regard, that it is an extension of the problem orientation, doing things for cause—"for cause, not because"—that it is all right, but what I object to is the extent of the profiling system and its overlaps. I would like to prune it considerably. Now, Dr. Henry says I have that opportunity. If he avails me of that opportunity, he is getting down to what I concluded in my remarks today: a few selective, carefully thought out and chosen tests, to get the thing going, then you can move from that point on if you find some fire.

Dr. Hicks: Dr. Henry, do you use panels for the examination of other fluids than serum, such as joint fluids, spinal fluids, et cetera? If so, would you please describe how?

Dr. Henry: To a small extent we have used other fluids with certain panels. We have one that I didn't discuss here. It is for our neurosurgical group, where obviously spinal fluid is examined. But for all practical purposes the panels are for blood and urine.

Dr. Overholt: I didn't answer my question completely. Our malpractice rate has not gone up. The cost of malpractice [insurance] has gone up, as it has everywhere, but the rate is not based upon our performance as a group, but on the performance of the country and the risk factors of the long duration of culpability of the physician if something happens *x* years from now.

Dr. Itskovitz: We are seeing the opposite. Sometimes when a test is applied which shouldn't be, we are getting malpractice suits. There was one, a Ph.D. in Milwaukee, who had had an arteriogram performed for renal hypertension which was not particularly indicated. He had a transverse myelitis and was paraplegic after that; the suit was for a million dollars. So suits can go for tests done as well as for tests not done.

Dr. Hicks: The next question for Dr. Overholt. How much $60 an hour physician time is utilized in limiting the work which needs to be done by the $6 per hour technologist?

Dr. Overholt: Well, I am not sure I can answer that except to state that if I am seeing the patient and doing an appropriate history and physical, it won't be nearly $60. It will take me about a half an hour, unless it is a very complex case, and that would be about $25 to $30, and if you include a couple of laboratory studies—such as the Pap. smear, stools for occult blood for people over forty—and if the patient hasn't had a chest film in five or six years or the SMA 12/60 battery in a few years, perhaps that might come to $60. But if I get everything, as I see often done, then it *will* be $60, for the laboratory work alone.

Now, the other thing that I would really support is Dr. Weed's comment about having the patient carry his own record. I am all for that. I see far too often a hospital not accepting the laboratory data or not knowing it is available. And even *knowing* it is available and not accepting it because it wasn't done in an institution or by a referring physician. The patient comes in with his barium enema and some lab. studies, and if you don't watch out somebody will repeat it just because *they* didn't do it. I think that has to be put into perspective, too. But if you have the record right there and there has been a decent physical and history and some reasonable laboratory studies, I don't see why you can't accept it as a valid part of the history and physical.

Dr. Itskovitz: I would also like to comment on that. I don't think the two are interchangeable. That is, I don't think that what you are getting out of the laboratory is necessarily the same as what you are getting out of the physician's evaluation. The two are complementary and you cannot substitute one for the other.

Dr. Hicks: Dr. Overholt, how do you feel about the claimed necessity of monitoring drug therapy by blood-level determinations. Also, what about daily glucose determinations on diabetics? At our institution many are ordered. Are physicians poor regulators of diabetes? Should they be using the urine dip-stick?

Dr. Overholt: In the first place, the physician is *always* a poor regulator of a diabetic. It is the patient who makes it easy. If he is an adult-onset type mature diabetic a modest amount of insulin will usually take care of him handsomely. You can follow his [case by checking his] urine very effectively. If I have been successful as a therapist in getting him to his ideal weight, and he feels well, is not having symptoms of hypoglycemia, I don't get blood sugars done.

But, if he is labile, fragile, brittle, I don't know anybody that knows how to do that kind of case well, and I guess we need the laboratory to help us out of our confusion.

In short then, the standard simple diabetic doesn't need many blood sugars done, if any. Obviously, diabetic acidosis is a different problem. We need that information to get the patient out of the acidosis. What makes diabetes difficult is that we are not the pancreas, we can't toss out the insulin on an as-needed basis depending on the sugar levels, and at best, as demonstrated by the slice-of-life studies at Mayo Clinic, we really don't do much with the swings or the fluctuations of the blood sugar with our therapy. We cut down the extent of it, but they still swing away like mad, and you are kidding yourselves [to believe otherwise]. It is part of our compulsiveness to overemphasize blood sugars in the management of a diabetic.

Dr. Hicks: What about drug monitoring?

Dr. Overholt: Well, for what? I am an internist. I don't have many people on drugs that require monitoring. I do use the prothrombin time. Anticoagulants are the most dangerous drugs known to man, and as I grow older and, I hope, wiser, I use them less and less. I have become so compulsive about this that I have not accepted a diagnosis of thrombophlebitis on the basis of calf tenderness and pain. You would be amazed at how few people with the diagnosis of phlebitis would actually have this diagnosis proved directly by venogram. I guarantee you that a high percentage of your patients [alleged to have thrombophlebitis] don't have it. Out the door they go. They don't have to be in the hospital; you put them on crutches. I don't know why they have calf pain sometimes, but they don't have deep thrombophlebitis, and if I can obviate seven days of hospitalization and anticoagulant therapy, and then the three months or six months they are going to be off work, depending upon your prejudices, I really save a lot of money by a simple $4 test.

The other one might be digoxin measurement. We don't monitor digoxin on an outpatient basis. We use the clinical parameters, and the only time I think a clinician should measure digoxin is when he has a complex problem of multi-system disease and the patient has nausea, vomiting, and so on. The physician wants to know if the drug concentrations are exceedingly high or does the patient have azotemia that might be causing it, and so forth—he is trying to sort this out. Other than that, I don't treat epileptics very often.

I suppose Dilantin would be the other area. I see neurologists use it.

Dr. Hicks: We have actually shown in our hospital that by measuring aminophylline in asthmatic children, we have greatly cut down on the hospitalization of children, certainly in intensive-care units, on the respirator, so I think doing monitoring tests actually maybe does reduce costs sometimes.

Dr. Itskovitz, perhaps you would like to repeat what you said about increased catecholamines?

Dr. Itskovitz: I said that about 90% or more of the time, when a patient comes to see us with a high VMA, it is a false positive. What we have to do in that circumstance is put them in the hospital, put them on the proper diet, and make sure they are not taking medications that can interfere with the test. Then if it is high it is meaningful, but in most situations something else has made it falsely high.

Dr. Hicks: And VMA is probably one of the most difficult tests to get clear of interferences, particularly in the Children's Hospital, where children love things that are full of vanilla.

Also, Dr. Itskovitz, did you exclude the screen for Cushing's disease on purpose?

Dr. Itskovitz: I didn't get into that, but the best screen there is a physical examination. One wants to know what the patient looks like.

Dr. Overholt: Yes. I agree.

Dr. Hicks: Dr. Henry, what type of instrumentation do you use in your laboratory to run the various panels?

Dr. Henry: We have a centrifugal analyzer, a Dupont *aca*, and a Technicon 6/60, and that is about it.

Dr. Hicks: Dr. Henry, how do you differentiate between general-health panel and nutritional status survey panel?

Dr. Henry: The metabolic panel and the general-health panel are similar, but the health survey was put together primarily for our general internists for primary care in the medical clinic arena and was to pull in some different things that we didn't have on the medical or the metabolic panel.

Dr. Hicks: Could you elaborate on the usefulness of free water clearance indicated in your renal panel?

Dr. Weisberg: Well, I think it is obvious that the people who are interested in free water clearance and in osmolar clearance are the nephrologists. They are the ones dealing with that aspect. It is now used by some physicians as a criterion for the diagnosis of incipient acute renal failure. There was a recent review on that with a critique with one of the check samples of the ASCP by Beck and his group. On the other hand, the nephrologists don't think that is a good idea. They want to have the filtered excretion fraction of sodium, which in essence is the ratio of the clearance of sodium over the creatinine clearance, and now we have a new one recently in a Letter to the Editor of *The New England Journal of Medicine*, where the authors talk about the RFI, renal failure index, which is the same ratio of the sodium clearance to the creatinine clearance, only you don't use the determination of the plasma sodium. I can't see the difference. The difference is about 40%, but they are both under one and over one on those individual bases, so it is a rather esoteric test, but where it *is* useful it can be extremely useful.

Dr. Henry: I would like to include it as sort of a vehicle for further education. I don't think that, aside from the nephrologist, most physicians do appreciate the free water clearance, and it gives us an opportunity to further clarify its utilization. If we look at our electrolyte problems, we do a lot of electrolyte measurements and generally, as you well know, there are fluid—i.e., water-problems as well as electrolyte problems, and certainly by having these together and looking at the free water clearance you can pretty well delineate your problem in terms of either solute excretion or depletion when you add the urine sodium measurement and you have a free water clearance to go with it. You can really help identify and certainly monitor the water-balance aspect as well as the solute aspect, which you just can't get a handle on otherwise.

Dr. Hicks: Dr. Overholt, if a nurse practitioner does a hemoglobin, blood pressure, and chest x-ray, and these are normal, does the physician need to give the patient a physical? Secondly, do you believe in complete physical examinations and history, even aspects unrelated to the primary complaint?

Dr. Overholt: In our particular clinic the nurse clinician is an emerging concept only in pediatrics. I can't speak then from any direct experience.

I, selfishly speaking, have no problems with dealing with my patients in doing their histories and physicals and enjoying it when I see healthy people. Indeed, I think it is the best therapeutic act I can perform, so I am not looking for any particular short-cuts in this 20-minute or half-hour dialogue that ensues.

Now, could a nurse do it as well? Most of the things that we do I am sure [a nurse could] do quite effectively.

I tell my residents, when I am making rounds and they present this 72- or 82-year-old person, I am not interested in their family history. They have already beat the system, so I take histories and physicals that are relevant and that are in context. I do give a complete physical exam, but I do it in a very simple manner. I am in the room, I watch the patient move about, the males especially, they dress right in front of me; that itself is a neurological exam essentially. A lot of sophistication goes on in that maneuver. A few other things can be done. I do everything else, including a testicular exam, rectal exam, pelvic exam, and so on. They can be done very efficiently and effectively while you are doing part of your history and physical, give your patient a few minutes of dialogue and they are off; it is all done in 15 minutes, 30 at the most. Then you can sit down and talk to your patient when it is all done, and that is the most fun part.

Dr. Hicks: Dr. Overholt, since you try hard to cut down on tests ordered, what is the hospital's policy relative to ordering of tests from the emergency room? Are all emergency room tests treated as stat; are the house staff taught that stat tests are more costly because of their special handling?

Dr. Overholt: They certainly are. They have to be ordered as stat if the problem is stat. They are taught this. The same cost-accountability or thought process goes on in the emergency ward, because as our emergency ward has evolved, it is a night clinic. It is a misnomer, emergency ward. Cut it out. They are night clinics. Some emergencies come in there, but in our American system, for the most part, hospitals are running night clinics to serve people because they can't get served any other way and so put your best clinicians down there. [*Ed. note: cf. The New England Journal of Medicine* **298,** 940, 1978, for a discussion of this.] At least that is what we have done, and sure, there is always some abuse of the emergency ordering of tests or stat testing and it has to be constantly worked at.

Dr. Hicks: Dr. Itskovitz, have you had any experience with the measurement of dopamine-beta-hydroxylase?

Dr. Itskovitz: Yes. In fact, my colleagues here from the pathology department do measure it with us and we have been measuring it as a possible indication of sympathetic nervous activity in hypertensive states and at this point have not found any great relationship between hypertension and DBH activity in blood.

Dr. Hicks: What added information does the intravenous pyelogram give you in suspected cases of renovascular hypertension when you are quite sure you are going to need to do angiography anyway?

Dr. Itskovitz: We aren't quite sure of that. If a patient has relatively severe hypertension, that is not necessarily an indication for angiography, but before we take that step, which is expensive and has great morbidity, we need further proof that there may be a localizing lesion. When we do an IVP we look for certain indications of a localizing lesion—one kidney smaller than another, one kidney taking up the dye at a different rate than the other, or one kidney holding the dye and concentrating it longer than the other. When we see that kind of a picture, then we think there may be a renal vascular lesion, and that is an indication for angiography. Otherwise, we would probably not do angiography.

Dr. Hicks: Dr. Henry, there are several questions relating to your choice of tests in the thyroid panel. One of them relates to the fact that since there is no clinical problem with distinguishing between hypo- and hyperthyroidism, why include TSH. Another is that since the incidence of T3 thyrotoxicosis is so low, why include the T3? Finally, isn't T3 more valuable than T4?

Dr. Henry: It is true that hyper- and hypothyroidism are dissimilar in terms of their clinical picture, but one of the problems appears to be in the very subtle manifestations in hypothyroidism; namely, there there are degrees of blatancy or obviousness of the clinical picture in this condition and there is no question but that TSH is more sensitive in identifying hypothyroidism.

As for T3 in thyrotoxicosis, I think the next question helped clarify this. The response is that all the answers are not in yet regarding sensitivity. There is some suggestion that maybe T3 will give us everything we got with the T4 before, and if that proves to be true, we obviously won't need to measure T4, but we are not convinced of that, so we are still doing both.

Dr. Hicks: What are the disadvantages of organ profiling?

Dr. Itskovitz: Well, I guess it is against my basic philosophy of life. If you think the most important function we have as physicians or as health-care people is to find disease, then I think there is a very good reason to do organ profiling. On the other hand, if you are concerned that perhaps our function in life is to help a person have a rich and rewarding life, I don't know if organ profiling adds to that.

Very often what it does is to point out something that we don't then do anything about anyhow, and for the rest of that person's life he is concerned with that little thing which may or may not be there, and may never be active to the same extent or enjoy life to the same extent. So that if basically you simply want to find disease, I say yes, let's do profiling, but if you are more interested in what the patient is going to get out of life, I don't know that I want to do it.

Dr. Weisberg: This is analogous to the situation with automated admission chemistry. In several hospitals I have been at in the past, they didn't have that. The physician had to order what he wanted himself. The other side of the coin is the aspect of reducing everything to mediocrity, where we all do the same thing. But this takes us back many years. Those of you in the audience who are physicians or those of you who have been in the laboratory long enough, know that each physician has his own pet orders. The nursing station would have Dr. X's routine and Dr. Y's routine and so on. This is equally true of orders for postoperative fluid. Dr. A wants 2 liters of saline; Dr. B wants 2.5 liters; and so there is difficulty in getting these various clinicians to agree with one another. I think what John Henry has said is that there is more uniformity in a state institution, where you have a little more clout. But when you deal with prima donnas, you never forget that the laboratory is an ancillary service. I am perfectly willing to work and to serve the patient through the doctor, as Dr. Overholt was talking about. But you have to now look at your own local community, and I think this is my major objection to what Dr. Weed was talking about. It is almost impossible to get all areas of the country to agree on what we want to do as a least common denominator.

Dr. Overholt: As I implied earlier, restraint is the key word as far as I am concerned. I think that if a patient has a physical problem, a complaint, or shows a few appropriate changes in his physical exam, I would go back to the selective process of ordering the most productive laboratory studies to help me resolve the patient's problem and manage it. The physician needs to think, to bear the responsibility of carefully selecting the diagnostic and therapeutic studies, get back again to that level from the cookbook approach, and not become enamored with a whole group of interesting but not very meaningful laboratory studies.

Dr. Hicks: Dr. Henry, do you want to comment?

Dr. Henry: First of all, I want to respond to Dr. Weisberg's comment. I don't have much clout in my institution and if you don't believe me, just ask any of my colleagues there, whether it is the house staff or any other. We do have a lot of prima donnas in the State University of New York Upstate Medical Center, and they are not bashful about commenting about their specific needs, but when you have elicited the response and support of the super specialists within their own discipline, particularly in internal medicine, on what you have put together, it takes the sting out of it, I think, within their own setting.

The disadvantage of organ panels that I see, and obviously I see very few, is the confusion that may exist in the minds of some people in terms of the emphasis of their application on diagnosis versus management. I think when they are used as management tools and they are basically diagnosis oriented that they are superfluous, they are excessive, and they are inappropriate, and I think the central question is whether or not we have to go to separate management panels for almost every organ. We are beginning to feel our way along in this area as well. There seems to be much more emphasis on management, particularly in chronic diseases, and as I have heard here the simplicity of this is much different than it is in the diagnostic application. I think we should be able to do something different here and not confuse people, but rather clarify the utilization of organ panels as either diagnostic or management.

Dr. Hicks: Dr. Henry, do you have evidence that diagnoses are made more rapidly and more efficiently with the panel approach than without, and do you ever write any interpretative comment on the clinical laboratory report?

Dr. Henry: The answer to the first is that I don't have good hard data other than with the initial study that we reported in 1971. At that time there was a little more distinction than there is now in rapidity of the diagnosis. Clearly, one can get back more information more quickly with the panel approach than with the sequential ordering pattern that preceded it.

Yes, we do write on those organ-panel report sheets. You will note that there is space below each assembly of panels and in these one can comment on the interpretation or translation, and make recommendations or suggestions for other measurements and examination.

Dr. Hicks: One final question for Dr. Itskovitz. What do you do if a physician has a patient on an anti-hypertensive drug and wants to measure catecholamines and the anti-hypertensive drug interferes with the test and the clinician will not take the patient off the drug?

Dr. Itskovitz: Well, if he wants an accurate answer he generally has to take the patient off the drug and you just have to explain to him what the problems are, and if worst comes to the worst, you can always hospitalize that patient, take him off the drug, and do it.

IV

RESPONSE TO ABNORMAL TEST DATA

The Clinical Usefulness of Multiple Result Ranges for a Laboratory Test, as Exemplified by Serum Urea Nitrogen

G. William Cole, M.D.

Division of Clinical Pathology, Department of Pathology, University of Alabama in Birmingham Medical Center, Birmingham, Ala. 35203

The usefulness of a laboratory test to detect, confirm, exclude or monitor a clinical problem or treatment is a matter of definition. It is customary to define results of tests as normal or abnormal. Such a definition ignores the quantitative nature of most tests performed in the clinical laboratory. In this era when accuracy and precision of laboratory tests are high, a re-examination of a test's medical usefulness is in order.

Much confusion focuses on the use of the term "normality." Medically defined, normality means "free of disease" *(1)*. Statistically defined, normality usually is interpreted as a range of results that is plus or minus two standard deviations from the mean value *(2, 3)*. Such definitions imply that the control population under study is representative of the entire population at large, as well as that each test measurement will contribute to a gaussian distribution curve. Both assumptions are often false *(4–6)*.

Many persons have recognized this problem and have proposed alternative methods for expressing a test result. In particular, a variety of statistical methods, such as percentile ranking, have been suggested as alternatives. These also miss the mark

This paper has been published in *Pathologist* 32, 101–104 (1978), and is republished here, in slightly altered form, by permission of the College of American Pathologists.

because they fail to take into account the medical purpose or decision-making powers of a test result.

Here I hope to demonstrate that test results have multiple overlapping ranges that alter medical decisions. The serum urea nitrogen[1] has been chosen as an example. I suggest that the following test-result ranges be defined for each laboratory test.

(*a*) Physiologic range
(*b*) Problem detection range
(*c*) Problem confirmation range
(*d*) Problem exclusion range
(*e*) Specific or toxic range
(*f*) Therapeutic range
  1. Significant deviation between results
  2. Significant trend between results

Methods

The serum urea nitrogen determination[1] is performed routinely on all patients over 12 years old upon hospital admission as part of a panel of laboratory tests. The medical records of 996 patients were reviewed retrospectively. These represent the number of patients admitted during a one-month period and represent over 90% of the hospital admissions during that time. No effort was made to pre-select patient categories. The hospital is a referral institution and serves the State of Alabama. Cardiovascular disease cases are common; relatively few women at term and children are admitted. Medicine, surgery, gynecology and psychiatric services are all active, and the hospital is classified as an acute disease—short stay institution.

Blood samples were obtained within the first 12 h after admission, and an almost equal number were collected while the subjects were in the fasting and non-fasting state. Samples were processed and tests performed within 4 h of collection, care being taken to prevent any sample evaporation. The urea nitrogen analyses were done by continuous-flow analysis (a Technicon SMA-660), by a modification of the urea nitrogen method developed by Marsh et al. *(7)*.

[1] *Ed. note:* Urea nitrogen × 2.14 = urea. The normal range listed by the Massachusetts General Hospital [*N. Engl. J. Med.* **298,** 37 (1978)] is 8–25 mg/dl (80-250 mg/l).

The patients' medical records were reviewed, and each patient's problems at the time of admission were tabulated. Patients were sorted on the basis of the results for urea nitrogen, from the lowest to the highest values. After this sort was made, it became apparent that there were certain clinical problems common to the group with low urea nitrogen values, and a different set of problems in the high-value group. From this sort, three subpopulations of patients were identified.

Group A (High Urea Nitrogen Group)

Subset One (Pre-renal Azotemia)

1. Urine relative density (specific gravity) >1.022 (no glycosuria)
2. Congestive heart failure, decompensated (pulmonary or hepatic congestion and (or) peripheral edema)
3. Acute bleeding in the upper gastrointestinal tract

Subset Two

1. Chronic diastolic hypertension
2. Renal or urologic disease (excluding urinary tract infection alone)
3. Renal dialysis patients

Group B (Low Urea Nitrogen Group)

1. Urine relative density (specific gravity) <1.005
2. Intravenous fluid administration, no Group A problem
3. Pregnancy, third trimester
4. Acute alcoholic intoxication
5. Recent gastric lavage

Group C—All patients not having problems listed for either Group A or Group B

Diabetes mellitus, vascular disease in general, and diuretic use were often simultaneously present in Group A patients. No independent correlation was found between these problems and values found for serum urea nitrogen.

Likewise, some patients in the low urea nitrogen group were vomiting, but many of them were receiving intravenous fluids at the time of specimen collection.

Results and Conclusions

Figure 1 depicts the histogram obtained for males (470) and females (526). There is considerable overlap between each of the

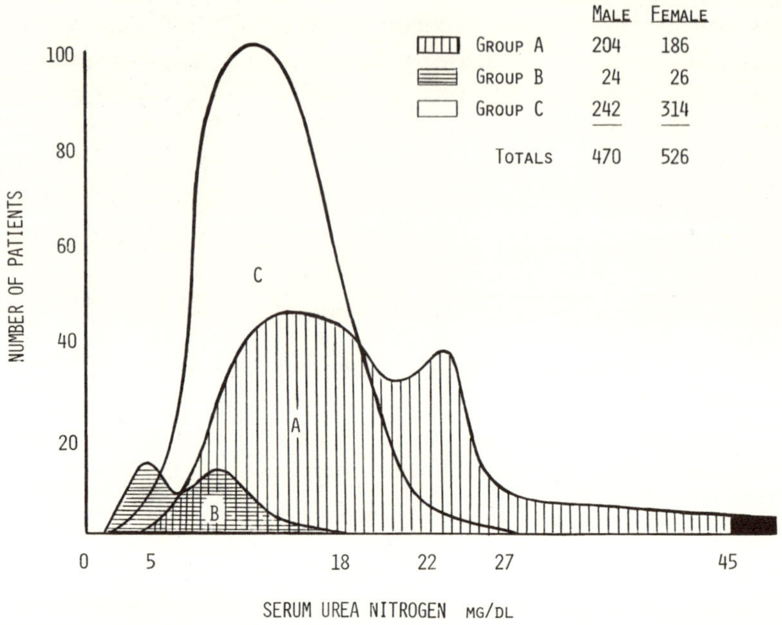

Fig. 1. Hospital Population Subgroups for Serum Urea Nitrogen

three groups. Only in the upper and lower ranges is discrimination possible. What is clear, however, is that the greater the urea nitrogen value, the more likely it is that a Group A problem is present, and the smaller the value, the greater the probability that a Group B problem is present.

Table 1 shows the number of patients falling into each test-result range, and in parentheses the probability or predictive value that one of the three problem sets (groups) is present in each range.

Range One, low result range, is the detection range for Group B problems and, at the same time, is the exclusion range for Group A problems.

Range Two, the normal result range, is the physiologic range. There is no clear discrimination between Group A, B, or C problems.

Range Three, the borderline high-result range, provides no distinction between Group A or C problems, but is the exclusion range for Group B problems. There is an equal chance that a Group A or C problem exists.

Table 1. **Clinically Significant Ranges for Serum Urea Nitrogen**

Range	*Serum urea nitrogen result range, mg/dl*	*Predictive value for clinical problem sets*			*Clinically significant ranges*	
		Group A	*Group B*	*Group C*	*Group A*	*Group B*
1	2–5	1/25 (0.05)	16/25 (0.63)	8/25 (0.32)	Exclusion range	Detection range
2	6–17	183/684 (0.27)	34/684 (0.5)	467/684 (0.68)	Indeterminant or physiologic range	
3	18–21 ♀ 18–22 ♂	72/142 (0.51)	0/142 (0.00)	70/142 (0.49)	Detection	Exclusion range
4	22–27 ♀ 23–27 ♂	60/70 (0.86)	0/70 (0.00)	10/70 (0.14)	Ranges	
5	28–44	42/43 (0.98)	0/43 (0.00)	1/43 (0.02)	Confirmatory range	
6	>44	32/32 (1.00)	0/32 (0.00)	0/32 (0.00)	Specific range	
7	Totals	390/996 (0.39)	50/996 (0.05)	556/996 (0.56)	All ranges helpful in therapeutic monitoring	

Range Four is the detection range for a Group A problem. A large proportion (0.86) of patients with urea nitrogen values at this level will have one or more of the problems of Group A. Further patient evaluation is required. Not always will a correlative problem be found.

Range Five is the confirmatory range. Virtually all patients with a urea nitrogen value this high will have a Group A problem. If one of the Group A problems is not clinically apparent, the validity of the result should be questioned. If the result is verified, then rarer causes for increased urea should be looked for.

Range Six is a specific range for severely compromised renal function. All patients in this range had creatinine values exceeding 17 mg/liter and also had a clinically apparent cause for renal disease. Though some of these patients had prerenal azotemia in addition, none had pre-renal azotemia alone.

Range Seven, the therapeutic or problem monitor range, encompasses all of the six ranges. Thus, for patients anticipated to have one of the conditions or treatments identified during their hospital stay, assay of urea nitrogen could provide useful information. Lacking, and beyond the scope of this study, is the degree of change between consecutive tests that are therapeutically significant and useful.

Discussion

A laboratory test's meaning and usefulness in the medical care of hospitalized patients is broader in scope than is often appreciated. A test's total medical value is more readily seen when test ranges are identified that relate to the *use* of the test. Although disease detection is an important reason for laboratory testing, the test result should not be expressed in detection terms alone. Thus, the terms "normal" and "abnormal" do not sufficiently define the usefulness of a test.[2] The decision-making properties of a test that measures a physiologic constituent are related to the quantitative result and are often independent of the "normal" range.

Serum urea nitrogen measurement has no decision-making powers unless defined in a medical context. The justification for its use must be defined in terms of its multiple clinical purposes. I have shown that there are at least seven overlapping test-result ranges of medical significance for this physiologic constituent. Each result range implies a somewhat different clinical interpretation and should result in a somewhat altered clinical action.

The reliability of these ranges for clinical decision-making primarily depend on the overall quality of the laboratory, the precision of the measurement, and a knowledge of the prevalence *(8–10)* of clinical problems in the population being evaluated.

Group A and Group B problems were found in 44% of our hospital population. Not included are those patients in whom

---

[2] *Ed. note: cf.* Miller et al., *Clin. Chem.* **14,** 836 (1968). Abstract.

some change in renal blood flow or function was anticipated rather than actually present at the time of admission. Thus, many medical and surgical therapies were applied after admission that are known to influence renal function. In this sense, the test is used as a therapeutic monitor.

The frequency with which repetitive testing is required or useful for therapeutic monitoring has not been defined, nor has the range of difference between results that are medically significant. Independent of the lack of such documented evidence, clinicians do make decisions based upon test differences defined by their own experience.

A test such as serum urea nitrogen measurement expressed only in terms of normality would have very limited clinical value. Viewed from the perspective of multiple different ranges and different clinical conclusions, it is easy to understand why this test is requested frequently.

It should also be understood that these ranges may vary according to the patient population studied. Thus, the actual numerical ranges might vary between institutions and laboratories, but this does not detract from the importance of defining multiple interpretative ranges.

References

1. *Webster's New World Dictionary,* Prentice-Hall, Englewood Cliffs, N.J.
2. Wooton, I. D. P., King, E. J., and Smith, J. M., The quantitative approach to hospital biochemistry. *Br. Med. Bull.* **7,** 307 (1957).
3. Collen, M. F., Siegelaub, A. B., Cutler, J. L., and Goldberg, R., Aspects of normal values in medicine. *Ann. N. Y. Acad. Sci.* **161,** 572 (1969).
4. Murphy, E. A., and Abbey, H., The normal range—A common misuse. *J. Chron. Dis.* **20,** 79 (1967).
5. Elveback, L. R., Guillier, C. L., and Keating, F. R., Jr., Health, normality, and the ghost of Gauss. *J. Am. Med. Assoc.* **211,** 69 (1970).
6. Burnett, R. W., Accurate estimation of standard deviations for qualitative methods in clinical chemistry. *Clin. Chem.* **21,** 1935 (1975).
7. Marsh, W. H., Fingerhut, B., and Miller, H., Automated and manual direct methods for the determination of blood urea. *Clin. Chem.* **11,** 624 (1965).

8. Vecchio, T. J., Predictive value of a single diagnostic test in unselected populations. *N. Engl. J. Med.* **274,** 1171 (1966).
9. Werner, M., Brooks, S., and Wette, R., Strategy for cost-effective laboratory testing. *Human Pathol.* **4,** 17 (1973).
10. Galen, R., and Gambino, R., *Beyond Normality,* John Wiley and Sons, New York, N. Y., 1975.

The Physician's Response to Unsolicited Tests

Lawrence J. Schneiderman, M.D.

Department of Community Medicine, S–004, University of California, San Diego, La Jolla, Calif. 92093

Since the introduction of automated laboratory testing, a number of investigators have attempted to evaluate the impact of this technology on medical care. By now, three findings have been common to studies in the patient-care setting:[1] *(a)* the larger number of abnormal results, *(b)* the small yield of clinically useful results, and *(c)* the scanty attention physicians pay to these results.

The first finding was easily predictable. Because the notion of "normal" emerged from statistical theory, rather than concepts of health and illness, it came to represent any value lying within an arbitrary distance of the mean—usually graphically depicted in the standard laboratory printout. Inevitably, therefore, the larger the screening panel, the larger the number of abnormalities *(2)*. Studies have shown such abnormalities to appear in anywhere from 20 to 55% of patients subjected to an automated chemical profile, depending on the number of tests in the panel *(3–8)*. Variations from this prediction depend either on the extent to which a particular test curve deviates from a symmetrical bell-shaped distribution *(9)* or arbitrary modifications adopted by the laboratory *(3, 7, 10)*.

The second finding—the low clinical yield—naturally was disappointing. Only a few studies have been done to trace the

[1] This review will limit itself to studies of automated laboratory testing applied to *case finding* [Sackett and Holland *(1)*], i.e., where the encounter is initiated by the patient and the purpose of the testing is the comprehensive assessment of health.

outcome of these abnormalities, most of which were unanticipated by the physician. In these, the majority of unexpected deviations from normal—whether slight or considerable—either remained abnormal or reverted to normal with no explanation for either phenomemon *(11–13)* (Table 1). Unsolicited laboratory-detected abnormalities that proved to be diagnostically or therapeutically useful usually were present in no more than 10% of adult patients *(3–7, 11)*.

It would be tempting to conclude that there is a logical connection between the first two findings—the tendency of automated laboratories to generate a profusion of abnormal results of little clinical significance—and the third—the tendency for physicians to ignore them *(7, 11, 12, 14)*. It is this issue which we will explore further.

Studies of the Physician's Response

Even before the advent of automated laboratory testing, Williamson et al. *(15)* reported the tendency of doctors to ignore unexpected abnormalities in routine screening tests. Only three rudimentary tests made up their screen (urinalysis, fasting blood glucose, and hemoglobin) and yet the investigators found that physicians "revealed no apparent response of any kind to approximately two-thirds of these test abnormalities" *(15)* (Table 2). An intensive education program produced some slight improvement, but most responses never met their "minimum adequate" criteria.

In a chart review of 547 patients subjected to a 12-channel

Table 1. **Outcome of Chemical Screening Battery Abnormalities[a] (Schneiderman et al.[11])**

	Alk phos[b]	*(> 12 KA units*	*Bili-rubin*	*(> 1.0 mg/100 ml)*	*AST*	*(> 50 U)*	*Urea*	*(> 20 mg/100 ml)*
Total abnormal tests	53	(13–35)	20	(1.1–3.0)	7	(59–100)	51	(21–51)
Diagnosis made	7	(14–23)	3	(1.3–3.0)	3	(60–90)	6	(21–51)
No diagnosis	46	(13–31)	17	(1.1–1.7)	4	(59–100)	45	(21–31)
Test repeated:								
becomes normal	5	(13–31)	3	(1.1–1.7)	1	(59)	13	(21–26)
remains abnormal	2	(13–23)	1	(1.5)	. . .		3	(21–22)

[a] Range of values for each category in parentheses.

[b] "Alk phos" signifies alkaline phosphatase; KA, King-Armstrong; AST, aspartate aminotransferase (EC 2.6.1.1); U, international (IUB) units; and urea, serum urea nitrogen.

biochemical screening as part of their general medical work-up, we found that physicians tended to ignore—i.e., neither underline or circle or record any comment on—about half of the abnormal results *(11)* (Table 3).

In a follow-up of over 8000 ambulatory patients subjected to a panel of 20 chemical tests under the Kaiser Health Plan, Friedman et al. *(8)* found that physicians ordered re-tests of laboratory abnormalities no more than a third of the time.

Similarly, Bates and Yellin *(14)* found that the 417 physicians they studied made little effort to follow up abnormalities obtained by multi-phasic screening (Table 4). Indeed, "in only three test categories—blood glucose level, urinary tract bacteria, and hemoglobin levels/hematocrit reading—did they check the screening results even half the time." Pursuing this matter by

Table 2. **Response of Physicians to Unexpected Abnormalities[a] (Williamson et al.[15])**

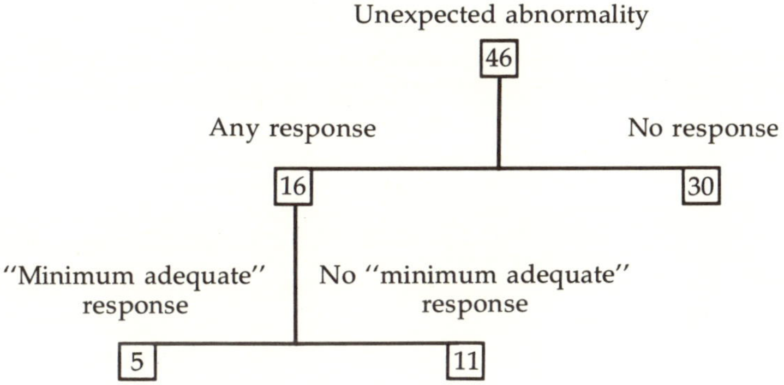

[a] Urinalysis, fasting blood glucose, and hemoglobin.

Table 3. **Physician Response to Abnormal Tests[a] (Schneiderman et al.[11])**

	Alk phos[a]	*(> 12 KA units*	*Bili-rubin*	*(> 1.0 mg/ 100 ml)*	*AST*	*(> 50 U)*	*Urea*	*(> 20 mg/ 100 ml)*
Total abnormal tests	53	(13–35)	20	(1.1–3.0)	7	(59–100)	51	(21–51)
Diagnosis made	7	(14–23)	3	(1.3–3.0)	3	(60–90)	6	(21–51)
No diagnosis	46	(13–31)	17	(1.1–1.7)	4	(59–100)	45	(21–31)
Interpreted as not clinically significant	22	(13/35)	7	(1.1–1.7)	. . .		16	(21–28)
No interpretation	17	(13–27)	7	(1.2–1.3)	3	(60–100)	16	(21–27)
No note	7	(13–14)	3	(1.1–1.7)	1	(59)	13	(21–26)

[a] Range of values for each category in parentheses
[b] Abbreviations as in Table 1

Table 4. **Percentages of Physicians Who Took Action to Confirm Previously Unknown Test Abnormalities (Bates & Yellin,[14])**

Test[a]	*No. of abnormalities not previously known*	*% Confirming action taken*
Hemoglobin and (or) hematocrit reading	91	50.5
Glucose (>1.7 g/l)	228	56.1
Cholesterol	140	12.1
Total lipids	129	13.2
Protein-bound iodine and T_4	65	29.2
Calcium	108	13.0
Potassium	118	19.5
Uric acid	96	30.2
Alkaline phosphatase	108	24.1
AST	105	25.7

[a] T_4 signifies thyroxine; AST, aspartate aminotransferase.

questionnaire, the authors elicited a wide range of reasons for the lack of follow-up *(14)* (Table 5).

Other Studies

Several studies employing different measurements of medicare processes provide insight into physicians' responses to unexpected abnormalities. Whitehead and Wooton *(5)*, in a carefully designed experiment, found that the number of general-practice patients recalled to a doctor's surgery increased as a consequence of the introduction of a routine biochemical profile, while the number of patients referred from the general practice to the hospital decreased. In the hospital setting, there were a reduction in the number of discretionary biochemical tests ordered, but no decrease in the workload of the hematology and the diagnostic radiology departments and no significant effect on the duration of hospitalization or patient outcome. They concluded that "the introduction of biochemical profiling led to dramatic changes in treatment for only a small proportion of patients."

In a hospital for children a similar experiment revealed a significant increase in the total number of tests requests with the introduction of a routine biochemical profile, with no effect on

Table 5. **Reasons Given by Physicians for Not Confirming Abnormal Screening Test**[a] **(Bates & Yellin,**[14]**)**

Reason	%
Result borderline or only slightly abnormal	15.3
Previously known	14.9
Referred or deferred management to another physician	12.2
Symptoms or signs absent or insignificant	11.6
Result explained by other data	9.9
Patient not seen since screening	9.2
Result relatively unimportant in context, would not affect management	6.1
Patient was not or is no longer under the physician's care	5.0
No other supporting laboratory data with which to correlate finding	3.7
Physician questions testing methods	2.9
Patient declined or was thought uncooperative	2.1
Physician accepted report as sufficient	1.9
Previously suspected	0.6
Follow-up unnecessary, reason unspecified	1.6
Miscellaneous, unclassifiable	3.2

[a] Figures indicate average percentages across all tests.

duration of hospitalization *(6)*. These authors concluded that "the profile made only a small contribution to the overall care of the patients."

In another controlled experiment, investigators found that the use of a routine screening profile, including biochemical tests as well as tests of cardiopulmonary function, color vision, and urinary bacteriology, increased the total costs of hospital care by 5% "without associated evidence of benefit to the patient" *(16)*.

Problems in Interpreting the Unexpected Abnormal Laboratory Result

Brief comments on a few selected laboratory tests will provide a sense of the wide variety of factors that physicians must contend with in judging the significance of the unexpected abnormality. In all these situations a major consideration is the effect of medications. There are few people today seeking help from the physician who are not already taking one or more drugs, and the number of biochemical deviations caused by these exceeds the physician's memory capacity—indeed a bibliography of this information now fills an entire volume *(17)*. The likelihood of polydrug ingestion alone to affect results of continuous-flow analysis is an impressive exercise in probability function.

Serum calcium. The concept of chemical screening first gained popularity with the report that routine calcium determinations disclosed hitherto unsuspected hyperparathyroidism *(18)*. In a program involving nearly 12 000 patients, nine new cases of hyperparathyroidism were detected, eight of which were confirmed at surgery. This technique was then vigorously promoted as a means of uncovering surgically curable asymptomatic parathyroid disease *(19–21)*. At the same time, however, critics began to point out that the natural history of such "biochemical hyperparathyroidism" is not well-known, and that evidence for the benefit of this approach to patients is controversial both on pathological and surgical grounds *(22, 23)*. Hence at this time there is no clearcut course for the physician to follow in the case of an elevated serum calcium value in a patient who lacks other evidence of damage related to parathyroid hyperactivity. Moreover, before calling the surgeon, and even before obtaining parathyrin (parathormone) values, the physician would do well to consider a variety of other factors affecting serum calcium, ranging from blood-drawing technique to common drugs such as steroids, antacids, contraceptives, and diuretics.

Glucose. The physician facing an above-normal value for blood glucose, whether obtained after an elaborate ritual of dietary preparation and oral glucose ingestion, or as a simple casual specimen, must evaluate this information in light of the current controversy that is raging about diabetic control. One group of diabetologists, basing their findings principally on animal experimentation, urge the careful regulation of glucose as a means of preventing the development of diabetic vascular disease *(24)*. Another group, equally expert, maintains that the clinical evidence in humans that strict regulation of blood glucose prevents diabetic vascular disease remains inconclusive *(25, 26)*. And once again, before choosing sides, the physician must consider that the high glucose value may have nothing to do with diabetes, because over a dozen common drugs might be responsible, including steroids, vitamin C, aspirin, Dilantin, diuretics, oral contraceptives and tranquilizers.

Serum cholesterol and triglycerides. It is well-known that serum cholesterol is an important factor in heart-attack risk in young men, an effect which appears to evaporate by age 65 *(27)*. Recent evidence indicates more powerful discrimination when high-density lipoprotein (HDL) cholesterol and low-density lipoprotein (LDL) cholesterol are determined, the risk being low in the former and high in the latter. Yet in the fever of enthusiasm the physician must still remember that the *fact* of association of heart disease with hypercholesterolemia does not necessarily prove the *premise* that medical intervention will be beneficial. How far should the physician pursue an elevated cholesterol value beyond simple advice to reduce risk factors such as weight, smoking, hypertension, etc.? Medical opinion in this area is in constant tempest. Those who are in the low-pressure zone would sweep in with angiograms and by-pass procedures even in the asymptomatic patient *(28)*. They are held off by a small but sturdy high-pressure front of critics who urge caution *(29)*. Adding to the tempest is the breezy enthusiasm of The New Aerobics *(30)* whose benefits have yet to be evaluated seriously. All this the physician with an elevated cholesterol value in hand must face once again, after considering possible spurious effects caused by over a dozen common medicines.

Bilirubin. Isolated bilirubin elevation unaccompanied by any evidence of liver or hematologic abnormalities has generally led to the diagnosis of an entity called Gilbert's disease. What is Gilbert's disease? Is it even a disease? A recent survey of some 23 000 men and women suggests that it is only a statistical illusion and that people who are now diagnosed as having Gilbert's disease may simply constitute the upper end of the normal range *(9)*. Physicians who up to now were unable to cope with this dire scourge will be grateful to learn that automated laboratory screening has eliminated it. They are left now only to decide whether any of the 40 drugs ranging from A to Z (aspirin to Zyloprim) has caused their patient's elevation.

Serum thyroxine (T4). The automated routine reporting of serum T4 seems to offer optimal benefits to the physician and patient. Both hyper- and hypothyroidism can be difficult to detect clinically, can be severe if overlooked too long, and are easily treatable when recognized. Nevertheless in about a half of such patients recently studied, the above-normal serum thyroxine seemed to be a consequence of non-thyroidal illness *(31)*. Thus the clinical evaluation of T4 in the patient presenting with symptoms of undiagnosed illness is uncertain. Indeed, one writer now recommends that routine screening of thyroid function be deferred in the presence of various intercurrent illnesses *(32)*—a recommendation that would obviate much of its diagnostic utility.

Discussion

If we attempt to justify the lack of response of physicians to unsolicited laboratory abnormalities by the low yield of these results, we come up against one of the fundamental rubrics of medical education: the routine medical work-up. Most of us continue to pay homage to this diagnostic formula despite the low yield of many of its parts. Medical students are still taught to record childhood exanthems, remote surgeries, diseases of ancestors, foreign travels, numbers of cocktails, and the like. After that they are enjoined to carry out a detailed examination that includes peering into ears, eyes, mouth and nostrils, putting fingers to the thyroid gland, a stethoscope to the chest, hands on the abdomen, rubber hammer to the tendons, and so on, with fingers, hands, and instruments down the rest of the body—carefully probing for the spleen tip and duly noting arcus senilis—never pausing to question the yield of these procedures. The goal is to be complete, to miss nothing. It is taken for granted that a great effort is being expended for a fractional benefit. Is this not the criteria by which automated laboratory testing should be judged? The answer probably is: yes—for the present. But when greater cost accounting is brought to bear on the health care industry the answer may be somewhat different.

One feature of modern medical life that has magnified routine examinations we are all well aware of is the threat of malpractice litigation. It is acknowledged that physicians today often order a barrage of tests—barium enemas, IVP's, upper GI series, gallbladder series—not for clinical reasons, but simply to escape potential liability. So, while we scrupulously debate the merits of the biochemical profile, the "radio-opaque profile" has already become the standard diagnostic approach to the dyspeptic patient whose physician knows better but fears worse.

There is an important difference, of course between the abnormality resulting from a test a physician orders in search of a disease and one which appears gratuitously. It is quite unlikely physicians would ignore the former, and when they do, they quite clearly are in legal peril. With regard to the latter, however, I do not think we have fully felt the malpractice implications of their behavior. The large assortment of abnormalities emerging from routine laboratory testing provides a rich harvest for litigation. (This problem, in fact, probably is compounded by printouts that offer helpful diagnostic suggestions to the physician *(33)*). What will a jury decide about the doctor who recorded no response to a modestly increased alkaline phosphatase in a patient who simply came for a check-up but five years later demonstrates metastatic malignancy? What will be the effect of the jury's decision on future medical practice? Will physicians go to the trouble of recording their reasoning behind every laboratory abnormality they believe requires no further work-up? Will they take the time—and face the difficulty—of explaining their actions to the patient? Will they begin to limit their use of the biochemical profile to avoid such contingencies? Will the results of such a review as this, indicating that unexpected biochemical abnormalities rarely yield significant diagnostic or therapeutic clues, be employed in the physician's defense? These are questions which we may begin to see answered in the near future. We can only hope that the answers chosen will be in the direction of careful and compassionate medical care.

References

1. Sackett, D. L., and Holland, W. W., Controversy in the detection of disease. *Lancet* **ii,** 357–359 (1975).

2. Davey, P. W., Letts, H. W., and Anderson, P. A., Experience with a hospital laboratory admission screen. *Can. Med. Assoc. J.* **103,** 140–146 (1970).

3. Bryan, D. J., Wearne, J. L., Viau, A., et al., Profile of admission chemical data by multichannel automation: An evaluative experiment. *Clin. Chem.* **12,** 137–143 (1966).

4. Carmalt, M. H. B., Biochemical screening—the results. In *Fifth Symposium on Advanced Medicine*, Proceedings of a Conference held at the Royal College of Physicians of London, 1969, R. Williams, Ed., Pitman Med. Pub. Co. Ltd., 1969, pp 268–278.

5. Whitehead, T. P., and Wootton, E. D. P., Biochemical profiles for hospital patients. *Lancet* **ii,** 1439–1443 (1974).
6. Leonard, J. V., Clayton, B. E., and Colley, J. R. T., Use of biochemical profile in children's hospital: Results of two controlled trials. *Br. Med. J.* **ii,** 662–665 (1975).
7. Korvin, C. C., Pearce, R. H., and Stanley, J., Admissions screening: Clinical benefits. *Ann. Intern. Med.* **83,** 197–203 (1975).
8. Friedman, G. D., Goldberg, M., Ahuja, J. N., et al., Biochemical screening tests. *Arch. Intern. Med.* **129,** 91–97 (1972).
9. Bailey, A., Robinson, D., and Dawson, A. M., Does Gilbert's disease exist? *Lancet* **i,** 931–933 (1977).
10. Sackett, D. L., The usefulness of laboratory tests in health screening programs. *Clin. Chem.* **19,** 366–372 (1973).
11. Schneiderman, L. J., DeSalvo, L., Baylor, S., and Wolf, P. L., The "abnormal" screening laboratory results. *Arch. Intern. Med.* **129,** 88–90 (1972).
12. Fletcher, S. W., Sourkes, M., Rabzel, M., and Fletcher, R. H., Multiphasic screening: Case-finding tool in a teaching hospital medical clinic. *J. Am. Med. Assoc.* **237,** 887–891 (1977).
13. Bradwell, A. R., Carmalt, M. H. B., and Whitehead, T. P., Explaining the unexpected abnormal results of biochemical profile investigations. *Lancet* **ii,** 1071–1074 (1974).
14. Bates, B., and Yellin, J. A., The yield of multiphasic screening. *J. Am. Med. Assoc.* **222,** 74–78 (1972).
15. Williamson, J. W., Alexander, M., and Miller, G. E., Continuing education and patient care research. *J. Am. Med. Assoc* **201,** 118–122 (1967).
16. Durbridge, T. C., Edwards, F., Edwards, R. G., and Atkinson, M., Evaluation of benefits of screening tests done immediately on admission to hospital. *Clin. Chem.* **22,** 968–971 (1976).
17. Young, D. S., Pestaner, L. C., and Gibberman, V., *Effects of Drugs on Clinical Laboratory Tests. Clin. Chem.* **21**(5) 1D-432D (1975). Special issue.
18. Boonstra, C. E., and Jackson, C. E., The clinical value of routine serum calcium analysis. *Ann. Intern. Med.* **57,** 963–969 (1962).
19. Haff, R. C., Black, W. C., and Ballinger, W. F., Primary hyperparathyroidism: Changing clinical, surgical and pathologic aspects. *Ann. Surg.* **171,** 85–92 (1970).
20. Purnell, D. C., Scholz, D. A., Smith, L. H., et al., Treatment of primary hyperparathyroidism. *Am. J. Med.* **56,** 800–809 (1974).
21. Attie, J. N., and Khafif, R. A., Surgical exploration in asymptomatic hypercalcemia: Early treatment of hyperparathyroidism. *Am. J. Surg.* **132,** 449–452 (1976).

22. Ahlvin, R. C., Biochemical screening—a critique. *N. Engl. J. Med.* **283,** 1084–1086 (1970).
23. Williamson, E., and Van Peenen, H. J., Patient benefit in discovering occult hyperparathyroidism. *Arch. Intern. Med.* **133,** 430–431 (1974).
24. Cahill, G. F., Etzwiler, D. D., and Freinkel, N., "Control" and diabetes. *N. Engl. J. Med.* **294,** 1004 (1976).
25. Siperstein, M. D., Foster, D. W., Knowles, H. C., Levine, R., Madison, L. L., and Roth, J., Control of blood glucose and diabetic vascular disease. *N. Engl. J. Med.* **296,** 1060–1063 (1977).
26. Ingelfinger, F. J., Debates on diabetes. *N. Engl. J. Med.* **296,** 1228–1230 (1977).
27. Gofman, J. W., Young, W., and Tandy, R., Ischemic heart disease, atherosclerosis and longevity. *Circulation* **34,** 679–697 (1966).
28. Vismara, L. A., Miller, R. R., Price, J. E., et al., Improved longevity due to reduction of sudden death by aortocoronary bypass in coronary atherosclerosis. *Am. J. Cardiol.* **39,** 919–924 (1977).
29. Selzer, A., and Cohn, K., Asymptomatic coronary artery disease and coronary bypass surgery. *Am. J. Cardiol.* **39,** 614–616 (1977).
30. Cooper, K. H., *The New Aerobics,* Bantam, New York, N.Y., 1970.
31. Birkhauser, M., Busset, R., Burer, Th., and Burger, A., Diagnosis of hyperthyroidism when serum-thyroxine alone is raised. *Lancet* **ii,** 53–56 (1977).
32. Editorial, *Lancet* **ii,** 74–75 (1977).
33. Wolf, P. L., *Interpretation of Biochemical Multitest Profiles,* Masson Pub., New York, N.Y., 1977.

The Usefulness of Laboratory Information in Medical Monitoring

C. H. Altshuler, M.D.

St. Joseph's Hospital, 5000 West Chambers St., Milwaukee, Wis. 53202

It is axiomatic that cost-effective medicine must provide for fast, accurate diagnoses and for appropriate therapy, appropriately administered. Discussion as to the best means to attain these goals have been difficult (and heated) because inadequate information is available as to the cost of making a diagnosis, the incidence of missed or inappropriate diagnoses, the cost of missing a diagnosis, and the relative effectiveness of various therapies. In this presentation, it will be shown that certain previously described laboratory systems provide an opportunity to obtain some of this necessary information and may facilitate the correction of some of the deficiencies found.

In an effort to exploit developments in laboratory automation and information handling, two systems were developed. Both have been described previously. One of the systems, the PALI (*p*rogrammed *a*ccelerated *l*aboratory *i*nvestigation), provides for a pathologist interface between the original screening data and the follow-on discriminatory or confirmatory tests (Figure 1). The technique, which entails no additional charge to the patient, has been well received by physicians (Figure 2) and offers some significant cost benefits (Figure 3). Other advantages of the PALI are listed in Figure 4.

The numerical data generated by the PALI and some other procedures have been stored on laboratory computer tapes for several years, and it was in an effort to determine the diagnostic content of given patterns of findings that we initiated these studies. In this effort, the laboratory information was transmit-

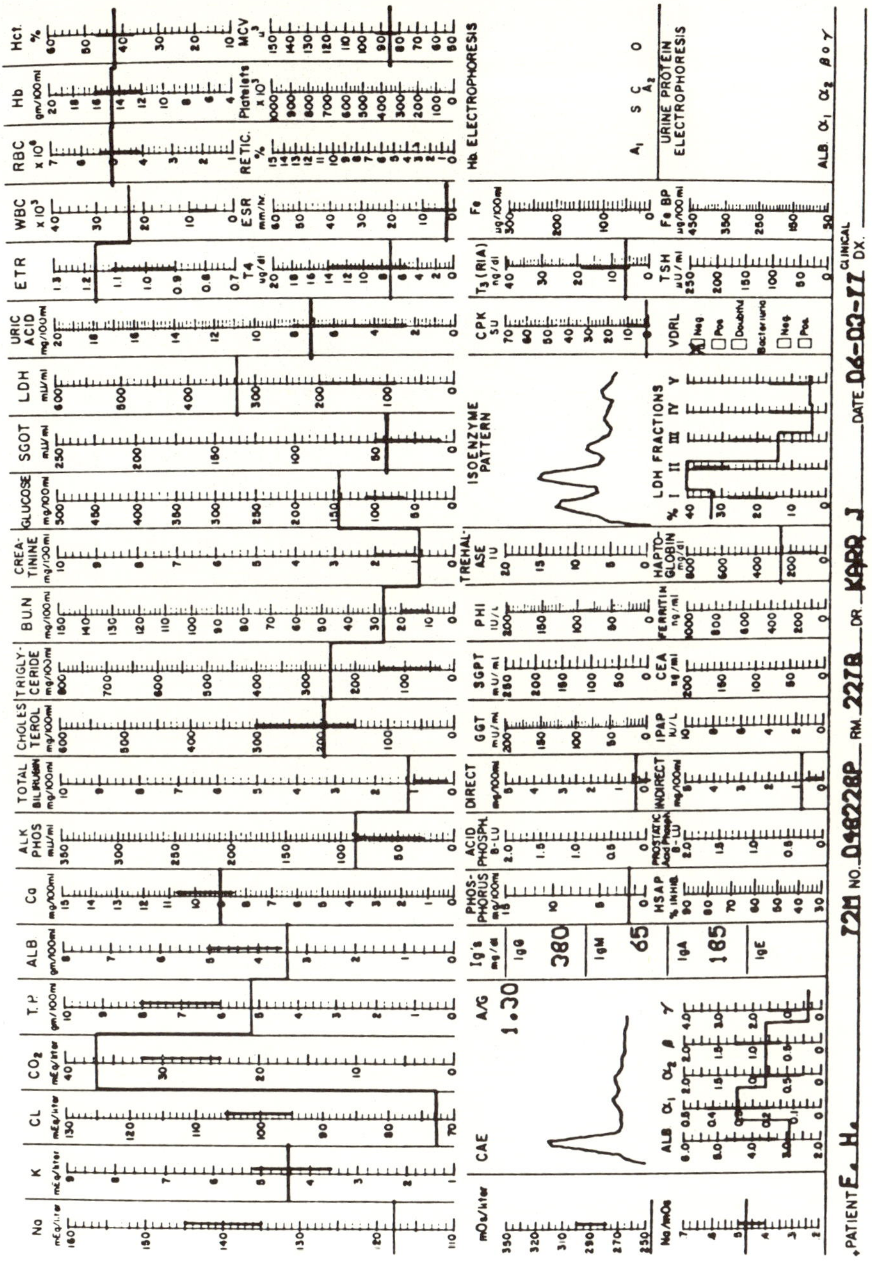
Na
K
CL
CO2
T.P.
ALB
Ca
ALK PHOS
TOTAL BILIRUBIN
CHOLESTEROL
TRIGLYCERIDE
B.U.N
CREATININE
GLUCOSE
SGOT
LDH
URIC ACID
ETR
WBC x 10³
RBC x 10⁶
Hb
Hct. %
ESR
T4
RETIC. %
Platelets x 10³
MCV
Hb. ELECTROPHORESIS
A1 S C A2 O
URINE PROTEIN ELECTROPHORESIS
ALB α1 α2 β γ
mOs/liter
Na/mOs
CAE
A/G
1.30
Ig's
IgG
380
IgM
65
IgA
185
IgE
PHOSPHORUS
ACID PHOSPH.
DIRECT
GGT
SGPT
PHI
TREHALASE
ISOENZYME PATTERN
CPK
T3 (RIA)
Fe
HSAP
PROSTATIC ACID PHOSPH.
INDIRECT
IPAP
CEA
FERRITIN
HAPTOGLOBIN
LDH FRACTIONS
VDRL
TSH
Fe BP
PATIENT E. H.
72H NO. 048228P
RM. 227B
DR. KARR J
DATE 06-03-77
CLINICAL DX.

Fig. 1

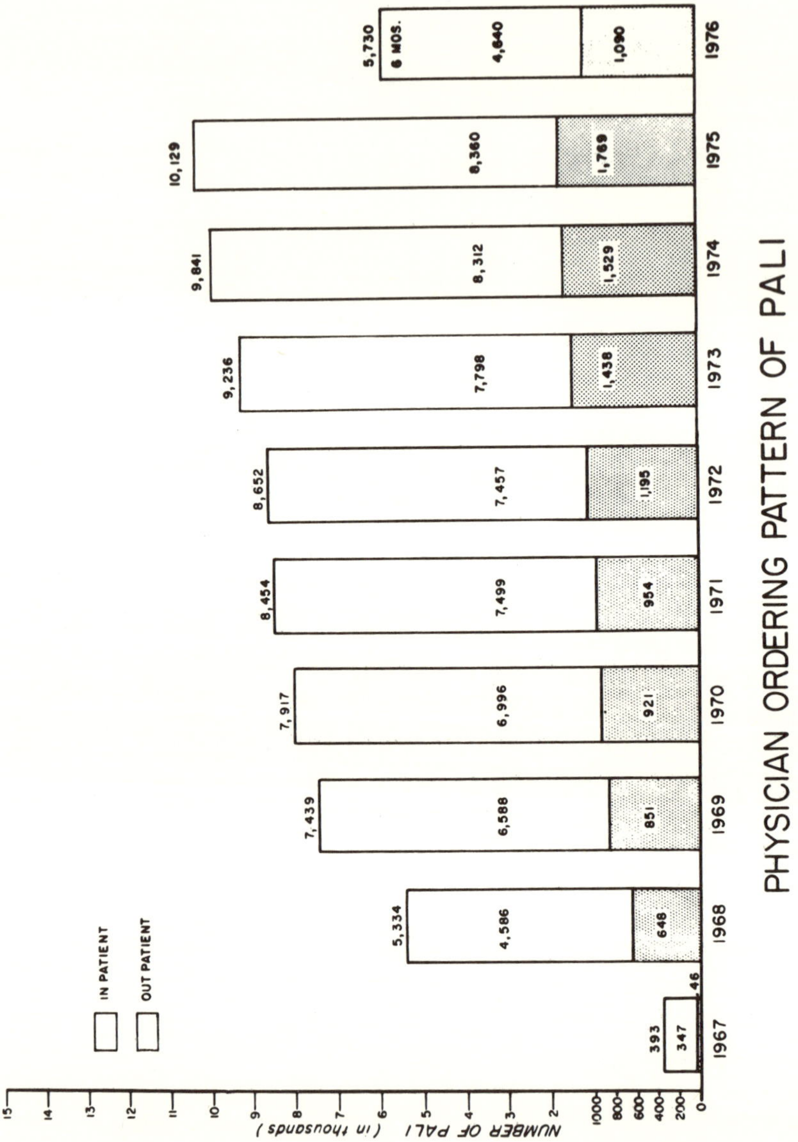

Fig. 2

ted to the hospital mainframe, where they were merged with the discharge diagnoses obtained from appropriate PAS tapes (Figure 5). (Later, drug profiles were also added to the data base.) As a matter of convenience, we created two separate files—one containing the records of inpatients discharged in 1975 and one for inpatients discharged in 1976. (The hospital discharged ap-

Fig. 3. *Cost Benefits of the PALI*

1) Minimizes costs of obtaining the specimen and associated paper work.
2) Maximizes the utilization of consumables.
3) Makes for efficient utilization of laboratory instrumentation.
4) Minimizes cost per test for those tests that require an expensive calibration curve.
5) Minimizes the expensive "stat" order.

Fig. 4. *Advantages of the PALI*

1) Is economically and technically feasible
2) Makes efficient use of the patient's serum and urine
3) Makes efficient use of the laboratory equipment
4) Shortens the patient's time in the hospital before a diagnosis is available
5) Is suitable for screening "normal" individuals
6) Is of value in establishing normal values
7) Facilitates introduction of new laboratory tests
8) Enhances the role of the pathologist and gives him better control of the quality of the laboratory output
9) Facilitates interpretation of the laboratory data
10) Minimizes iatrogenic illness of laboratory origin, and
11) Facilitates laboratory development planning

proximately 24 000 patients each year, and each year approximately 8500 PALI's had been done on inpatients.)

In early studies it became apparent that the technique could not be employed for the purpose of data evaluation because of several shortcomings. These limiting factors are listed in Figure 6. For example, when patients with the pattern of high IgG, low IgA, and low IgM were isolated, the diagnosis of multiple myeloma was not always made, even though some of the patients were shown to have a monoclonal peak on both cellulose acetate electrophoresis and immunoelectrophoresis and a positive test for Bence Jones protein in the urine (Figure 7). Similarly, in patients with other clinical problems (thyroid, anemia, parathyroid, etc.) the incidence of unrecognized disease in the presence of characteristic laboratory data was distressingly high (Figure 8).

For some of these diseases, the physicians involved were

Fig. 5.

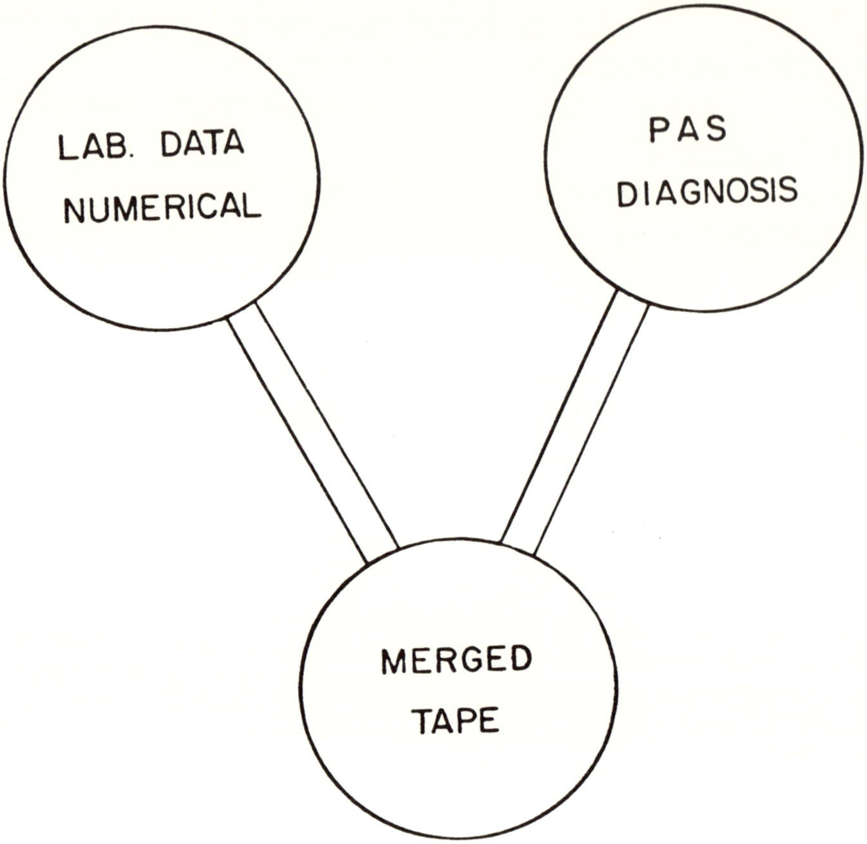

Fig. 6. Deficiencies of the System *Relative to Diagnostic Content Studies*

1) Diagnosis not recorded
   a) Missed
   b) Known diagnosis but patient in hospital for some other problem
   c) Diagnosis suspected but could not be unequivocably established by the time of discharge
   d) Many pathophysiologic states not considered to be a diagnosis
2) Inherent shortcomings of HICD-A coding
3) Some findings characteristic of only a brief period in the course of disease. Sampling time inappropriate
4) Significant test not obtained
5) Some test values the net result of conflicting pathophysiologic factors
6) Therapeutic effects obscure the "characteristic" test results

```
A12R08                                              ST. JOSEPH'S HOSPITAL            5/27/77
                                             PATIENT DIAGNOSIS AND SELECTIVE TEST REPORT

STUDY # 76152   BASE CONDITION:
                AND CONDITIONS:              TEST 052/01 IGG      RANGE: 1801.00 - 9999.00
                                             TEST 052/02 IGA      RANGE:    0.00 -   89.00
                                             TEST 052/03 IGM      RANGE:    0.00 -   59.00

PAT NO   NAME                                  VST AGE SEX RACE    ADM DATE  DSCH DATE
      DIAGNOSIS INFO                               TEST TITLE    DATE      VALUE  DATE

                                                 2  74Y  M  WHITE      6/01/76      7/05/76
   FINAL DX: Y11.6  HX-LYMPHATIC MALIGNANCY      052/01 IGG            6/02 2100.00
   OTHER DX: Y10.4  HX-PROSTATIC MALIGNANCY      052/02 IGA            6/02   70.00
   OTHER DX: 427.9  CONGESTIVE HEART FAILURE     052/03 IGM            6/02   20.00

                                                 1  83Y  F  WHITE      8/12/76      8/15/76
   FINAL DX: 776.1  EPISTAXIS                    052/01 IGG            8/12 3180.00
   OTHER DX: 440.9  ARTERIOSCLEROSIS NOS         052/02 IGA            8/12   10.00
   OTHER DX: 444.9  ARTERIAL EMBOLISM NEC        052/03 IGM            8/12   15.00
   OTHER DX: 715.0  ARTHRITIS NOS

                                                 1  67Y  F  WHITE      7/26/76      8/28/76
   FINAL DX: Y11.6  HX-LYMPHATIC MALIGNANCY      052/01 IGG            7/27 2240.00
   OTHER DX: 496.0  CHR OBSTRUCTIVE LUNG DIS     052/02 IGA            7/27   30.00
   OTHER DX: 412.9  CHR ISCHEMIC HRT DIS NEC     052/03 IGM            7/27   15.00
   OTHER DX: 595.9  CYSTITIS NOS

                                                 2  68Y  F  WHITE      10/11/76     10/15/76
   FINAL DX: Y11.6  HX-LYMPHATIC MALIGNANCY      052/01 IGG            10/12 2160.00
   OTHER DX: 412.9  CHR ISCHEMIC HRT DIS NEC     052/02 IGA            10/12   35.00
   OTHER DX: 723.0  OSTEOPOROSIS                 052/03 IGM            10/12   15.00

                                                 1  75Y  F  WHITE      4/22/76      7/08/76
   FINAL DX: 203.0  MULTIPLE MYELOMA             052/01 IGG            4/23 2325.00
   OTHER DX: 288.1  LEUKOPENIC STATE NEC         052/02 IGA            4/23   40.00
   OTHER DX: 782.0  GI BLEEDING NEC              052/03 IGM            4/23   20.00
   OTHER DX: 776.1  EPISTAXIS
   OTHER DX: 805.4  FRACT LUMBAR CLOSED
   OTHER DX: 427.9  CONGESTIVE HEART FAILURE
   OTHER DX: 306.3  PARANOID SCHIZOPHRENIA

                                                 1  75Y  M  WHITE      1/14/76      1/24/76
   FINAL DX: 440.2  ARTERIOSCLER-EXTREMITY       052/01 IGG            1/15 2940.00
   OTHER DX: 432.0  PRECEREBRAL OCCLUSION        052/02 IGA            1/15   50.00
   OTHER DX: 437.0  CEREBRAL ISCHEMIA NOS        052/03 IGM            1/15   25.00
   OTHER DX: 412.9  CHR ISCHEMIC HRT DIS NEC
   OTHER DX: 427.9  CONGESTIVE HEART FAILURE
   OTHER DX: 496.0  CHR OBSTRUCTIVE LUNG DIS
   OTHER DX: 203.0  MULTIPLE MYELOMA

                                                 1  42Y  M  WHITE      12/06/76     12/20/76
   FINAL DX: 203.0  MULTIPLE MYELOMA             052/01 IGG            12/07 7500.00
   OTHER DX: 573.9  LIVER DISEASE NEC & NOS      052/02 IGA            12/07    6.00
   OTHER DX: 286.9  COAGULATION DEFECT NEC       052/03 IGM            12/07   14.00
```

Fig. 7

predominantly orthopedic surgeons, urologists, and opthalmologists, who were treating the patients for other readily apparent disorders—fractured leg, cataracts, prostatic hyperplasia, etc. Thus, these specialist physicians were confronted with data which revealed a problem outside their area of expertise.

For other conditions, however, this distribution of physician types did not pertain. For example, in a study done by an internist on iron-deficiency states for the medical audit committee, it was determined that of 40 undiagnosed cases of iron deficiency, 12 patients had been cared for by general surgeons,

A12R08 ST. JOSEPH'S HOSPITAL 10/19/76
PATIENT DIAGNOSIS AND SELECTIVE TEST REPORT

PAT NO NAME / DIAGNOSIS INFO	VST	AGE	SEX	RACE	ADM DATE	DSCH DATE	TEST	TITLE	DATE	VALUE
	1	66Y	F	WHITE	1/24/75	1/28/75				
FINAL DX: 244.9 HYPOTHYROIDISM NOS							705/01	TSH	1/25	100.00
OTHER DX: 562.1 DIVERTICULOSIS COLON							706/01	T3-RIA	1/25	25.00
OTHER DX: 728.9 RADICULAR SYNDROME NEC							707/01	ETR	1/25	0.84
							732/01	BLTHYR	1/25	4.00
	1	82Y	F	WHITE	9/07/75	9/30/75				
FINAL DX: 731.1 SYNOVITIS OF SHOULDER							705/01	TSH	9/06	58.00
							706/01	T3-RIA	9/06	25.00
							707/01	ETR	9/06	0.83
							732/01	BLTHYR	9/06	4.00
	2	82Y	F	WHITE	10/31/75	12/23/75				
FINAL DX: 731.1 SYNOVITIS OF SHOULDER							705/01	TSH	11/01	40.00
							706/01	T3-RIA	11/01	25.00
							707/01	ETR	11/01	0.84
							732/01	BLTHYR	11/01	4.00
	1	83Y	F	WHITE	12/24/74	1/08/75				
FINAL DX: 812.2 FX HUMERUS SHAFT CLOSED							705/01	TSH	12/26	52.00
OTHER DX: E88.7 ACCIDENTAL FALL							706/01	T3-RIA	12/26	25.00
OTHER DX: 357.4 ARM NERVE DISEASE NEC							707/01	ETR	12/25	0.85
							732/01	BLTHYR	12/26	4.00
	1	87Y	M	WHITE	1/01/75	2/19/75				
FINAL DX: 153.8 MALIG NEO LG BOWEL NOS							705/01	TSH	1/02	80.00
OTHER DX: 244.9 HYPOTHYROIDISM NOS							706/01	T3-RIA	1/02	25.00
OTHER DX: 999.6 TRANSFUSION REACTION							707/01	ETR	1/01	0.85
OTHER DX: E93.1 SURG OR MED COMPLICATION							732/01	BLTHYR	1/07	4.00
	1	35Y	F	WHITE	6/24/75	7/01/75				
FINAL DX: 244.9 HYPOTHYROIDISM NOS							705/01	TSH	6/25	100.00
OTHER DX: 678.0 PUERPERAL LACTATION DIS							706/01	T3-RIA	6/25	25.00
OTHER DX: 574.0 GALLBLADDER CALCULUS							707/01	ETR	6/25	0.86
							732/01	BLTHYR	6/25	4.00

Fig. 8

11 by internists, 11 by obstetricians, and the remainder by others of the medical staff.

It had previously been thought that systematic investigation of the patient by the pathologist might generate information with which the clinician might be unfamiliar. For that reason, a microfiche information-transfer system had been designed and was made available at the points of need (nursing stations, doctors' offices, emergency rooms, etc.) This system, AIDE (*a*ccessible *i*nformation for *d*iagnosis and *e*valuation), has also been described in detail and has been used on the floors for at least six years. Some of the advantages of the AIDE system are listed in Figure 9. Since the lapses in diagnosis previously described occurred in the presence of available microfiche, it is apparent that the AIDE system was not being appropriately utilized.

It is apparent that some procedure(s) must be developed to

Fig. 9. *ADVANTAGES OF THE AIDE SYSTEM*

1) Provides a readily retrievable and easily accessible data base at the point of need.
2) Provides required graphic information and color illustrations.
3) Is economical.
4) Allows for associative indexing.
5) Allows the information to be presented at any level of sophistication desired and to be modified for any specific audience.
6) Couples the problem with the solution.
7) Encourages a logic flow for the user.
8) Equipment can be used for other informational needs.
   a) Administrative—nursing procedures, ward clerks, e.g.
   b) PSRO
   c) Education
      1) Purpose
         a) Self-learning and self-testing
         b) Lecture residual
         c) Seminar residual
      2) Advantages
         a) Facilitates discussion at ward rounds
         b) Allows "educatory" information to be retrieved when there is a need to know
   d) Operational information—laboratory statistics, test evaluation.

correct these deficiencies in practice. It would appear that the following would be helpful:

- on admission, an efficient comprehensive systematic laboratory investigation be carried out,
- some effective means of communication be developed to improve information transfer between the laboratory and the clinician, and
- some means of monitoring the system be available

The acquisition of the comprehensive data base is essential because without it there is no convenient and effective means to determine whether the diagnosis is missed, whether effective concentrations of drugs were attained in the blood, and so forth.

The communication problem is difficult, for one must be assured that the data is neither overlooked nor inappropriately interpreted. Since telephoning the results and interpretation is not a practical solution except in isolated cases or when only small numbers of abnormal results are generated, other means of communication must be developed and available forms of information-transfer enhanced.

One solution to the educational aspect of the communication problem, we believe, may be the improved utilization of microfiche. For this purpose, we have proposed that they be employed in the continuing medical education (CME) efforts. For this purpose, we have written a series of topic-oriented questionnaires (Figures 10 and 11) for use by the physician. The answers to the questions on any one questionnaire are available

Fig. 10. *QUESTIONNAIRE—PROTEINS-3–085 056*

Note: More than one answer may be correct

1) Alpha fetoprotein is said to be a diagnostic aid in instances of neonatal jaundice. It is said to be elevated in
   (a) neonatal hepatitis
   (b) biliary atresia
   (c) erythroblastosis
2) Immunoglobulins are thought to have two chief regions—a variable region and a constant region. The constant region, in turn, is divided into three areas. It is believed that certain of the constant areas have specific functions. The area that is believed to be cytophilic is
   (a) CH1 (CH—constant region of heavy chain)
   (b) CH2
   (c) CH3
3) IgG is divided into four subclasses. IgG2 and IgG4 are believed to migrate differently from IgG1 and IgG3 or electrophoresis. IgG2 and IgG4 are believed to migrate
   (a) more anodal
   (b) more cathodal
4) The various IgG subclasses may be identified by Gm allotyping and by its reactivity with staphylococcal protein A. The IgG(s) that do not react with protein A is (are)
   (a) IgG1
   (b) IgG2
   (c) IgG3
   (d) IgG4
5) The IgG subtype(s) that binds to monocytes is (are)
   (a) IgG1
   (b) IgG2
   (c) IgG3
   (d) IgG4

on a specific fiche. Thus, a physician who wishes to brush up on tumor immunology or protein electrophoretic patterns, for example, could utilize the appropriate questionnaire and fiche. After the questionnaire has been completed, the physician is supplied with the answers (Figure 12) and the grid location of the pertinent information is indicated. We feel that there are unique advantages of microfiche in CME. These advantages are listed in Figure 13.

In addition to this educational effort, communication may be enhanced by the merger of the laboratory data with a diagnostic file and a drug profile. This merged data provides a practical opportunity for the creation of a computer-based message file. For example, if the laboratory data indicate a low ETR, low

Fig. 11. *QUESTIONNAIRE—HEMATOLOGY 025 015* 000 105

Note: More than one answer may be correct

1) Patients with atypical pneumonia sometimes show evidence of hemolysis. This finding is usually accompanied by
    - (a) a cold agglutinin
    - (b) an IgG antibody to I (big I)
    - (c) an IgM antibody to i (little i)
2) Patients with vitamin B12 deficient megaloblastic anemia may show gastric acid. This combination is seen in
    - (a) adult Addisonian pernicious anemia
    - (b) juvenile autoimmune pernicious anemia
    - (c) congenital intrinsic factor deficiency
    - (d) Grasbeck-Imerslund syndrome
3) The B12 binding protein is increased in pregnancy in megaloblastic anemia of pregnancy. The vitamin B12 levels are
    - (a) increased markedly
    - (b) increased slightly
    - (c) decreased markedly
    - (d) decreased slightly
4) Insulin or sucrose hemolysis tests are used for the diagnosis of
    - (a) pernicious anemia
    - (b) paroxysmal nocturnal hemoglobinuria
    - (c) thalassemia
5) Sickle cell disease with persistent F is characterized by
    - (a) severe crises and anemia
    - (b) occasional crises
    - (c) being essentially asymptomatic
    - (d) spherocytosis

Fig. 12. *ANSWERS—HEMATOLOGY 025 015* 000 105

			Grid Position
1)	(a)	A cold agglutinin The direct Coombs test is positive when using anti-complement antibody or when using a bivalent serum containing both anti IgC and anti-complement antibody. The most common specificity of such cold agglutinin is IgM anti I	D–3, B–3
2)	(c)	Congenital intrinsic factor deficiency	F–5
	(d)	Grasbeck-Imerslund syndrome	
3)	(d)	Decreased slightly	B–5
4)	(b)	Paroxysmal nocturnal hemoglobinuria	F–7
5)	(b)	Occasional crises	D–8
6)	(b)	Iron deficiency anemia	D–7
	(c)	Thalassemia	
	(d)	Hemoglobin C disease	
7)	(a)	High blood iron binding protein	B–7, C–4,
	(c)	Increase in free erythrocyte porphyrins	D–4
	(d)	Decrease in serum ferritin	
8)		All correct	D–1
	(a)	Microcytosis is characteristic of thalassemia and pyridoxine deficiency anemia	
	(b)	Burr cells may be indicative of microangiopathic hemolytic anemia	
	(c)	Macrocytes may be seen with reticulocytosis	
	(d)	Tear drop red cells may occur in myelofibrosis	
9)	(c)	Triose phosphate isomerase	E–6
0)		All correct	E–7
	(a)	Refractory anemia	
	(b)	Abnormal red cell enzymes	
	(c)	Evidence of iron overloading	
	(d)	Increased unconjugated bilirubin and	
	(e)	Morphologic abnormalities in the bone marrow	

Fig. 13. *Advantages of Fiche in CME*

1) The user would have a means of assessing his knowledge of a subject.
2) The user would learn something about the subject.
3) The user would have the information in a readily retrievable form when required sometime in the future.
4) The user would get CME credit.
5) Facilities would be provided for those users who have an interest in self-learning
6) Analysis of the answers would determine whether the material on the fiche is presented in a readily understandable manner.

R00040 ST. JOSEPH'S HOSPITAL 09/20/77 PAGE 005

MEDICAL MESSAGES

PATIENT NAME:

PATIENT NUMBER: AGE: 51 YEARS

ADMISSION DATE : 07/12/77 SEX: FEMALE

DISCHARGE DATE : 07/20/77 RACE: WHITE

THE LABORATORY FINDINGS SUGGEST HYPOTHYROIDISM. DOES THIS FIT THE CLINICAL PICTURE?

AIDE REFERENCE

012047 A13-D13
021057 A1-A13

*********************** CRITERIA ************************

TYPE	NUMBER	DESCRIPTION	RANGE	PATIENT VALUES
LAB TEST	707/01	EFF. THYROXINE RATIO(ETR)	0.00- 0.89	.86
LAB TEST	732/01	TOTAL THYROXINE(T4)	0.00- 5.40	3.60
LAB TEST	705/01	THYROID STIM. HORMONE(TSH)	11.00- 999.00	18.00
DIAGNOSIS			244.0 - 244.9	NONEXSIST

00001 00002 00020

Fig. 13

thyroxine, and a high thyrotropin, and no pertinent diagnosis had been made, a message is generated urging the physician to consider the diagnosis of hypothyroidism (Figure 14). This message file has already been developed partly, but has not as yet been implemented. At the present stage of development, the message file deals with data that are six to eight weeks old. Ultimately, of course, the goal would be to generate the messages concurrently.

In summary, comprehensive laboratory data, when merged with the discharge diagnoses and drug profiles, provides a superior means for the evaluation of medical care and for the study of utilization patterns of laboratory data. Future inclusion of cost information into the system may be helpful in obtaining the information badly needed for the intelligent development of cost-control mechanisms.

PANEL DISCUSSION

Dr. Uddin: Dr. Altshuler, what is the actual cost of a system such as you describe, the full computerized system?

Dr. Altshuler: Well, I really think that we probably all should be making those microfiche together. What is appropriate to us may not be entirely appropriate to all other institutions, and I think that one of the advantages of the microfiche system is that you could use people with special knowledge and special experience and make that experience available to everybody, so actually the only capital cost that you have would be the readers and the cost of the fiche themselves. That's all. I don't think it is desirable to have more than one or two data bases.

Those readers each cost about $150 to $250. Many hospitals already have them because they have been using microfiche for archival purposes. One of the disadvantages of the microfiche system is that you have to overcome the antipathy that physicians have for microfilm, because in the past they have microfilmed the charts and usually have done very poorly, and certainly if you can't read a guy's handwriting to start with, you still can't read it after you take a picture of it.

Dr. Uddin: If a patient has only hyperglycemia without glycosuria, don't you still think it of benefit to the obese patient to advise him to lose weight, to not smoke if he is a smoker, or to drink less well if he drinks?

Dr. Schneiderman: I think he should drink better!

Well, the renal threshold for glucose is about 170, so I assume we are talking about blood sugar between 100 and 170, and the patient shows no other reason for its excretion. I personally tell patients that they can enjoy life better if they lose weight, if they stop smoking, and if they drink for fun and not in excess, but in doing that I am expressing simple points of view about life, not my professional judgment as a clinician, in that my professionalism enters much less than my judgment as a human being and my relationship with the patient.

I think this is an important distinction: sometimes we talk to our patients as physicians and we try to give them facts and help them run their lives; we are technical consultants. But at other times we are quite clearly guides of another sort, and so I would put the answer to that question as being more in the second sort.

Dr. Uddin: What is the potential of malpractice suit if the lab finds an abnormality and does not report it, because the abnormal result was for a test not requested?

Dr. Schneiderman: I didn't know that you did things like that. Well, now I can see of course that you must do things like that, because you set up panels, and then, if you happen to do a serum potassium on the same panel and it wasn't requested, I can see that. But I don't know the answer.

Dr. Uddin: Dr. Altshuler, in your system, how do you avoid redundant PALI workups on patients who are readmitted by a physician that doesn't know the last workup was done?

Dr. Altshuler: Well, usually a PALI is only done on admission and we don't ordinarily see a repeat PALI except in a patient who has been in the hospital for a very long time, very complicated, and has had many things happening, problems with water, electrolyte balance, problems of renal failure, and so on. We never really had that kind of problem.

Dr. Cole: I don't know of anybody who goes into a hospital who doesn't have blood collected for some test. Now, the cost of collecting that sample, the cost of going through the whole process just to get one test done by the laboratory has a basic unit cost, and the question of how many you do relates to the question of how much do you increase the cost over and above what it otherwise would be, in relationship both to the population as a whole in the hospital, as well as otherwise.

Now, I think that is quite in contrast to an outpatient evaluation, and I think that sometimes it is very confusing in people's minds when you are doing multiphasic laboratory testing, whether you are doing it in a hospital or on an outpatient basis. It is just not the same. You can't compare the cost of one to the other. But I keep hearing this justification for a battery of tests in a hospitalized patient, who already has been sick, by definition, because he has been admitted; there must be something wrong or the patient would not be there. This is not the same kind of problem we face on an outpatient basis, where you may encounter the worried well. It is a totally different problem, but we keep confusing them.

Dr. Uddin: Do you, Dr. Cole, plan to use the method of looking at two or more tests in a way to eliminate or reduce tests, or to set up a sequence of testing?

Dr. Cole: I suppose the most notable example is that we have many different forms of discriminative analysis and different equations, just like the osmolality model that was referred to this morning. We have developed in our laboratory a mechanism to detect alcohol by measuring the BUN, glucose, and sodium, doing the osmolality, and then subtracting the difference and checking to see if ketones are present. We could [instead] determine whether or not alcohol was present, and whether we should go ahead and do a whole bunch of other alcohol studies. In the admission profile, the thing I latched on very quickly was that if the lactate dehydrogenase/aspartate aminotransferase ratio, by the methods used in the Technicon system, falls below 2/1 the probability is about 0.98 that the cause for those enzyme elevations is hepatic. If that ratio is above 10/1, the chances of those changes coming from the liver are remote by an equal probability. Therefore, to do isoenzymes or other tests and say is this aspartate aminotransferase elevated because of liver disease or not really reduces the frequency with which you need to do it as a diagnostic procedure. So here are two tests acting in synergism, giving you information each done singly would not.

Dr. Uddin: Much has been said about the tendency of physicians to ignore unsolicited abnormal results. Do they not also sometimes ignore the results of specifically solicited tests?

Dr. Schneiderman: That is true.

Dr. Cole: I think that if you are ordering the tests because you want to rule our something that you really don't believe is there, and the test result comes back and it shows that maybe what you wanted to rule out is there, it does present some problems of probability. When I review the charts on a large number of patients, often I discover several clinical reasons why that value should be disturbed, but I don't really know which of the reasons is the correct one. It is not so much trying to find in some cases *what* is wrong, as it is that you have several reasons, but don't know which is the correct one.

For example, if you have a patient who is on steroids and

diuretics, and has a low potassium value, you really are not quite sure whether that hypokalemia is due to the steroids or the diuretics. If you have a patient in congestive heart failure and who simultaneously has a myocardial infarction and you are looking at some of the enzyme activities, sometimes it is difficult to discriminate with a couple of them just what is going on. I think that much of the time I find that without some real discrimination, the abnormality is attributed to the wrong reason. It becomes a matter of how carefully you follow through on it and how important it is to know.

Dr. Altshuler: I think the behavior of physicians, to say the least, is a little baffling. First of all, when they order the PALI, they are going to get whatever we can give them, if we haven't overlooked it ourselves, relative to the cause of the test abnormality. It is not really an "unsolicited" abnormal laboratory finding.

A friend and I were having lunch in our hospital one day, sitting there with a group of surgeons, and I have to relate a very frightening conversation. I said, "Supposing 100 hypothyroids were admitted to the hospital—to any service, orthopedic, opthalmological, or anything else—and we gave you the information that would be required to make a diagnosis of hypothyroidism (in essence the study that we did). How many times do you think the diagnosis would be made?" Their guess was worse than the real result. One or two guys guessed about 50% and one orthopedic surgeon said 3% of the time, and then he added to his comment. He said, "I don't know anything about that enzyme jazz and all that sort of stuff." So one wonders why they order the PALI, really—but we have got to try and help them. We have got to help ourselves. Somebody has got to help them.

Dr. Uddin: Dr. Cole, in what form is your interpretation of lab data submitted to the physician or the patient's chart?

Dr. Cole: We have our own print-out that is independent of the cumulative computer print-out or the interim report, so that all the data that goes to the floor is reported in an unadulterated way. In addition, we develop a different print-out for the admission profile, and on that report there is space provided for our written comments relative to those data. We have a percentile ranking system for each of the tests in terms of the degree of abnormality present, and we have many different setups for calculating different ratios and different other things from the results. All this is really to provide assistance in data interpretation. That then is posted to the record and is a part of the permanent medical record in much the same way an x-ray report would be.

Dr. Uddin: Dr. Schneiderman, while the screening panel may benefit a rare patient whose treatable disease may have otherwise been overlooked, is it not possible that more patients may be harmed by morbidity generated by working up false positive results?

Dr. Schneiderman: Absolutely, and I think that is one of the questions we are asking, and I think efforts are being made in that direction. We are still continuing research in this area, and I am interested in the question of elevated T4 in a nonthyroidal disease. Particularly, we see it more commonly with people with liver disease, such as hepatitis—the most common increase in T4 is in patients with hepatitis. It comes through on the panel, and I have asked just that question: What happens to the patient whose T4 is abnormally high and whose physician then begins to react to it? How many follow-up tests are ordered? What is the cost of this? I don't have the information because my student hasn't finished with the project, but in any event those are the kind of questions I think we should be asking.

Dr. Altshuler: One of the things about doing an automatic follow-on test is the fact that you minimize this iatrogenic aspect, the impact of the iatrogenic abnormality, as we do with the ETR, we follow with T4 and the T3 if necessary. Similarly, if you had an increased T4 value and you thought it might really be an abnormality, you might want to do a TBG, or a T3 uptake, or a thyroid TSI, to determine whether or not the patient really has hypothyroidism.

Dr. Schneiderman: I agree that one could either pursue it through the laboratory or could pursue it by watching the patient. It is also quite possible to use the old expression "don't just do something, stand there." If your clinical judgment tells you that the patient is not suffering from whatever the problem appears to be or at least can afford to wait, then that is a reasonable clinical approach. It may well be that you just simply repeat the T4 and it is normal, and that takes care of that.

Dr. Uddin: Is it a practical approach to determine normal or baseline values for individuals? If so, when would such an approach be available so each individual would have his reference normal value?

Dr. Cole: I think Dr. Williams in San Francisco has been the major impetus behind trying to establish everybody's individual normal range. I don't see it [happening routinely] before the year 2000. Unless all clinicians suddenly decide that the laboratory will rule the world in medicine, we are unlikely to come up with that kind of approach.

I think that it is correct that indeed we all do have our own individual normal values; that in fact if you wanted to pick the right tests—whatever they are—to follow during a person's lifetime, to use in some sort of predictive way or to detect something early, most certainly that sounds as if it would be profitable—medically at least, but probably not economically.

Dr. Schneiderman: From only the little information that I showed you—where tests can either revert to a lower level, so-called normal, or remain abnormal—I conclude that at this point we can't yet predict what the long-range values of some individual will be. I do have a fantasy that there may be some advantage in recognizing a patient with, say, an elevated alkaline phosphatase activity, who has no present evidence of bone or liver disease. Knowing about hyper alkaline phosphatasemia may be of benefit. We have never thought of that. That is why I don't like the term "abnormality"; I think I have made that clear, that these are deviations from a range, which we say is an average range for that particular population that is used in producing your so-called normal range. I think the words you use are very important because you are very influential on the behavior of physicians. If you call it abnormal, what is the physician going to do? He has to do something—if not on ethical considerations, then legal ones. So I urge you to pay attention to the precise use of words: when you send back a laboratory result, what does it really convey?

Dr. Uddin: Dr. Schneiderman, do you think that by being rationally selective in our tests we can prove the diagnosis and the risk?

Dr. Schneiderman: There are some diseases you diagnose and treat through one set of perceptions, such as a physical exam, and there are others that come through by others and, obviously, both ways work. We are going to have to continue to use both ways, so the answer is "it depends."

Dr. Uddin: Does not the critical level of urea nitrogen vary with age? Secondly, wouldn't a different test, creatinine, be better than urea for [detecting] renal disease in your scheme?

Dr. Cole: I have made no decision at this time as to what test is best for what situation when we are dealing with this kind of physiological measurement. Each test is still under evaluation, being plotted out, the same kind of way as I showed for urea nitrogen. As we move along, then we will start doing our discriminate analyses to try to figure out what contribution each test makes to the problem-set as a whole.

Now, at times it is important to know that the BUN is increased as a cause of pre-renal azotemia in a patient, and that it gives you some reflection, perhaps, of the degree of dehydration; so it does have some usefulness. Likewise, I have some suspicions from looking at the data on creatinine and urea. In terms of renal function, it appears that BUN may yield information quicker or more frequently than creatinine measurement does, even though the specificity of the creatinine measurement for renal disease is much greater than that of the BUN. So there is a trade-off here of how you want to use the test. I think that the other ingredient that is so critical is the laboratory. As you all know, when we in the laboratory get a request, we don't have any clinical information to go with it so we don't know why the test has been ordered in this kind of way. We assume it could have been ordered for any of the following *x* number of reasons. Now, of those *x* reasons, what is it most likely to correlate with and how? To me, this is approaching the whole problem from a totally different direction than it is approached clinically.

We were taught rigidly in medical school that you first do a history, then you make a physical examination, and then you do the laboratory and x-ray work that is deemed appropriate. I think really a lot of the discussion that we are having relates to the fact that if we look at a battery of laboratory tests as an entity, that may be a [good] reverse approach from taking the history first, doing the physical second, and then doing laboratory tests. Conceivably one could look at it and say, "What if we just did all the laboratory tests and we didn't worry about the history and the physical, or we did only the history and the physical that were implied from the laboratory tests?" So there is more than one side to this coin. That poses lots of threats to our own background and training as physicians; much of it really comes as "future shock."

What we must do is study what we have, to determine the maximum usefulness of it. It is not a question of whether it does or doesn't have value the way it seemingly is quoted, or of what the medical-legal implications are. What we are trying to do is help the patient, whatever way we can within the system. I think the other questions are somewhat irrelevant, frankly.

Dr. Overholt: Dr. Uddin, I would like to ask the audience a question. If the number of times that you make a diagnosis of patients that go in the hospital when you have the data is at best 70%, what do you think the miss-rate must be when you *don't* have the data?

Quiet here, isn't it? Well, it is only a rhetorical question.

Dr. Uddin: Dr. Altshuler, in what manner are the costs for the PALI systems or additional tests borne? Are they spread out over all of the patients or third parties?

Dr. Altshuler: The cost for the PALI is $65, and there is no additional charge to the patient, and we do whatever we deem appropriate to the investigation of the case. We really haven't cost-accounted it because we haven't had the data. We will as soon as we get the rest of the thing out of the computer, because there really is no effective way to do it otherwise.

V

EDUCATION IN LABORATORY UTILIZATION

Current Practices in Laboratory Utilization in the Teaching Hospital: Possible Impact of Education Programs

Paul F. Griner, M.D.

Department of Medicine, University of Rochester School of Medicine and Dentistry, Rochester, N. Y. 14603

As those of you who are attending this conference are well aware, clinical teachers and laboratory directors have indicated concern over the liberal use of laboratory tests in health care. The rhetoric enveloping this issue has evolved not only around costs but is equally concerned with the need for physicians to place laboratory test results in perspective with their own clinical observations. Some seven years ago, in an effort to provide a quantitative background for discussions of these issues, we surveyed patterns of laboratory utilization on the medical service of Strong Memorial Hospital, Rochester, New York *(1)*. This is the parent teaching hospital of the University of Rochester School of Medicine and Dentistry. The purpose of the study was not only to quantitate patterns of laboratory use but to determine whether changing trends were accompanied by measurable improvements in the outcomes of patient care. Where such trends were not accompanied by evidence for improved outcomes, we were interested in identifying the factors which appeared to promote the use of the laboratory in excess of those necessary for optimal care.

The results of the study that I mentioned indicated a rather striking increase in the growth of laboratory tests ordered over the five-year period 1966–1970. For a few well-defined clinical conditions, the increases in laboratory tests applied in the diagnosis and management of the clinical problem could not be shown to be accompanied by improved outcomes of care. Finally, this study identified a number of obvious factors contributing to excessive use of the laboratory.

Our next step was to develop educational strategies designed to improve laboratory utilization by focusing primarily upon those factors which were considered to be remediable. These strategies were then implemented with the intent to eventually re-assess patterns of laboratory use to determine whether the educational inputs per se resulted in any change. I should like, first, to review briefly some of the initial data we collected in 1970; secondly, to indicate the factors promoting overutilization of the laboratory that were identified; thirdly, to review the educational strategies that were developed; and, finally, to present some data and conclusions from a recent analysis of patterns of laboratory use at our institution.

Materials and Methods

Data pertaining to laboratory studies were collected for all adult inpatients discharged from the medical service of the University of Rochester Medical Center during a three-month period, March 1, 1970, through May 31, 1970. Comparable data were obtained in 1971 and 1977. A random sample, comprising about 10% of discharges, was also obtained for the year 1976.

Laboratory tests were grouped according to the following categories: clinical chemistry, hematology, microbiology, radiology, and miscellaneous evaluations. For the purposes of these studies, test analyses performed by automated techniques (e.g., SMA-12, SMA-6, Coulter Counter) were considered to represent a single test.

Additional data collected included mean and median figures for both patient age and length of hospital stay and the distribution of patients by sex. The significance of any apparent differences was estimated by the application of Student's *t*-test or chi-square analysis, as appropriate.

For the years 1975–77, comparable data were obtained from the medical service of an affiliated teaching hospital where no defined educational program concerning laboratory utilization was in effect. Laboratory trends from this hospital were compared with those at Strong Memorial Hospital.

Results

Data Pertaining to Studies Performed in 1970

Table 1 indicates the mean number of tests performed per patient per hospitalization, grouped according to the categories indicated. The mean duration of hospitalization for these 855 patients was 13.8 days. Note that these data are expressed as mean rather than median numbers of tests per patient. Later in the discussion, I will refer back to these data in terms of median figures when comparing current utilization patterns.

Figure 1 indicates that for the year 1970, 95% more clinical chemistry tests were performed than during the reference year, 1965. This figure also indicates that the total laboratory budget (chemistry, hematology, microbiology, and radiology) increased at a rate almost double that of the total hospital budget during the five-year period.

The average number of specific tests performed per patient per hospitalization (data not shown) suggested that the frequency of follow-up laboratory determinations was excessive. For example, an average of 4.3 hematocrit determinations per patient per hospitalization was documented. Yet for those patients whose hematocrit was normal on admission (representing the majority of all admissions), fewer than one in six subsequently developed an abnormal value.

For two well-defined acute clinical conditions, we could demonstrate no evidence for improved patient outcomes despite a significant increase in the application of laboratory tests. For example, there was no difference in mortality or length of hospitalization among patients admitted with diabetic ketoacidosis in 1966 versus 1969 despite a 25% increase in numbers of tests ordered in the latter group. Likewise, a 530% increase in the number of arterial blood-gas measurements performed upon patients admitted with acute pulmonary edema during the two years after the opening of our Medical Intensive Care Unit was not accompanied by any difference in mortality, duration of symptoms, or length of hospitalization when these patients were compared with a retrospective control group.

We thus concluded that the use of laboratory tests was excessive and we proceeded to examine the possible factors contributing to overutilization.

Table 1. **Overall Laboratory Use According to the Type of Test (n = 855; mean L.O.S. = 13.8 Days)[a]**

	No. tests per patient	
Laboratory	*Av.*	*Maximum*
Chemistry	38.26	620
Hematology	18.45	209
Bacteriology	6.11	122
Radiology	3.40	35
Miscellaneous detns.	2.67	49
Total	68.89	

[a]From ref. *1*. L.O.S., length of stay

Factors Contributing to Excessive Laboratory Utilization, and Educational Strategies Designed to Remedy

Table 2 represents a list of those factors we identified as contributing to excessive use of the laboratory in the teaching hospital. Some of these, of course, can be expected to apply to the non-teaching hospital as well and (or) the office practice of medicine. For the sake of efficiency, I should like to combine the review of these factors with the educational strategies we developed as they pertain to the specific factors indicated.

Level of training. From data subsequently collected and analyzed, we have shown that the use of laboratory tests by third-year medical residents is consistently less than that of first-year residents and more closely approximates the "ideal." We assume this to reflect the increased confidence of the senior residents in their own clinical (i.e., bedside) observations. Traditionally, the first-year resident assumes the responsibility for writing orders upon hospitalized patients and is intimately supervised by a second-year resident. This year, for the first time, we have restructured our training program such that some elective time is provided for second-year residents, and third-year residents return to the inpatient floors for three-month periods to supervise first-year residents. We are anxious to determine whether this curricular innovation is accompanied by more judicious use of the laboratory by the first-year resident.

Lack of awareness of test characteristics. Through both anecdotal examples and systematic study of our residents, it is clear that they are not sufficiently well informed about the specificity, sensitivity, expected rate of change, and reproducibility of labo-

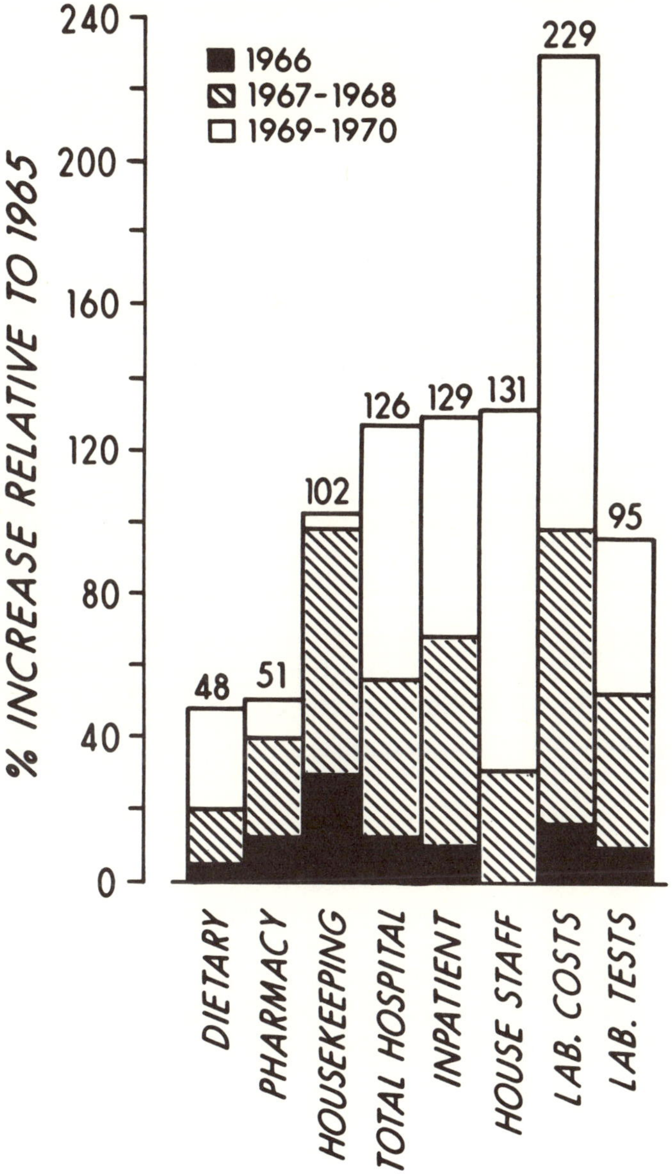

Fig. 1. Percentage increase in selected hospital costs, 1966 to 1970, relative to 1965. (From ref. *1*)

Table 2. **Factors Contributing to Overutilization of the Laboratory**

1. Level of *training*
2. Lack of awareness of *test characteristics*
3. *Uniform interpretation* of test results regardless of prior probabilities
4. Diagnostic *overkill*
5. Tests to *blanket* differential diagnostic possibilities
6. Test *packages:* a) Developed by lab.
b) "Developed" by house staff
7. Medical practice *routines*
8. *Consultative* care
9. Lack of awareness of test *interferences*
10. *Inappropriate* testing
11. Logistical *conveniences*
12. Laboratory *inefficiencies*
13. *Economic* considerations: a) Hospital b) Office practice
14. Need for patient *reassurance*
15. Desire to uncover *occult* disease
16. *Malpractice* considerations
17. Healthy *curiosity*

ratory tests. An obvious example is the ordering of an ASLO titer each day for five successive days in a patient suspected of a recent streptococcal infection.

Uniform interpretation of test results regardless of prior probabilities. It is equally clear that many physicians (house staff and attendings alike) do not interpret test results in relation to the expected prevalence of the condition being sought. Thus, an abnormal test result obtained in the evaluation of an apparently healthy patient who presents for a routine examination is often interpreted in the same light as when the test was used to confirm a suspected diagnosis.

Our educational strategies concerning the need for better knowledge of test characteristics and for differential interpretation of test results according to the population at risk have been multiple. We conduct regular seminars with our residents on such subjects as decision analysis, efficacy of screening laboratory tests, and epidemiologic principles. More recently, plans are being developed for collaboratorive teaching of both medical students and house staff with our colleagues in the clinical laboratory. This teaching will take place both on a case by case basis as well as through joint conferences.

Diagnostic overkill and tests used to blanket the differential diagnosis. Only one or two examples are necessary to highlight the nature of this problem. In the "diagnostic overkill" category fall such phenomena as ordering a serum osmolality in addition to an SMA-6 in a hyponatremic patient when the results of the latter are quite sufficient for precisely estimating the osmolality. A more common factor in overutilization is the tendency, at the time of admission, to order a complete line of secondary and tertiary tests, regardless of the clues provided by a set of abnormal primary data. An obvious example is the ordering of serum vitamin B_{12} and folic acid concentrations, iron and iron-binding capacity, and a Coomb's test when confronted with a newly diagnosed anemic patient regardless of the results of the erythrocyte indices and blood smear.

Our strategies concerning these components of overutilization have also been multiple. Each new first-year resident receives a set of guidelines concerning the use of laboratory resources. These guidelines reinforce the need for judicious ordering of laboratory tests and the application of sequential logic in the workup of a diagnostic problem. The attending physician as well as second- and third-year residents are requested to take an active role in monitoring the use of laboratory tests in supervising and assessing the performance of the first-year residents. Thirdly, we have introduced weekly conferences revolving around cases that represent either difficult or classical problems in diagnosis or management, or both. At these conferences, discussions of the efficacy of various laboratory procedures holds a high priority.

Test packages and medical practice routines. In addition to automated test packages offered by the laboratory, the day-to-day life of the residents is made less complicated by one resident group handing down to the next a variety of convenient test "packages" to be applied in given clinical circumstances. Such packages rapidly become traditions. Two recent such examples on our medical service are: *(a)* the routine ordering of both a prothrombin time and partial thromboplastin time in patients presenting with gastrointestinal hemorrhage but without either any additional bleeding site or a prior bleeding history; and *(b)* the simultaneous ordering of assays for thyroxine, triiodothyronine and thyrotropin in the process of ruling out hyperthyroidism. In numerical terms, a more overt problem is the establishment of routines in various hospital environments that are not really optimally effective. The routine of daily sputum cultures and chest roentgenograms among all patients hospitalized on the Respiratory Intensive Care Unit represents a case in point. Many other such examples could be enumerated that result in frequent follow-up determinations that often are of little or no value for patient management.

Some of the educational strategies already described apply to these factors contributing to overutilization. In addition, however, we have developed operational guidelines and recommended policies on practice concerning the use of laboratory tests for high-intensity areas of the hospital such as the Intensive Care Unit and the Emergency Department. In addition, we have regularly distributed to the residents a random sample of the cumulative hospital bills of patients they are currently managing. This practice has been most informative by both quantitating the use of the laboratory and indicating the charges incurred.

I will touch only briefly on the remaining items listed in Table 2. In the tertiary care setting, where difficult diagnostic and management issues are common, the consultant will often add to the problem of excessive use of laboratory resources. Current laboratory technology is such that in vitro test interferences appear to be less of a problem than in years past. The failure to integrate all existing clinical data before ordering subsequent tests is a frequent cause of inappropriate testing. A common occurrence in our hospital is the ordering of hemoglobin electrophoresis to confirm the diagnosis of homozygous sickle cell disease in a patient recently transfused with normal blood. The convenience of having phlebotomy technicians draw blood samples rather than their being drawn by either medical students or residents also promotes liberal use of the laboratory. For example, the single most striking increment in laboratory tests noted at Strong Memorial Hospital over the past decade occurred during six months after a phlebotomy team began work. The modern centralized laboratory with its rapid turn-around time for test results will obviously reduce some test excesses. In New York State, cost-containment mechanisms and reimbursement formulas that are based upon costs has resulted in a more equitable balance between costs and charges for laboratory procedures. This further reduces the tendency for superfluous or excess numbers of tests. The remaining items on the list are rather obvious and are not easily subject to educational remedies.

In addition to the strategies that I have already enumerated, we have found that a continuing analysis of health practices on the medical inpatient service and the sharing of our findings with both house officers and attendings has further facilitated our objectives of improved laboratory utilization. For example, Table 3 indicates the predictive value of information on the mean corpuscular volume for the diagnosis of deficiencies of either folic acid or vitamin B_{12} among anemic patients hospitalized at Strong Memorial Hospital, in relation to the prevalence of these conditions on the medical service and the specificity and sensitivity of the mean corpuscular volume for anemia due to these deficiency states. Because the probability was found to be 1 in 854 for either low folic acid or vitamin B_{12} in anemic patients whose mean corpuscular volume is less than 95, automatic ordering of such tests on newly diagnosed anemic patients, regardless of the results of the indices, is of extremely limited value. Likewise, the probability of only 18% for a deficiency of either vitamin concomitant with an elevated mean corpuscular volume highlights the numerous other reasons for large erythrocytes among anemic hospitalized patients.

Current Patterns of Laboratory Utilization

The following data will examine the extent, if any, to which patterns of laboratory utilization on the medical service of our hospital have changed over the seven-year period since the educational strategies I described were effected. Figure 2 indicates the trends in the utilization of clinical chemistry, hematology,

Table 3. **Probability of Either Low Folic Acid or Vitamin B_{12} in Serum of Anemic Hospitalized Patients, in Relation to Mean Corpuscular Volume (MCV) Values**

	MCV, fl[a]	
	<95	*≥95*
Probability of low folic acid or B_{12}	0.0011	0.18
% of anemic patients	87	13

[a] fl (femtoliters) is the SI unit that eventually will replace the unit now commonly used, cu. μ or μm^3

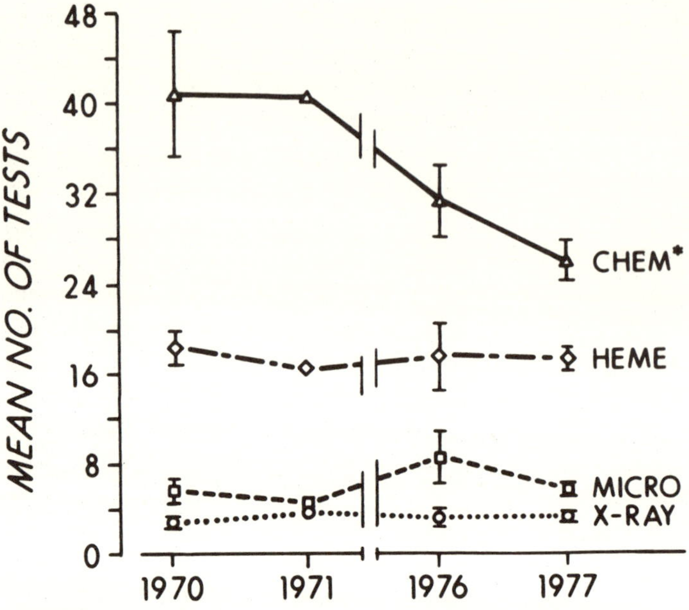

	1970	1971	1976	1977
NO. OF PTS.	93	947	85	768
LENGTH OF STAY (DAYS)	13.8	12.8	12.1	10.8

*p 0.001 FOR SIGNIFICANCE OF DIFFERENCE BETWEEN MEAN VALUES FOR 1977 vs. 1970

Fig. 2. Mean numbers of tests per patient per hospitalization according to test categories indicated (± SEM)

microbiology, and diagnostic radiology tests. The data are expressed as mean numbers of tests in each category per patient per hospitalization. Because the data collected in 1970 and 1971 did not permit an estimate of standard errors, the data presented in this figure for the year 1970 represent roughly a 10% sample of the original data, collected in such a way that these estimates could be calculated. The data for 1971 were almost identical with those of the previous year, so no standard errors were calculated. You will note that there has been a decline in the mean number of clinical chemistry tests performed per patient per hospitalization, from 40.9 in 1970 to 26.6 in 1977. This difference is highly significant ($p < 0.001$). There has been essentially no change in the numbers of hematology, microbiology, and diagnostic radiology tests ordered. The apparent increase in numbers of microbiology tests observed for the year 1976 is not statistically significant; it appears to be attributable to sampling error. For the year 1977, the mean numbers of tests ordered in the three categories indicated were 17.7, 5.9, and 3.8, respectively.

Figure 2 also indicates the trend in mean length of hospitalization during this seven-year period. Note the steady reduction in length of stay, from 13.8 days in 1970 to 10.8 days in 1977. In order to correct the 1970 data for a comparable length of stay relative to 1977, a 15% sample of the original 1970 data was obtained, which matched the 1977 sample with respect to distribution of length of hospitalization. Comparison of these means show a reduction in number of clinical chemistry tests per patient per hospitalization, from 34 in 1970 to 26.2, the figure previously noted, for 1977. This difference is significant at the 0.05 level. Once again, no significant differences were noted in numbers of hematology, microbiology, or diagnostic radiology tests.

Figure 3 indicates the distribution of the number of clinical chemistry tests performed per patient per hospitalization for the year 1977. Approximately 43% of all patients had 10 or fewer such tests ordered. On the other hand, 9.5% of the total sample had 50 to 264 chemistry tests ordered per hospitalization. This skewed (but expected) distribution pattern has obvious implications for expressing utilization trends in terms of mean values. For this reason, the data were re-examined and displayed in terms of median numbers of tests per patient per hospitalization. Figure 4 indicates these trends. For clinical chemistry tests, a median figure of 13 is noted for the year 1977, as contrasted with 25 for the year 1970. A decline from 15 in 1970 to 11 in 1977 is observed for hematology tests. For microbiology, the respective values are 4 and 2. Diagnostic radiology remained constant at 2. It is necessary to point out, once again, that the comparisons on this figure have not included an adjustment for the trend to briefer hospital stay observed between 1970 and 1977. Using the same technique described for the calculation of mean numbers of tests for comparable lengths of stay, we found that the median figures for numbers of clinical chemistry tests declined from 18

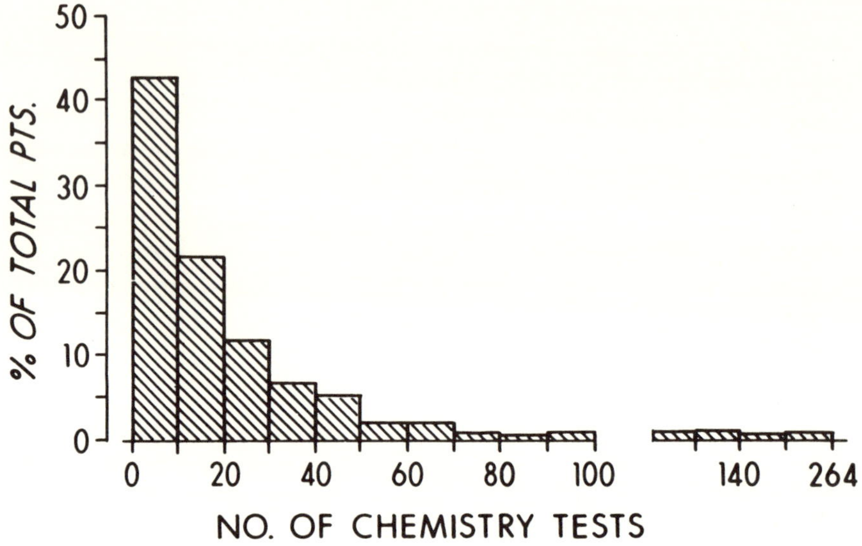

Fig. 3. Distribution of number of chemistry tests performed per patient per hospitalization in 1977. (n = 768)

in 1970 to the previously noted value of 13 in 1977. The adjusted median numbers of hematology, microbiology, and diagnostic radiology tests remained virtually constant.

These data thus show a significant decline in utilization of clinical chemistry tests among hospitalized medical inpatients over the seven-year period after adjusting for comparable length of stay, whether expressed as mean or median numbers of tests. No significant change was noted in the numbers of hematology, microbiology, or diagnostic radiology tests. These data are in contrast to the previously described 95% increase in numbers of clinical chemistry tests ordered between the years 1966 and 1970.

In an effort to determine whether the educational strategies employed were responsible for the observed reduction in numbers of clinical chemistry tests ordered and the stability in the other test categories, a comparison (Figure 5) was made between utilization trends observed on the medical service at Strong Memorial Hospital with similar data from an affiliated community hospital that had no such educational program. There were differences in the definition of tests between the two hospitals. For example, in the affiliated hospital, an SMA-12 determination

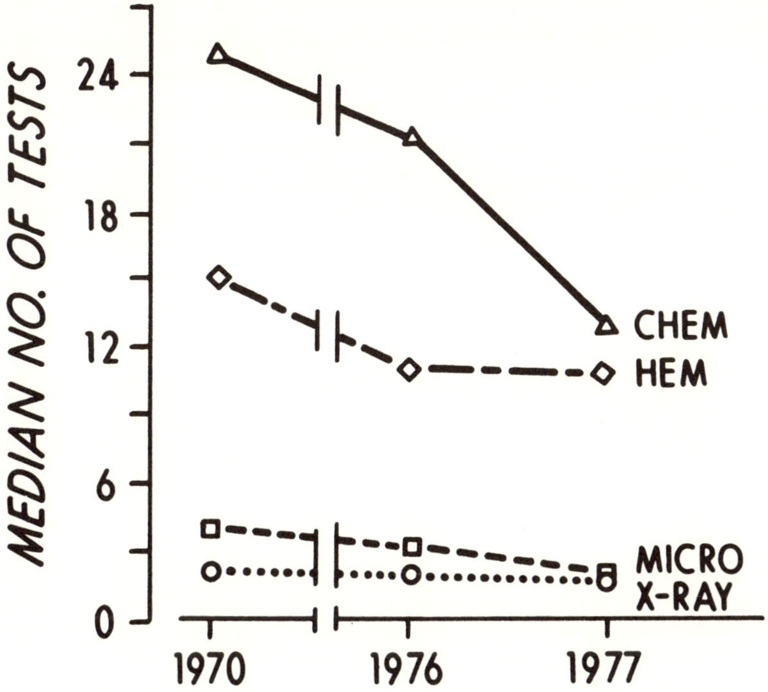

Fig. 4. Median numbers of tests per patient per hospitalization according to test categories indicated

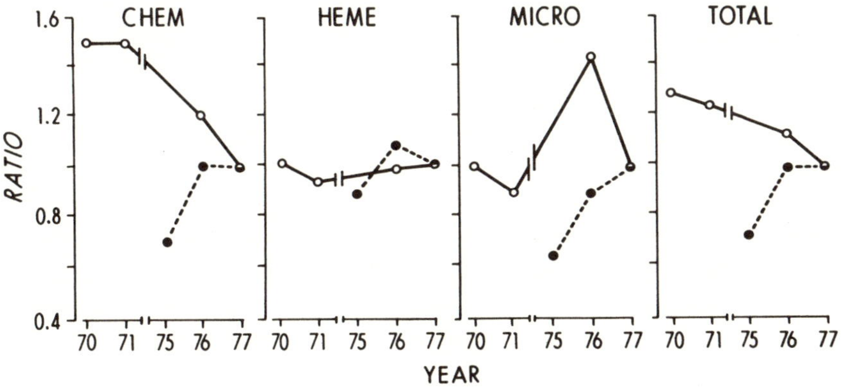

Fig. 5. Ratio of tests relative to 1977 in hospital with (○) and without (•) educational programs relating to laboratory utilization.

was recorded as 12 tests, rather than as a single test as at Strong Memorial. For this reason, the data in this figure are displayed as the ratio of number of tests per patient per hospitalization relative to the year 1977. For the affiliated hospital, data were available only for the years 1975–1977. The previously noted decline in clinical chemistry tests on the medical service at Strong Memorial Hospital contrast with the increase in such tests at the affiliated hospital. The differences in numbers of hematology tests between the two hospitals are not significant. An increase in microbiology tests at the affiliated hospital is noted, comparable to the increased utilization of clinical chemistry tests. No data were available to make a comparison between numbers of diagnostic radiology tests between the medical services of the two institutions. The final box in Figure 5 indicates the cumulative trends when chemistry, hematology, and microbiology tests are combined. The difference between decreasing utilization on the medical service at Strong Memorial Hospital and increased utilization at the affiliated hospital is clear. The data from the affiliated hospital does suggest, however, that a stable trend may be developing some years behind that noted at Strong Memorial Hospital. Further observations over the next few years will be necessary to confirm this possibility.

Discussion

The data I have presented thus suggest that the educational strategies we have used during the past seven years have been accompanied by a distinct trend in the direction of more optimal laboratory use. This trend is characterized by a significant decline in the numbers of clinical chemistry tests ordered per patient per hospitalization and no growth in numbers of hematology, microbiology, and diagnostic radiology tests. Further studies are needed to answer some important questions posed by the data presented in this table. For example, since the numbers of specific tests available in 1977 are greater than those in 1970, do the constant figures noted for hematology, microbiology, and diagnostic radiology actually represent a decrease in the utilization of individual tests? Data bearing on this question are currently being collected. Secondly, with a medical inpatient census consistently in excess of 90% of capacity at

Strong Memorial Hospital and an increasing population base, it is likely that the average severity of sickness among hospitalized inpatients is greater in 1977 than was the case in 1970. If this is so, the trends in terms of reduced utilization of the laboratories are even more impressive, but additional data will be required to confirm this speculation. Thirdly, how do these trends in laboratory utilization apply to groups of patients with sharply defined clinical conditions, such as acute myocardial infarction and diabetic ketoacidosis? Can we demonstrate decreased reliance upon laboratory observations for these clear-cut types of illnesses? Fourthly, are the outcomes of care for hospitalized patients at least as good as those noted in 1970 despite a reduction in the use of laboratory tests? Finally, what are the implications of these trends insofar as hospital costs are concerned and, specifically, the contribution of laboratory costs to total hospital costs? Additional studies are currently underway that address these specific questions.

I believe that the data just presented permit the conclusion that the educational strategies used during the past seven years have been accompanied by a clear trend toward more judicious application of laboratory tests on the medical inpatient service of our hospital. This conclusion should reinforce the importance of the more highly organized approaches to teaching in this arena that are presented further on in this book.

References

1. Griner, P. F., and Liptzin, B., Use of the laboratory in a teaching hospital. Implications for patient care, education, and hospital costs. *Ann. Intern. Med.* **75,** 157 (1971).

Use of a Computer to Issue Interpretive Reports: Do Clinicians Find This Form of Reporting Useful?

Richard L. Reece, M.D.* and
Russell K. Hobbie, Ph. D.**

**Luftin Medical Laboratories, P.O. Box 629, Minneapolis, Minn. 55440, and Metropolitan Medical Center, Minneapolis, Minn. 55404; and **School of Physics and Astronomy, University of Minnesota, Minneapolis, Minn. 55455*

This paper has three purposes: 1) to describe an interpretive report that our laboratory uses routinely; 2) to analyze the results of a questionnaire recently sent to physicians about the usefulness of report; and 3) to discuss the accuracy of interpretive reporting in a series of 450 hospital patients.

Background

For 10 years, we have been building and using a comprehensive computer reporting system that presents both raw test results and interpretive information *(1)*. For seven years, mostly before audiences of pathologists, one of us has been writing and speaking about the concept of interpretive pathology and interpretive reporting *(2)*. Presently, we are in the process of expanding our program to include all tests that we perform or refer. We are, in short, committed to interpretive reporting and have a great deal of experience with it.

The Interpretive Report

Our interpretive report consists of two pages. The first presents demographic information and numerical test data;

the second lists interpretive information (Figures 1–4). This format is intended to answer four questions the clinician might have: 1) What tests are abnormal? 2) How abnormal are they? 3) What might the abnormal tests mean? 4) What are logical follow-up tests?

We feel that an interpretive report does five things: 1) calculates derived values (e.g., free thyroxine index); 2) compares test results with a standard of health (normal or reference range); 3) reminds clinicians of possible causes of a single abnormal result; 4) suggests differential lists of possibilities based on multiple abnormals; and 5) suggests (and possibly initiates) further laboratory tests.

The Questionnaire

Do clinicians find this form of interpretive reporting useful? In August 1977, we sent questionnaires to clinicians who use our interpretive services to try to answer that question. We asked them what parts of our report they found useful and what they liked or disliked about it. Many of these clinicians had been receiving our report for as long as six years, long enough to form a firm opinion.

By September 1 we received 140 completed questionnaires, representing the opinions of about 10% of the clinicians who use the program, and we began to analyze the results. No one, to our knowledge, has systematically investigated how clinicians respond to interpretive reports. Others have evaluated our program *(3–5)*, but not from the clinicians' point of view.

We received responses from the following specialties: family medicine (65), internal medicine (31), surgery (22), others (22). "Others" included the director of an alcoholic treatment center, psychiatrists, pediatricians, industrial medicine physicians, clinical pathologists, and nephrologists. This specialty distribution is a realistic cross-section of the clinicians we serve. Of the tests we do, 90% are performed on outpatients, mostly in the private-office or large-clinic setting. Because 65% of University of Minnesota medical graduates are now choosing to specialize in family medicine, the proportion of family medicine specialists in our physician population in Minnesota is growing. The relatively large percentage of family practice physicians may be one

LUFKIN MEDICAL LABORATORIES (612) 335 6521 77/08/26. 13.04.32.
PO BOX 629 MINNEAPOLIS, MINN 55440 UNIPORT, COPYRIGHT 1973, 1977

8/26/77 ID 173

AGE 52 MALE WALKING

L27

DR THOMPSON

CHEMISTRY TESTS			NORMAL RANGE FOR
NORMAL TESTS	VALUE		AGE, SEX, POSTURE
GLUCOSE	103.	MG/DL	70. - 120.
CALCIUM	10.4	MG/DL	8.7 - 10.5
IONIZED CALCIUM-EST.	4.5	MG/DL	4.0 - 4.8
PHOSPHORUS	2.6	MG/DL	2.0 - 4.3
TOTAL PROTEIN	7.9	G/DL	6.3 - 8.1
ALBUMIN	4.8	G/DL	3.8 - 5.0
GLOBULIN	3.1	G/DL	2.0 - 3.6
ALBUMIN/GLOB RATIO	1.5/1		1.1 - 2.3
CHOLESTEROL	234.	MG/DL	150. - 290.
TRIGLYCERIDES	105.	MG/DL	0 - 250.
BLOOD UREA NITROGEN	17.	MG/DL	9. - 31.
CREATININE	1.4	MG/DL	.7 - 1.8
BUN/CREAT RATIO	12.1/1		6.0 - 20.0
SODIUM	141.	MEQ/L	135. - 148.
POTASSIUM	3.7	MEQ/L	3.5 - 5.4
CHLORIDE	100.	MEQ/L	95. - 111.
CARBON DIOXIDE	31.0	MEQ/L	22.0 - 33.0
BILIRUBIN (TOTAL)	.8	MG/DL	0 - 1.3
LDH	179.	U/L	0 - 220.
SGOT (AST)	17.	U/L	0 - 40.
CPK (CK)	61.	U/L	0 - 200.
ALKALINE PHOSPHATASE	88.	U/L	0 - 120.
IRON	118.	MCG/DL	45. - 210.

ABNORMAL TESTS	VALUE		NORMAL RANGE	ST DEV BEYOND NORMAL LIMIT
HIGH SERUM URIC ACID	10.6	MG/DL	3.0 - 9.0	2

Fig. 1. The first page of a sample report, showing the patient demographic information and normal and abnormal test results.

The second page is shown in Figure 2.

reason our program has been accepted so well. Table 1 shows the overall response to the questionnaire and a breakdown by specialty.

The response confirms our impression that physicians like to have normal results separated from abnormal results. Having all the abnormal results in one place saves the physician time; internists may not like this feature as well, because the results are not grouped by organ system. Age-, sex-, and posture-dependent normal ranges are well accepted. Our method for determining these has been discussed *(1)*.

We list abnormal tests by descending degree of abnormality,

```
                                                                  08/26/77

COMMENTS

DIAGNOSTIC POSSIBILITIES FOR HIGH URIC ACID

        THIAZIDE DIURETICS
        HYPERLIPOPROTEINEMIA (ESP. TYPE III AND
         IV)
        VASCULAR DISEASE ASSOCIATED WITH
        DIABETES, ARTERIAL HYPERTEN-
        SION, OBESITY, ATHEROSCLEROSIS
        ALCOHOL INGESTION (WILL BE HIGH IN 25
         PCT OF THOSE WHO HAVE IMBIBED 48 HOURS
         BEFORE TEST. ACCOUNTS FOR ABOUT 10 PCT
         OF ELEVATED URIC ACIDS)
        GOUT
        PRE-RENAL AZOTEMIA
        EXCESSIVE CELL DESTRUCTION
        ASPIRIN
        TOXEMIA OF PREGNANCY (IN APPROPRIATE
         INDIVIDUALS)
        DIABETIC ACIDOSIS
        CERTAIN CASES OF HYPERTENSION
        VIOLENT MUSCULAR EXERCISE
        DIET OR WEIGHT LOSS
        X-LINKED PRIMARY HYPERURICEMIA
        DEHYDRATION
        COFFEE
        VOMITING WITH ACIDOSIS


SUGGESTIONS

POSSIBILITIES BASED SOLELY ON CHEMISTRY TESTS. LIST MAY NOT BE COMPLETE.
                                                              8/26/77
```

Fig. 2. This second page of the report shown in Figure 1 contains interpretive information.

In this case there is a list of diagnostic possibilities but no comments or followup suggestions.

to highlight the very abnormal tests. This is not a foolproof procedure because standard deviations means different things for different tests. This is reflected by the smaller percentage of physicians who like this feature.

Still less useful is the statement "All results are borderline and may be clinically insignificant." This statement was put into the program several years ago when multiple tests were first requested, to remind clinicians of the fact that some abnormal results in healthy patients are statistically inevitable. With time, clinicians have become educated to this fact, and the statement is no longer useful.

LUFKIN MEDICAL LABORATORIES (612) 335 6521 77/09/12. 15.10.30.
PO BOX 629 MINNEAPOLIS, MINN 55440 UNIPORT, COPYRIGHT 1973, 1977

9/ 9/77 ID 245

AGE 57 MALE WALKING

L29PCB

DR

CHEMISTRY TESTS			NORMAL RANGE FOR
NORMAL TESTS	VALUE		AGE, SEX, POSTURE
GLUCOSE	109.	MG/DL	70. - 120.
CALCIUM	9.4	MG/DL	8.7 - 10.5
IONIZED CALCIUM-EST.	4.5	MG/DL	4.0 - 4.8
PHOSPHORUS	3.5	MG/DL	2.0 - 4.3
TOTAL PROTEIN	6.6	G/DL	6.3 - 8.1
ALBUMIN	4.1	G/DL	3.8 - 5.0
GLOBULIN	2.5	G/DL	2.0 - 3.6
ALBUMIN/GLOB RATIO	1.6/1		1.1 - 2.3
SERUM URIC ACID	7.7	MG/DL	3.0 - 9.0
CHOLESTEROL	256.	MG/DL	150. - 290.
TRIGLYCERIDES	133.	MG/DL	0 - 250.
BLOOD UREA NITROGEN	21.	MG/DL	10. - 32.
CREATININE	1.1	MG/DL	.7 - 1.9
SODIUM	144.	MEQ/L	135. - 148.
POTASSIUM	4.7	MEQ/L	3.5 - 5.4
CHLORIDE	104.	MEQ/L	95. - 111.
CARBON DIOXIDE	29.0	MEQ/L	22.0 - 33.0
BILIRUBIN (TOTAL)	.4	MG/DL	0 - 1.3
SGOT (AST)	21.	U/L	0 - 40.
ALKALINE PHOSPHATASE	72.	U/L	0 - 120.
TBG (EST.)	18.	MCG/DL	12. - 21.
IRON	123.	MCG/DL	45. - 210.

ABNORMAL TESTS	VALUE		NORMAL RANGE	ST DEV BEYOND NORMAL LIMIT
HIGH CPK (CK)	312.	U/L	0 - 200.	3
LOW THYROXINE (T4 RIA)	.8	MCG/DL	4.5 - 11.0	3
HIGH LDH	313.	U/L	0 - 220.	2
LOW FREE THYROXINE INDEX	19.		113. - 385.	2
LOW T3 RESIN UPTAKE	24.	PERCENT	25. - 35.	1

Fig. 3. The first page of another sample report.

The second page is shown in Figure 4.

Nearly all physicians like to have the results grouped on one page.

There is more controversy about calculated values. The estimated lipoprotein phenotype *(6)* is useful to about 80% of the respondents. The free-thyroxine index is next most useful. Since we have introduced these statements into the report, complaints and confusion about the meaning of thyroid and lipid tests have almost completely subsided. Other calculated results (globulin, albumin/globulin ratio, estimated thyroxine-binding globulin,

```
                                                              09/12/77

COMMENTS
   ACCORDING TO FREDRICKSON, LIPOPROTEIN PHENOTYPE MAY BE
   ESTIMATED FROM CHOLESTEROL AND TRIGLYCERIDES:
   ALTHOUGH CHOLESTEROL AND TRIGLYCERIDES ARE WITHIN NORMAL LIMITS
     AT LEAST ONE EXCEEDS IDEAL LIMITS (250 AND 150 RESPECTIVELY)
   CONSISTENT WITH HYPERLIPOPROTEINEMIA TYPE 2A
     OCCASIONALLY DUE TO ESTROGENS.  SEVERE FORM IS USUALLY GENETIC.
     A LOW CHOLESTEROL DIET IS OFTEN EFFECTIVE.

POSSIBILITIES                                                          7


  HYPOTHYROIDISM                                                       7
     LOW:  FREE THYROXINE INDEX
     HIGH: LDH                    CPK (CK)

  ALCOHOL INGESTION OR ALCOHOLIC HEPATITIS                             5
     HIGH: LDH                    CPK (CK)

  MYOCARDIAL INFARCTION                                                5
     HIGH: LDH                    CPK (CK)

  SEIZURES                                                             5
     HIGH: LDH                    CPK (CK)

  TISSUE NECROSIS, IE ABSCESS,GANGRENE,PERITONITIS,ETC                 5
     HIGH: LDH                    CPK (CK)

  CEREBRAL DAMAGE (EG STROKE)                                          5
     HIGH: LDH                    CPK (CK)

  MUSCLE DISEASE(DYSTROPHY,DERMATOMYOSITIS,IM INJECT)                  5
     HIGH: LDH                    CPK (CK)

  PANCREATITIS WITH CHOLECYSTITIS                                      5
     HIGH: LDH                    CPK (CK)


SUGGESTIONS
  CONSIDER TSH TO CONFIRM HYPOTHYROIDISM OR THYROIDITIS
     PROFILE (T3 RIA, TSH, ANTITHYROGLOBULIN ANTIBODY, ANTIMICROSOMAL ANTIBODY
     SUSPECT HASHIMOTOS THYROIDITIS.
  CONSIDER ISOENZYMES IF CAUSE OF ELEVATED LDH IS UNCLEAR


POSSIBILITIES BASED SOLELY ON CHEMISTRY TESTS. LIST MAY NOT BE COMPLETE.
                                                         9/ 9/77
```

Fig. 4. The second page of the report shown in Figure 3.

Comments are made about lipids, even though the cholesterol and triglycerides values are normal. The patient has hypothyroidism. Followup suggestions are made for the thyroid problem and for the supranormal value for LDH (lactate dehydrogenase, EC 1.1.1.27).

estimated ionized calcium, anion gap, chloride/phosphate ratio and serum urea nitrogen/creatinine ratio) are less well received, probably because these results are less reliable or are pertinent for only a few diseases.

The list of diagnostic possibilities (a simple list or "diagnote" for one abnormal; the results of a pattern-matching program for multiple abnormals) is the most controversial section of the report in the minds of laboratory directors who consider imple-

Table 1. **Results of the Survey of Physicians Using the Program (Results Are Expressed in Per Cent)**

Feature	*Overall (n = 140)*			*Family medicine (n = 65)*			*Internal medicine (n = 31)*			*Surgery (n = 22)*			*Others (n = 22)*		
	U	*I*	*D*	*U*	*I*	*D*	*U*	*I*	*D*	*U*	*I*	*D*	*U*	*I*	*D*
Separated normals and abnormals	95	2	3	98	0	2	87	6	6	100	0	0	90	5	5
Age- and sex-dependent normal ranges	95	4	1	95	3	2	97	3	0	90	10	0	95	5	0
Listing of abnormal results in order of standard deviations from normal	80	16	4	79	16	5	81	16	3	90	10	0	75	20	5
Statement that borderline results may be insignificant	70	23	6	75	20	5	58	32	10	80	15	5	65	30	5
Results grouped on one page	94	5	1	96	2	2	95	5	0	90	10	0	90	10	0
Lipoprotein phenotypes	81	16	3	96	2	2	72	28	0	78	22	0	91	9	0
Free-thyroxine index	75	20	5	76	20	4	68	21	11	87	13	0	72	22	6
Other calculations[a]	63	33	4	67	27	5	55	26	19	59	41	0	65	35	0
Comments[b]	82	15	3	85	12	3	65	27	8	82	18	0	94	0	6
"Diagnotes"	84	12	5	93	5	2	60	27	13	89	11	0	82	12	6
Diagnostic possibilities	87	9	2	95	3	2	70	17	13	89	11	0	84	16	0

[a] Globulin, albumin/globulin ratio, estimated ionized calcium, thyroxine binding globulin.
[b] Comments include statements about dehydration, serum urea nitrogen/creatinine ratio, and lipid abnormalities.
U, useful; **I**, indifferent; **D**, distracting

menting our system. Yet Table 1 shows that it is liked by physicians more than some other features of the report.

We have placed all interpretive information on the second page of the report for four reasons: *(a)* there is often not room on the first page; *(b)* it makes the diagnostic information less obtrusive; *(c)* the second page can be discarded if the physician wishes to do so; and *(d)* the second page can be programmed out of existence if a client does not want it.

Table 2 shows the response to certain questions about the second page. Nearly all physicians like it. One-third discard it. We also found that 35% of them always look at it, but 5% never look at it. On the average, the physician looks at it 60% of the time. The respondents report that the second page helps them manage their patients 38% of the time. Not surprisingly, internists found it least helpful. Fewer than 2% found the second page misleading or a hindrance to patient management.

The Accuracy of the Diagnostic Possibilities

It is difficult to characterize the diagnostic accuracy of the report. When the tests are few in number and are done for a specific reason (such as thyroid, lipid, anemia, or rheumatoid

Table 2. **Responses to Questions About the Second Page**

Question	Yes	No
1. Do you like the second page?	90	10
2. Do you see medical-legal implications in the second page?	35	65
3. Do you discard it?	34	66
4. Would you like to see more follow-up suggestions?	78	22

batteries) the diagnostic accuracy is high. The accuracy of a 12- or 20-test profile on the general population is much worse.

The program now contains "diagnotes" (lists of possibilities for single tests) for about 200 tests, including serum chemical values, creatinine clearance, hematology, serology, and immunology. When two or more tests are abnormal, the program performs a pattern-matching exercise among 103 diagnostic possibilities. We do not claim to make a diagnosis, but rather to remind the physician of certain possibilities which should be considered.[1]

The association of certain abnormalities with each diagnostic possibility was made originally by personal experience and from reports in the literature. Over the last several years, we have attempted to put these correlations on a more systematic basis. We have studied hospital charts for patients with a given discharge diagnosis. When the pathologist[2] reviewed the chart and considered the diagnosis to be correct and the case was not complicated by other problems which could alter the chemistry pattern, we recorded the values for an SMA-12 battery done at admission. For a few diseases, other results were also recorded. A total of 1459 patients in the 34 disease categories indicated in Table 3 have been collected. These were used to determine the patterns, such as that shown by the example in Table 4.

The pattern-matching program assigns a score to each disease, based on which abnormal tests are present and how abnormal

[1] Dr. David Jacobs, Provident—St. Margaret's Hospital, Kansas City, Kansas, who has adapted our program for his own use, approaches the diagnostic problem by simply presenting a detailed list of the diagnostic possibilities for the most abnormal result. He says that the proper diagnosis is almost always present in that list.

[2] We have enjoyed the collaboration of P. C. J. Ward, M.D., and M. D. Burke, M.D., of Mt. Sinai Hospital, Minneapolis, in this study.

Table 3. **Diseases for Which We Have Determined the Pattern of Abnormalities for an SMA-12 Battery**

Bronchial asthma
Bronchopneumonia
Bone metastases
Carcinoma
Acute cholecystitis
Cirrhosis
Degenerative vascular disease
Diabetes mellitus
Gastrointestinal bleeding or peptic ulcer
Hemolysis (venipuncture, serum on clot, hemolytic anemia, disseminated intravascular coagulation)
Acute or chronic viral hepatitis
Hepatic or biliary damage
Hyperparathyroidism
Hyperthyroidism
Hypothyroidism
Infectious mononucleosis
Intestinal obstruction
Leukemia
Lipemic serum
Lymphoma
Metabolic acidosis
Metastatic tumor in liver
Myocardial infarction
Obstruction of common duct
Paget's disease
Pancreatitis with cholecystitis
Pernicious, macrocytic, or hemolytic anemia
Plasma cell myeloma
Pre-renal azotemia
Regional enteritis
Renal insufficiency
Rheumatoid arthritis
Tissue necrosis (abscess, gangrene, peritonitis, etc.)

they are. These scores are further modified by specific values of some of the tests *(1)*.

When these data on 1459 patients were run back through the program, we found that the correct diagnostic possibility was listed first in 30% of the cases. If we print the five possibilities with the highest scores, we find that the correct possibility is listed 68% of the time. Printing more than five possibilities does not improve the percentage.

To confirm the accuracy of the program independently of the patients used to determine the patterns, we examined the hospital charts of 450 patients who were admitted consecutively to Metropolitan Medical Center. Of these patients, several had diagnoses of trauma or psychiatric problems for which no SMA-12 pattern would be expected. There were also patients in this group with multiple diagnoses.

These patients had many different diagnoses. In many cases there were only one or two patients with a given diagnosis. If we eliminated those categories for which there were fewer than five patients, we found that the overall success rate was 22% listed in first place and 50% listed in the top five. If we restricted ourselves further to those categories for which we had studied the

Table 4. **Determination of the Pattern for Renal Insufficiency**[a]

	Percent high			*Percent low*	
Test	*RI*	*Cont*	*Pattern*	*RI*	*Cont*
Calcium	9	9	L	22	7
Ionized calcium (estimated)	9	7	L	12	4
Phosphorous	61	15	H	1	2
Glucose	38	30	H	0	1
Total protein	12	8	L	16	9
Albumin	3	10	L	48	21
Globulin	33	17	H	4	3
A/G ratio	0	0	L	40	18
Uric acid	57	12	H	0	2
Cholesterol	20	20	H	1	3
BUN	96	11	H[b]	0	8
Total bilirubin	10	17	—	0	0
LDH	26	27	H	0	0
AST	21	41	H	0	0
APT	20	22	H	0	0

[a] Abbreviations: RI: 77 patients with renal insufficiency; Cont: 1459 patients with various other diseases.

AST: aspartate transaminase (EC 2.6.1.1); APT: alkaline phosphatase (EC 3.1.3.1); A/G, albumin/globulin; BUN, blood (serum) urea nitrogen; LDH, lactate dehydrogenase (EC 1.1.1.27).

[b] Essential.

patterns as described above, then the rates were 34% and 67%, respectively. (These figures do not address the question of false positives. Because five categories are usually listed, the false-positive rate would be at least 80%. It is not a useful concept in terms of our goal of presenting reminders.)

The program does well for certain diseases (Table 5). For example, we have instructions in the program to elevate common-duct obstruction to a high position if the total bilirubin and the alkaline phosphatase values are a certain amount above normal. This makes the program do relatively well for common-duct obstruction. Diabetes mellitus is promoted to a high position if the glucose value exceeds 2000 mg/liter. The same type of strategy is used with renal insufficiency, if the serum urea nitrogen (BUN) exceeds 400 mg/liter, or the BUN and creatinine are both above normal.

We have written the program to always list diabetes in first place if the blood sugar value exceeds 2000 mg/liter. The fact that it was listed in first place only 56% of the time indicates that

Table 5. **Results, by Diagnosis, for 450 Consecutive Patients Admitted to Metropolitan Medical Center**

	Number of patients				*Percent*	
Disease	*Admitted*	*With no abnormals*	*With one abnormal*	*With two or more abnormals*	*First*	*Listed at all*
Common-duct obstr.	5	0	0	5	20	100
Diabetes	36	2	4	30	56	86
Renal insufficiency	36	0	4	32	50	84
Tissue necrosis	6	1	0	5	40	80
Cirrhosis	9	0	0	9	55	77
Myocardial infarction	10	0	0	10	40	70
Alcohol ingestion	44	7	11	26	19	57
Thiazide diuretics	11	1	0	10	0	50
Dehydrated pt. or nothing by mouth for 12 h	14	0	1	13	36	54
Liver disease (e.g., fatty metamorphosis)	6	0	1	5	0	40
Bone metastases	8	2	1	5	20	40
Acute or chronic infect.	11	3	1	7	14	42
Degenerative vascular dis.	74	4	13	57	15	38
Cerebral damage	6	0	0	6	0	33
Intravenous fluids	10	1	1	8	12	25
Gastrointestinal bleeding	13	1	5	7	0	14
Carcinoma	18	1	3	14	0	14
Patient dehydrated	26	4	4	18	5	5
Percent correct, using only diseases for which there were five or more patients					22	50
Percent correct, using only diseases for which we studied the patterns					34	67

many of the patients who had diabetes were under control and did not have an increased blood sugar. This is an example of another problem with evaluating a program like this: a patient may have a disease, such as diabetes or hypothyroidism, that is under treatment, so that the patient may show no chemical abnormality. Should such patients be excluded in evaluating a program such as this? This problem has also been noted by Andersen et al. *(5)*.

In summary, for six years we have used a computer program to report test data and interpretive information on a current total of about one million patients. The report contains laboratory results on the first page and interpretation of them on the second page. We sent a questionnaire to the physicians who regularly receive the report. The response was generally favorable, certain features of the report being more liked than others.

When 450 consecutive hospital patients were studied, the correct diagnosis was found to be in the top five listed, on the basis of an admitting chemistry profile, about 67% of the time. The program's diagnostic ability varies widely with the disease.

References

1. Hobbie, R. K., and Reece, R. L., Computer interpretation of laboratory test results. *Fed. Proc. Fed. Am. Soc. Exp. Biol.* **34,** 2152 (1975), and papers cited therein.
2. Reece, R. L., and Hobbie, R. K., Letters, notes and computers: The evolution of a laboratory information system. *Conn. Med.* **36,** 324 (1972).
3. Button, K. F., and Gambino, S. R., Laboratory diagnosis by computer. *Comput. Biol. Med.* **3,** 131 (1973).
4. Tanner, F. H., et al. Computer assisted analysis of abnormal laboratory tests—a report of local experience. Parts I and II. *Nebr. Med. J.* **59,** 419 and 466 (1974).
5. Anderson, C. T., Cembrowski, G. S., and Toren, E. C., Computer-assisted differential diagnosis of laboratory abnormalities and followup testing—an evaluation. *Am. J. Clin. Pathol.* **65,** 234 (1976).
6. Frederickson, D. S., A physician's guide to hyperlipidemia. *Mod. Concepts Cardiovasc. Dis.* **16,** 31 (1972).

Teaching Proper Laboratory Use to Medical Student and Physicians

Patrick C. J. Ward, M.D., Charles A. Horwitz, M.D., and M. Desmond Burke, M.D.

Mount Sinai Hospital, Minneapolis, Minn. 55401

From the interest shown in our recent communication on this topic *(1)*, from discussions with faculty members at other institutions, and from the notable dearth of literature on the subject, we conclude that few medical schools have offered systematic instruction in interpretative aspects of laboratory medicine during the senior (clinical) years, either as core or elective instruction. On the other hand, most core curricula include some instruction in basic laboratory techniques in the junior years, such as making blood smears and examining urine and feces. Such instruction, while to an extent technical, is designed to give the students some insight into the role of the laboratory in arriving at a clinical diagnosis. However useful this exercise, it must be considered introductory, analogous to general pathology as a prelude to systemic pathology. Students subsequently going on to senior electives in clinical subjects are ill-equipped to make rational choices among the many laboratory tests available when faced with a sick patient. Although they pick up snatches of "appropriate tests" from clinicians on ward rounds, the information gleaned is random and to a large extent depends on the patient population, although such factors as instructor interest and philosophy may also operate. It is partly because of insecurity in laboratory utilization that "blanket" testing was evolved. Hence, as widely acknowledged, there is improper utilization of laboratory services, at enormous cost to the patient.

Other factors, however, must be considered as contributing

to improper utilization. They are summarized in the following diagram (Figure 1).

The net effect on clinicians of the continued rapid technical growth is confusion. Test batteries are frequently dictated by the instrumentation at hand rather than by logic. In the absence of adequate training in laboratory utilization, clinicians become slaves to technology. Attempts to follow simple investigational sequences are often thwarted by lack of availability of test(s), inordinate amounts of time taken to complete tests, or incorporation of tests into immutable batteries that are dictated by inflexible instruments or procedures, or both.

Systematic Instruction in Laboratory Interpretation and Utilization

In 1968, we began our current efforts to train senior medical students and pathology residents at the University of Minnesota in interpretation and utilization of laboratory data. We added one more elective to the roughly 200 senior electives available to classes ranging from 200 to 250 students. Only one student applied in 1968. The exponential growth of applications thereafter is shown in the next diagram (Figure 2).

Since 1974–75, we have accepted a maximum of 150 students per year in three groups of 50 each. In the 1977–78 academic

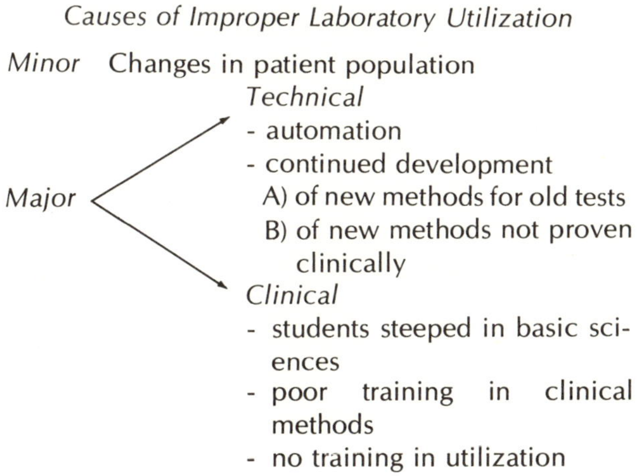

Fig. 1

1968:	1
1969:	10–12
1970:	18–24
1971:	25–30
1972:	75 (25 × 3)
1973:	100 (33 × 3)
1974:	145 (45, 50, 50)
1975:	150 (50, 50, 50)
1976:	150 (50, 50, 50)
1977:	150 (50, 50, 50)

Fig. 2

year, 320 students from the two senior classes applied for the 150 positions available. However, we continue to accept only 150, due to limitations in optimal group size (50) and in time available for tutoring.

The instructors focus on four basic questions:

1. What test(s) should be ordered in a given clinical situation?
2. What should be ordered first, second, etc.?
3. How should the test results be interpreted?
4. What are the common pitfalls in interpretation?

The course material is appended in summary form (Table 1).

The course work is in part didactic, in part Socratic, the latter effected by problem-solving by use of mini-histories, with lab. data from our hospital population. None of these are contrived or taken from the literature. The factors relating to the success of the course are (in our opinion):

1. A small faculty (3–4).
2. Strong faculty interest in teaching (supported by handout material, bibliographies, good photographic illustrations).
3. Prior ward experience on the part of the students, preferably with patient responsibility in medicine or pediatrics, or both.

We have surveyed hundreds of student critiques over the past eight years. The overwhelming majority of students favor taking the course *after* rather than *before* ward experience. The students state that early ward experience dramatizes what they need to know of laboratory lore, thus making our course all the more relevant at a later date.

Table 1. **Summary of Course Material**

Instructor 1

Practical Principles of Immunology
-Plasma proteins (ELP, IEP)
-Dysproteinemias
-Coombs test
-Haptoglobins
-Warm AIHA
-Cold AIHA

Laboratory diagnosis of SLE, including drug-induced syndromes
Laboratory diagnosis of rheumatoid arthritis
-Rheumatoid factors
-Immunologic aspects of joint fluids

Serologic Diagnosis of
-Hashimoto's disease,
-Primary biliary cirrhosis,
-Chronic active hepatitis

Diagnosis of pregnancy
Diagnosis of heterophil-positive I.M.
Diagnosis of heterophil-negative I.M.
Serologic tests for syphilis
Diagnosis of gonococcal infection
C-reactive protein, sed. r., and ASO.

General Principles of Blood Banking
Component therapy

Coagulation Disorders
-Basic cascade
-Screening tests
-Anticoagulation therapy & monitoring
-DIC
-Congenital disorders
-Platelet function tests

Principles of Detection of Drugs of Abuse

Instructor 2

Laboratory Decision Making
-Quality control
-Predictive value and effectiveness
-Biochemical screening profiles
-Cost/benefit analysis
-Examples and case studies (uric acid, carcinoembryonic antigen, pheochromocytoma, renovascular hypertension)
Clinical Enzymology & Liver Function
-Acute myocardial infarction
-Acute pancreatitis
-Prostatic carcinoma
-Muscle disease
-Malignancy
-Hematologic disease
-Liver function tests
-Case studies
Calcium, PO_4, & Metabolic Bone Diseases
-Serum proteins
-Ca, PO_4, alk. p'tase. in bone diseases
-Primary hyperparathyroidism
-Hypoparathyroidism
-Case studies
Laboratory Tests and Hypertension
-Pheochromocytoma
-Renovascular hypertension
-Primary aldosteronism
-Case studies
Electrolytes and Blood Gases
-Hyponatremic and hypernatremic states
-Metabolic acid-base disturbances
-Potassium disorders
-Hyperventilation, ventilatory failure, hypoxemia & hypoxia
-Case studies
Laboratory Endocrinology
-Thyroid function
-Adrenocortical function
-Hypothalamic-pituitary-gonadal function
-Endocrine syndromes in gastroenterology
-Z-E syndrome, insulinoma, diabetes
-WDHA syndrome
-Case studies

Instructor 3

Hematology I
-Cytoplasmic maturation defects
-Nuclear maturation defects
-Poikilocytosis I
-Poikilocytosis II

Hematology II
-Normochromic normocytic anemias
-Case studies I
-Case studies II
-Case studies III

Hematology III
-Leukopenias & leukocytoses
-Thrombocytopenias & thrombocytoses

Miscellaneous
-Synovial fluid analysis
-Cerebrospinal fluid analysis
-Lipoproteins

Renal
-Glomerular function
-Proteinuria
-Urinary sediment

Chemical Profiles (SMA 12)
-Fractional
-Fractional
-Total
-Total
-Total

Day 29: Comprehensive exam. by each instructor

Day 30: Review of exam. grades.

Teaching Methods and Examples

Hematology: In residency, as in office or hospital practice, the investigation of an anemia begins with a peripheral blood smear and indices. Morphologic variations in the erythrocyte are first reviewed. Then the students are led through a series of problem cases in which mini-histories are presented with indices alone. They are then asked to predict morphology (if possible). The smear is then briefly shown on a screen. From both indices and smear the students are then asked to suggest and defend further definitive tests if any. This exercise is designed: 1) to convince the students that both smear and indices should be examined in concert at the beginning, and 2) to make them approach test ordering in a logical (and economic) sequence. Coulter "S" indices are used for several reasons: *(a)* they are standard in 80% of all U.S. hospitals with 300 beds or more, *(b)* they are accurate and highly informative, *(c)* they readily characterize the majority of anemias encountered in practice, i.e. iron-deficiency, chronic disease, and B_{12}/folate deficiency. A typical case mini-history is used to demonstrate one approach to test sequencing.

Example: Indices in the diagnosis of anemia (Figure 3).

Question 1: How would you characterize this anemia?

Answer 1: Mild, normochromic normocytic.

Question 2: Does this smear substantiate that interpretation (slide)?

Answer 2: Yes. Polychromasia does not appear to be increased.

Question 3: What are the diagnostic possibilities?

73 y/o w/f (E.L.) admitted 9/77 for evaluation of low-back pain of acute onset (days). Past history of mastectomy in 1953, ASHD since 1970, mild anemia not responding to iron since 1974, fractured lumbar vertebra in 1976. On admission, the sedimentation rate was noted to be 110 mm/h.

Coulter 'S':	RBC	3.29
	Hgb	10.1
	Hct	30.4
	MCV	89
	MCH	30.6
	MCHC	33.5
WBC: 7400.		

Fig. 3. Data supplied from history.

Answer 3: The most likely five causes:

(1) Early iron deficiency.

(2) Anemia of chronic disease.
(3) Anemia of renal disease (creatinine normal).
(4) Hypoplastic marrow (leukocytes and platelets normal).
(5) Infiltrated marrow (x-rays negative).

Question 4: What should be ordered next?

Answer 4: Serum iron (Fe) and iron-binding capacity (TIBC). The serum iron was 200 μg/l, TIBC 2.25 mg/l.

Question 5: Are these data diagnostic?

Answer 5: No. The per cent saturation is 8.8. There is a significant overlap between the per cent saturation in iron deficiency and in the anemia resulting from chronic disease.

Question 6: What would you order next?

Answer 6: Normally in this setting a serum ferritin assay should be ordered for quantitation of iron stores *(2)*. This costs $25 and is non-invasive whereas a bone-marrow study costs $72 and *is* invasive. In this case, however, with the possibility of occult malignancy, a bone-marrow study might be considered the investigation of choice, because it would serve both to quantitate iron stores and detect metastatic disease. Actually, a serum ferritin was ordered first. It was markedly elevated.

Question 7: Are these data diagnostic?

Answer 7: They certainly rule out iron deficiency and strongly favor the anemia of chronic disease.

Sequel: Although the total proteins were normal, immunoelectrophoresis was performed. This showed a slight increase in the IgA fraction, later quantitated at 4.00 g/liter. A bone-marrow study was done later. This showed no evidence of either myeloma or tumor. Iron stores were increased. The back pain was ultimately ascribed to the patient's collapsed vertebra. The nature of the chronic disease was not elucidated. Iron therapy was stopped and the patient was discharged.

Serology: Example. Diagnosis of systemic lupus erythematosus: It should be recognized that the diagnosis of SLE is largely a clinical one and that lab. tests are at best only supportive. Before seriously pursuing the diagnosis of SLE, multisystem disease should be documented. The first test to be used in screening is the fluorescent antinuclear antibody test (FANA). If negative, SLE is no longer a viable consideration. If positive, the titer should be quantitated. Measurement of serum complement may be helpful. Other diseases, notably rheumatoid arthritis (RA) and mixed connective tissue disease (MCTD), should be ruled out. The algorithm is simple (Figure 4).

The students are next shown the common errors in initial test ordering for the diagnosis of SLE (Table 2).

Endocrinology: Example. Diagnosis of Cushing's syndrome: Steroids confuse clinicians. Faced with the problem of an adrenal work-up, most physicians order laboratory tests in haphazard sequence. When the results are obtained, they realize their error, ask for help, and start all over again.

The best way to detect Cushing's syndrome and to diagnose a specific cause is as follows:

1. If the patient has the appearance typical of this syndrome, order a 24-h urine for measurement of free cortisol. This test is the most sensitive and specific test. If the result is normal and collection is adequate, exclude diagnosis.
2. If the urinary free cortisol is above normal, do a 1-mg dexamethasone suppression test. Values for 8 a.m. plasma cortisol greater than 100 μg/liter after the one-night suppression

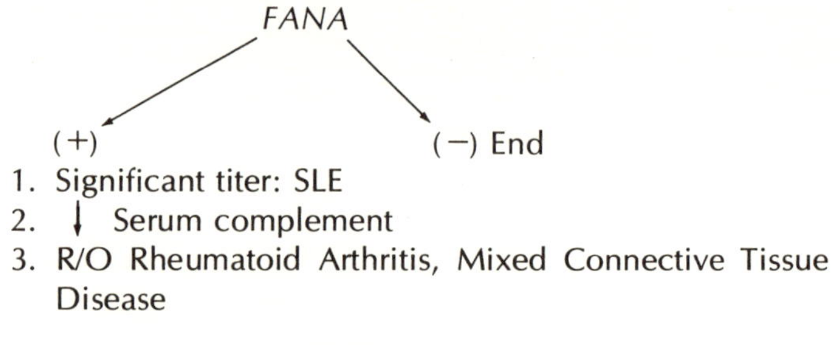

Fig. 4. *Evidence of Multisystem Disease*

Table 2. **Common Errors in Test Ordering for Lupus Erythematosus**

1. LE clot as screen (20–30% false negatives).
2. Anti-native DNA as screen (best use: to follow therapy).
3. Anti-single-stranded DNA as screen (not especially useful in diagnosis or follow-up).
4. Anti-ENA (primarily useful for diagnosis of Mixed Connective Tissue Disease.
5. Complement (may be supportive of diagnosis if low, and is useful for follow-up).

indicate Cushing's syndrome. Values less than 100 should be re-evaluated.

3. Two tests are used to determine the specific cause for Cushing's syndrome: Plasma corticotropin (ACTH) and the high-dose (8-mg) dexamethasone suppression test.
   (a) If plasma cortisol suppression exceeds 50%, the diagnosis is Cushing's disease (bilateral hyperplasia).
   (b) If suppression is less than 50%, measure plasma ACTH. If ACTH is suppressed, the diagnosis is adrenal tumor; if not, the diagnosis is ACTH-producing tumor.

Electrolytes. Example: Significant hyponatremia. Electrolyte determinations are virtually routine, but when significant changes from normal are found in the absence of obvious fluid and electrolyte losses, physicians, because they lack understanding of underlying pathophysiology, go berserk and indulge themselves in repeat test-ordering, particularly of urine electrolytes. In most instances, use of a few simple principles, if followed carefully will lead to a definite diagnosis, as in the following example.

1. Make sure the sample is not lipemic. If it is, measure the triglycerides: the triglyceride concentration in g/liter $\times$ 0.2 is a good approximation of the per cent decrease in sodium. Unless the triglycerides exceed 20 g/liter, lipemia is not the cause.

2. If the patient has congestive failure he has dilutional hyponatremia and does not need sodium.

3. If the patient has clinical evidence of dehydration he has depletional hyponatremia and needs sodium.

4. If there is no evidence of congestive failure, or dehydration, he has one of the following:
   (1) Vasopressin dysfunction: The syndrome of inappropriate antidiuretic hormone secretions (SIADH)
   (2) Hypothyroidism
   (3) A relatively subtle renal loss of sodium

Hypothyroidism is excluded by thyroid function tests. SIADH is distinguished from renal salt loss by demonstrating that urinary output of sodium exceeds intake in renal salt loss; the reverse is true in SIADH. Furthermore, SIADH can be confirmed by water-loading.

Multichannel screening: After first discussing the merits and cost/yield of multichannel screening, the students are tutored daily on such topics as uric acid metabolism (and its relevance to clinical diagnosis), the calcium–phosphate axis, and liver/heart enzymes. In the last week of the course, SMA-12 profiles are used to bring together widely disparate and seemingly unrelated chemical abnormalities. Like the Coulter "S," many hospitals use either the visual profile or computer-translated numerical data as part of an admission profile. Whether espoused or not as a screening device for metabolic, renal, or cardiohepatic disorders, the visual profile proffered by the SMA-12 has been extremely useful as a device for teaching clinical chemistry *(4)*.

Initially the students are led through a series of stylized partial SMA-12 data (slide). This slide shows increased values for calcium, uric acid, and alkaline phosphatase. Diagnostic considerations would include: breast carcinoma metastasizing to bone, bronchogenic carcinoma with parathyrin-like activity (with or without bony metastases), or primary hyperparathyroidism.

Another example of partial data is now shown (Figure 5). Here, the abnormalities are in the hepatic group. This profile implies infiltrative disease in the liver until proven otherwise. The pathophysiology of these and other stylized patterns is discussed along the lines previously described *(4)*.

Finally, the students are given real profiles from our teaching files and after study, asked to suggest and defend possible diagnoses. These profiles may be single or sequential, depending on the disease process involved. An example of the latter will now be considered (Figure 6*A*).

Question: At this stage, given the pancytopenia and back pain, what are the possibilities?

Answer: Metastatic tumor, myeloma, leukemia, etc., involving marrow. Somewhat against a diagnosis of myeloma is the normal total protein value.

Over the next six days, the patient had a variety of laboratory studies, including acid phosphatase and protein assays, immunoelectrophoreses, and a bone scan, all of which were normal. On Day 6, bilateral bone-marrow aspirates and trephine biopsies were obtained. Both aspirates and trephines showed massive marrow necrosis. The necrotic cells, however, were

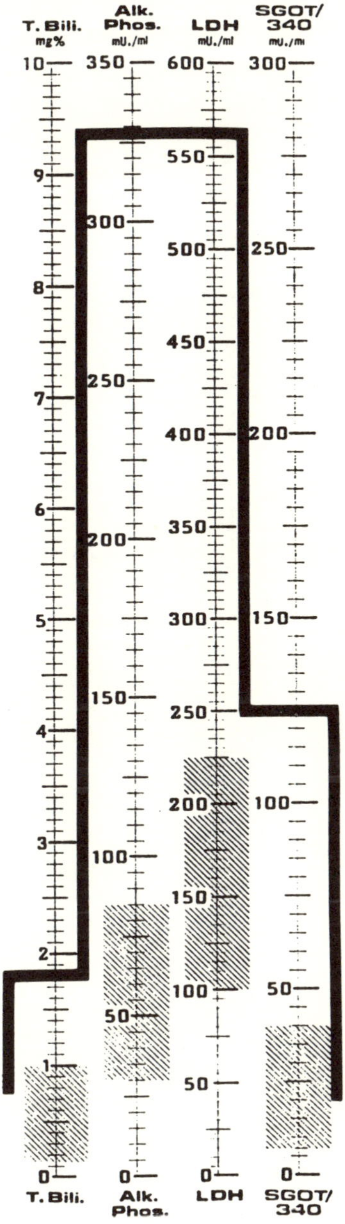
T. Bili.
mg%
Alk.
Phos.
mU./ml
LDH
mU./ml
SGOT/
340
mU./ml
10
9
8
7
6
5
4
3
2
1
0
350
300
250
200
150
100
50
0
600
550
500
450
400
350
300
250
200
150
100
50
0
300
250
200
150
100
50
0
T. Bili.
Alk.
Phos.
LDH
SGOT/
340

Fig. 5

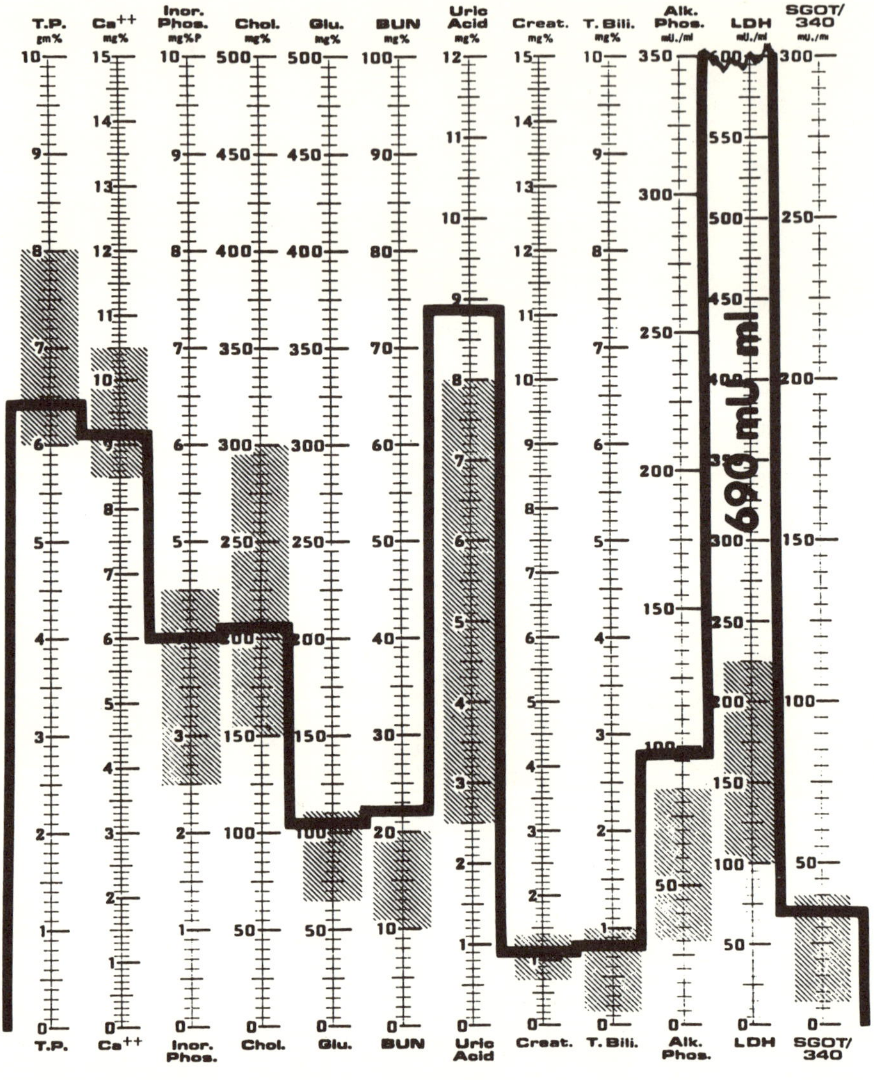

Fig. 6A

considered to be malignant and furthermore to reflect a hematopoietic malignancy. On Day 7, another SMA 12 was performed (Figure 6*B*). (The shaded areas reflect the increments in inorganic phosphorus, BUN, uric acid, creatinine and alkaline phosphatase. The LDH units are written in, since in both profiles levels were off the chart).

Question: At this stage, how would you interpret: *(a)* the

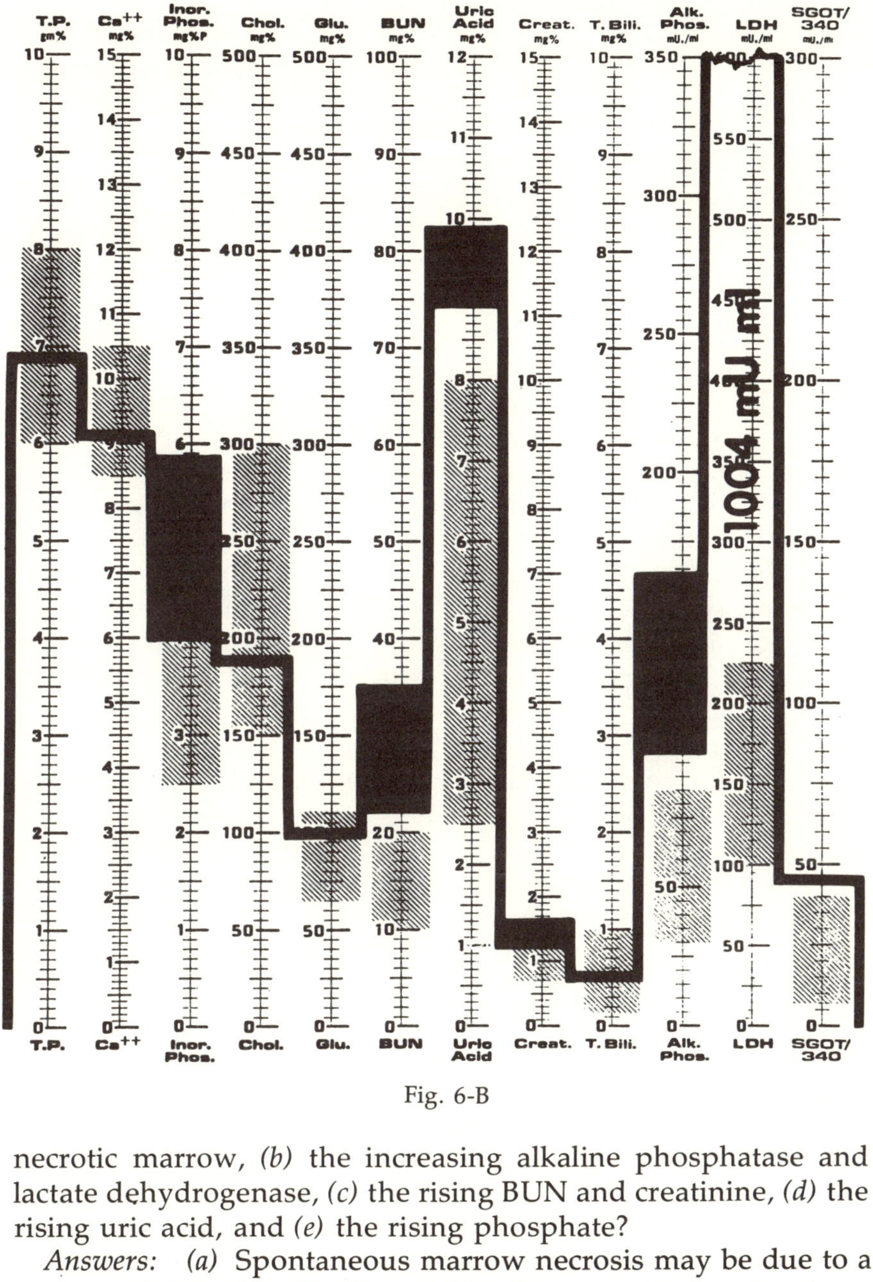

Fig. 6-B

necrotic marrow, *(b)* the increasing alkaline phosphatase and lactate dehydrogenase, *(c)* the rising BUN and creatinine, *(d)* the rising uric acid, and *(e)* the rising phosphate?

Answers: *(a)* Spontaneous marrow necrosis may be due to a variety of disorders *(5, 6)*, notably the acute leukemias, lymphomas, and metastatic tumor (when ionizing radiation, sepsis, and sickle cell disease have been excluded). Among the acute leukemias, necrosis is more common among lymphoblastic than myelogenous types.

(b) The rising alkaline phosphatase and LDH probably reflect progressing disease. The alkaline phosphatase could reflect active bone necrosis or progressive liver infiltration. The normal bilirubin value offers no clues as to origin. Isoenzyme fractionation of alkaline phosphatase would be helpful in distinguishing bone from liver origin. The LDH is probably of tumor origin, although damage to hepatic cells by infiltrating tumor might be a contributing factor. LDH fractionation showed a characteristic tumoral pattern (increased isoenzyme III).

(c) The asynchronous rise in BUN and creatinine probably reflect some pre-renal azotemia. The creatinine clearance (50 ml/min) is normal for this age decade and militates against significant renal disease at this time (although hyperuricemic nephropathy is an immediate risk in this clinical setting).

(d) The rising uric acid possibly reflects azotemia, in this case pre-renal. (Uric acid nephropathy would be accompanied by deteriorating renal function). More probably, however, nucleotides are being added to the central uric acid pool as a result of tumor necrosis.

(e) The rising phosphate is somewhat puzzling. When hemolysis (in vivo or in vitro) has been excluded, for practical purposes a high serum phosphate is due to: 1) increased intake of phosphates, 2) decreased renal excretion of phosphate, or 3) entry of phosphate into the extracellular fluid from tissue breakdown *(7)*. There was no evidence of increased phosphate ingestion. Phosphate should not increase in the serum when the glomerular filtration rate exceeds 25 ml/min *(8)*. The phosphate might therefore be coming from tissue breakdown or tumor necrosis, as in this case. Although hyperphosphatemia is well known to occur in patients undergoing chemotherapy for leukemia *(9)*, it has not been reported to occur spontaneously in leukemics.

Bone-marrow studies, repeated on Day 13, again showed necrotic marrow in iliac aspirates and trephines. A simultaneous sternal marrow aspirate, however, showed viable cells and the diagnosis of acute monoblastic leukemia was made (slides).

Summary

I have briefly reviewed an elective course in *Interpretation of Laboratory Data,* for senior medical students and pathology residents at the University of Minnesota. The factors responsible for its success are considered. Examples of teaching techniques are presented in the divisions of hematology, immunology, endocrinology, electrolytes, and multi-channel screening. In addition to highlighting interpretation of laboratory data, the course focuses on cost/yield considerations and attempts to teach students to use only the most informative laboratory tests in an effort to avoid overutilization of laboratory services and minimize cost to the patient.

References

1. Ward, P. C. J., Harris, I. B., Burke, M. D., et al., Instruction in interpretative aspects of laboratory medicine. *J. Med. Educ.* **51,** 648–656 (1976).
2. Jacobs, A., and Worwood, M., The clinical use of serum ferritin estimation (Annotation). *Br. J. Haematol.* **31,** 1–3 (1975).
3. Burke, M. D., Hyponatremia. *Check Sample CC-103, ASCP,* 1977.
4. Ward, P. C. J., Chemical profiles of disease. *Hum. Pathol.* **4,** 47–65 (1973).
5. Kiraly, J. F., III, and Wheby, M. S., Bone marrow necrosis. *Am. J. Med.* **60,** 361–368 (1976).
6. Carloss, H., Winslow, D., Kastan, L., et al., Bone marrow necrosis. Diagnosis and assessment of extent of involvement by radioisotope studies. *Arch. Int. Med.* **137,** 863–866 (1977).
7. Slatopolsky, E., Rutherford, W. E., Rosenbaum, R., et al., Hyperphosphatemia. *Clin. Nephrol.* **7,** 138–146 (1977).
8. Goldman, R., and Basset, S. H., Phosphorous excretion in renal failure. *J. Clin. Invest.* **33,** 1623 (1954).
9. Zusman, J., Brown, D. M., and Nesbit, M. E., Hyperphosphatemia, hyperphosphaturia and hypocalcemia in acute lymphoblastic leukemia. *N. Engl. J. Med.* **289,** 1335–1340 (1973).

Teaching Laboratory Utilization by Use of Computer-Based Clinical Simulations

Richard B. Friedman, M.D., and Donald R. Korst, M.D.

Department of Medicine, University of Wisconsin, 427 Lorch St., Madison, Wis. 53706

We are currently in the midst of a development in the practice of medicine that may well rank with Sydenham's redefinition of disease, Morgagni's use of necropsy to examine the interior of the body after death, or Virchow's introduction of microscopy in the study of disease. This development is the gradual emergence of the clinical laboratory data as an increasingly important source of information for the physician.

Interestingly, this major shift in health-care delivery has had very little effect on the curriculum of the nation's medical schools. Today as in the past, medical students are introduced to clinical medicine through a course in physical diagnosis, which concentrates on the skills necessary to take a complete medical history and perform a skillful physical examination. This course rarely includes any training in the use of clinical laboratory data. In learning to take a complete history, medical students are told that those data represent 90% of their diagnostic information. They are trained to get this information by taking long and involved histories. Yet as their experience increases, their histories become shorter and shorter, because they quickly learn that those data represent 90% of their diagnostic information. included in the one-sentence chief complaint, "my stomach hurts." The ability to perform a skillful physical examination is similarly stressed. Weeks are spent listening for heart murmurs, feeling for the kidneys, spleen, or liver and testing cranial nerves. Yet, what cardiovascular surgeon would operate without a cardiac catheterization or what internist initiate treatment for an enlarged liver, spleen, or kidney without the appropriate laboratory tests, scan, or roentgenogram?

Training in the use of the clinical laboratories is left to the student himself, who must learn by reading texts or observing how physicians (frequently housestaff officers with only a year or two additional training) perform this task. Factors such as cost of a test, its availability, the length of time it takes to get results, its reliability, drug interactions, alternative methods of obtaining the same data, interpretation of results, etc. are rarely formally discussed. The absence of a clearly defined training program in laboratory medicine undoubtedly has contributed to the current misuse of the clinical laboratory.

At a few medical schools across the country this is beginning to change and formal training in the use of the clinical laboratories is being included in the curriculum. Unfortunately, even at these institutions the training is usually didactic and only involves the interpretation of test results. Lecturers rarely approach questions of test cost, availability, reliability, interactions, etc.

At the University of Wisconsin, training in the use of the clinical laboratories is included as part of the third-year medical clerkship through the incorporation of computer-based clinical simulations in the teaching program. These clinical simulations permit students to practice ordering laboratory procedures on simulated patients, to review the results, and then to use this information in making diagnostic and therapeutic decisions. In this way the student is given an opportunity to see realistically how factors such as patient condition, time of day, cost, test reliability, and laboratory hours can all affect the value of the material he obtains from the laboratory.

Computer Simulation

The computer-based simulation of the patient-physician encounter utilized at the University of Wisconsin *(1, 2)* differs from most previous clinical simulations *(3–11)* in that it contains a "time-axis," which permits the patient's condition, test results, and test availability to vary as the case progresses. The Hospital Encounter Simulation (HECS) session begins with the presentation of a brief introductory statement describing the patient. The student is then informed of the time, day of the week, and is asked to request any tests or consultations necessary to determine the diagnosis. The following will serve as an example:

> BACKGROUND INTRODUCTORY REMARKS:
>
> A 37-year-old woman was admitted the the hospital because of fever. It is currently 7:00 a.m. Tuesday, on Day 1 of hospitalization. The patient has now been placed in the care of Student Dr. Hughes.

The student is then free to request any tests he desires, in any order he desires. All tests or consultations (> 400) available at a good general hospital are included in the model. The user may make a diagnosis at any time during the encounter. The user requests information either by typing into the computer the actual text of his request (for example, "electrolytes") or, if he is unable to type, by entering the code number designation for the text (for example, "224" for electrolytes). The student may request tests individually or in groups, and after each test request he is informed whether the test is available, how long it will take, and how much it will cost. He may then decide either to request the test or to reject it without penalty. In the example given above, the dialogue might continue in this manner:

> REQUEST?
> *history*
>
> HISTORY $9.00
> PROCEDURE REQUIRES 0.1 HOURS RESULTS DUE TUESDAY AT 7:06 A.M.
> ORDER THIS PROCEDURE? *yes*

When a test is ordered, the stated amount for each test is added to the patient's bill. When the results of the tests are requested, the time sequence ("clock") is moved ahead to the time the test would be available. If the user decides to order a number of tests at one time, the system will "stack" these requests and return the results when the "clock" reaches the date and hour the test would normally be completed. Test results are only available during that period of the day when those tests would normally be available in a functioning general hospital. If the user wishes to obtain a result on an emergency basis, he may order the test or consultations as a STAT at an additional cost. To illustrate, the sample program might resume as follows:

IT IS CURRENTLY 7:00 A.M. ON TUESDAY DAY 1 OF HOSPITALIZATION

REQUEST?
vital signs

VITAL SIGNS $0.00
PROCEDURE REQUIRES 0.1 HOURS RESULTS DUE TUESDAY AT 7:06 A.M.
ORDER THIS PROCEDURE? *yes*

REQUEST?
results

IT IS CURRENTLY 7:06 A.M. ON TUESDAY DAY 1 OF HOSPITALIZATION

RESULTS OF VITAL SIGNS
T–41.4 C. R–46/MIN P–120 BP–90/55

REQUEST?
electrolytes

ELECTROLYTES $7.00
RESULTS DUE TUESDAY AT 5:00 P.M.
ORDER THIS LAB TEST? *no*

REQUEST?
stat electrolytes

EMERGENCY ELECTROLYTES $7.00
LAB TEST REQUIRES 0.5 HOURS RESULTS DUE TUESDAY AT 7:36 A.M.
ORDER THIS LAB TEST? *yes*

Any time there is a change in the patient's condition, the program automatically generates "Nurse's Notes." These notes appear spontaneously and stop the movement of the clock so the student is free to reassess his test requests in light of this new development.

IT IS CURRENTLY 11:30 A.M. ON TUESDAY DAY 1 OF HOSPITALIZATION

REQUEST?
stat chest x-ray

EMERGENCY CHEST X-RAY $22.00
PROCEDURE REQUIRES 0.5 HOURS RESULTS DUE TUESDAY AT 12:00 P.M.
ORDER THIS PROCEDURE? *yes*

RESULTS?
results

IT IS CURRENTLY 12:00 P.M. ON TUESDAY DAY 1 OF HOSPITALIZATION

NURSE'S NOTES:

T–41.1 C. PATIENT IS TACHYPNEIC AND COMPLAINING OF CHEST PAIN

IT IS CURRENTLY 12:00 P.M. ON TUESDAY DAY 1 OF HOSPITALIZATION

RESULTS OF EMERGENCY CHEST X-RAY:

MARKED CONGESTION AT THE RIGHT BASE WITH A DIFFUSE INCREASE IN DENSITY THROUGHOUT THE LUNGS.

REQUEST?

The results of all tests in the system can vary hourly as a result of changes in the patient's condition or because of the effects of the medication prescribed by a "simulated" house staff. Once the student feels that he has enough information to make the correct diagnosis, he types the text of that diagnosis (or its code number) into the computer. If the student makes an incorrect diagnosis, he is penalized in time and money, an amount corresponding to how serious that misdiagnosis would be at this time in this particular case. If he is correct, he is given the total cost of the workup as well as the amount of time the patient has been in the hospital. The computer–physician dialog for the sample case might be as follows:

DO YOU WISH TO MAKE A DIAGNOSIS? *yes*
DIAGNOSIS = *pneumococcal pneumonia*

CORRECT DIAGNOSIS

This young woman suffered from pneumonia. Sputum cultures grew out pneumococcus. Treatment with penicillin, or erythromycin would result in resolution of the infiltrate in seven days. If diagnosed incorrectly the patient's condition gradually deteriorates and she expires on the tenth day.

NUMBER OF DAYS IN HOSPITAL	9
COST OF DIAGNOSTIC WORKUP	$ 527.00
COST OF HOSPITAL ROOM ($90/DAY)	$ 810.00
TOTAL COST OF HOSPITALIZATION	$1,337.00

The program can be used in conjunction with a wide range of audiovisual source materials. If the script writer wishes to incorporate such material into his case, he constructs it so that instead of printing the results (or the interpretation of these results) at the terminal the program informs the user which x-ray, EKG, biopsy slide, or other material he is to take from the nearby files and examine. In this way the program can still control the cost, availability, and results of each test, and yet permit the student to view the source material and make his own interpretation.

Cases are entered into the system via a special script-writing program. This program operates in an interactive mode so that the script writer need have no knowledge of computers or computer programming. Script writing is simplified by the fact that the model itself contains all information on test costs and test availability, as well as the normal results for each test. The script writer need only enter the results for those tests that are abnormal. For each case, this usually represents a 20- to 50-test subset of the more than 400 available tests.

The system is programmed using the MIIS dialect of MUMPS, a high-level interpretive time-shared computer language. The programs operate on a time-sharing Digital Equipment Computer PDP-15 located at the University of Wisconsin. The computer can handle up to 16 users simultaneously and interaction with the computer requires no peripheral devices other than a standard computer terminal connected to the computer via telephone lines. The average encounter lasts between 30 and 45 minutes.

Evaluation

At the end of the encounter the student's workup is evaluated by the program. Data for the evaluation program is generated by a committee of clinicians with recognized expertise in the area covered by the case material. This committee goes over the case and, for each of the over 400 possible tests, procedures, or consultations in the model, assigns each of them scores. These scores range from +3 (test which is essential to do) to −3 (a test which is essential to avoid). (See Figure 1.) When a student requests an evaluation, the program first goes through its file and prints all tests not ordered by the user which were essential (+3) or important to do (+2). It next prints out a list of all tests ordered by the candidate which were essential (−3) or important to avoid (−2). The program then computes an Efficiency, Proficiency, Errors of Omission, Errors of Commission, and Competence Index (see Figure 2) based on these previously

+3 = Test or procedure which is essential to do in this case.
+2 = Test or procedure which is important to do in this case.
+1 = Test or procedure which is useful to do in this case.
0 = Test or procedure which is optional to do in this case.
−1 = Test or procedure which is useless to do in this case.
−2 = Test or procedure which is harmful to do in this case.
−3 = Test or procedure which is essential to avoid in this case.

Fig. 1.

$$\text{Efficiency} \ldots\ldots = \frac{\text{number of positive items selected}}{\text{total number of items selected}}$$

$$\text{Proficiency} \ldots\ldots = \frac{\text{sum of all pos. and neg. items selected}}{\text{maximum possible score}}$$

$$\text{Errors of Omission} \ldots\ldots = \frac{\text{sum of the positive items NOT selected}}{\text{maximum possible score}}$$

$$\text{Errors of Commission} \ldots\ldots = \frac{\text{weighted sum of neg. items selected}}{\text{maximum possible score}}$$

$$\text{Competence Index} = \text{efficiency} \times \text{proficiency}$$

Fig. 2.

assigned scores. An evaluation of the above case might be as follows:

EVALUATION FOR CASE

TESTS ORDERED WHICH WERE ESSENTIAL TO AVOID
thoracotomy

TESTS ORDERED WHICH WERE HARMFUL TO THE PATIENT
none

TESTS NOT ORDERED WHICH WERE ESSENTIAL TO DO
sputum culture
urinalysis

TESTS NOT ORDERED WHICH WERE IMPORTANT TO DO
sputum stain for tuberculosis
feces guaiac
blood urea nitrogen
electrolytes

EFFICIENCY	75
PROFICIENCY	28
ERRORS OF OMISSION	30

ERRORS OF COMMISSION	5
COMPETENCE INDEX	21

Application

At the University of Wisconsin the computer-based clinical simulation has been used as an integral part of the teaching program for third-year medical students for the past five and one-half years. The model has been incorporated into the teaching program in the following manner:

1. To Permit Students to Gain Clinical Experience

All third-year students are told of the availability of the computer-based clinical simulation at the start of their medicine rotation. A computer terminal is available at each of the major teaching hospitals in the program. The students are given the opportunity to complete simulated encounters on their own as a means of gaining experience in case workups and diagnosis. About 20 different cases, representing a broad cross-section of internal medicine, are available.

Each year an increasing percentage of the third-year class has utilized the simulator for self-assessment. In 1972–73, the first full year of this program, third-year students completed 490 case encounters. In the current academic year they will complete over 2500 encounters.

2. As Part of a Series of Teaching Conferences

Each week during the three-month third-year medicine clerkship, a 1.5-h topic-oriented conference is held in which one area of medicine is considered in some detail (e.g., hypertension, anemia, abdominal pain, etc.). For each session, two students serve as the conference organizers and are responsible for planning the conference, developing handouts, and inviting a faculty expert to attend the conference. These conferences have met with high student acceptance and attendance is excellent.

Each group of students can plan the conference as they wish, utilizing lectures, audiovisual materials, paper-and-pencil case simulation, computer simulation, etc. Each quarter these student leaders elect to utilize computer simulations for about one-third of their conferences. The simulations are utilized in two different ways. In one format, the students are each requested to complete a simulated case encounter on the week's topic before the conference. The case is then discussed at the conference by the faculty expert. In the second format, a computer terminal is brought to the conference and a case dealing with that week's topic is worked on by the group. The group orders each test after a consensus is reached and the faculty expert serves as a consultant. In both cases the visiting "experts" discuss in some detail how they would go about working up the case. They indicate the relative strengths and weaknesses of various tests or procedures and indicate why they chose one test over another. Various aspects of the reliability, availability, cost, and informational content of the tests are discussed.

3. As Part of a Testing Program

All third-year students are given three unknown clinical simulations as part of their final examinations in medicine. They are given 2 h in which to complete the three cases, which count as 10% of their final grade in medicine. The student's performance on each case is graded with respect to: *(a)* whether the correct diagnosis was obtained; *(b)* the number of "simulated" days until correct diagnosis; *(c)* the patient's total hospital bill; *(d)* the total "actual" time spent by students on each case; and *(e)* the student's performance on each of the following parameters: proficiency, errors of omission, errors of commission, competence index.

Student Response

As an evaluation of the students' response to the simulation experience, a subgroup of students was sent a one-page questionnaire, soliciting their evaluation of the program to complete and return (unsigned). A total of 165 students returned the questionnaires. The students were asked to evaluate the experience with respect to *(a)* whether they enjoyed the experience, *(b)* whether they were technically able to use the model, *(c)* whether they found the experience generally useful, and *(d)* whether they believed it was helpful for self-assessment purposes. They were asked to indicate how accurately their performance on the model reflected their assessment of their medical knowledge or their performance on the ward, and whether they would be interested in using the simulation model in the future for self-evaluation, gaining experience, or testing. The results of the questionnaire (see Figure 3) revealed a high degree of user enjoyment, a general consensus that the model was easy to use, and a belief that the simulated encounter sessions were helpful. The students indicated they would continue to use the model for self-evaluation, but some had reservations as to whether performance on the model reflected either their true medical knowledge or performance on the ward. A number had serious reservations about using the model as a testing device.

The student evaluation scores presented here represent a composite of those received over the entire five years the program has been used at Wisconsin. It is interesting to note that the scores received on the computer simulation examination were

1. *I found the Simulated Patient Encounter (SPE):*		*Average score*
0 1 2 3 4x 5 Terrible	Enjoyable	4.32
0 1 2 3x 4 5 Difficult to Use	Easy to Use	3.20
0 1 2 3x 4 5 A Waste of Time	Useful	3.70
0 1 2 3x 4 5 Useless for Self-Evaluation	Helpful for Self-Evaluation	3.60
2. *I believe my performance on the SPE:*		
0 1 2 3x 4 5 Did Not Reflect My Medical Knowledge	Accurately Reflected My Medical Knowledge	3.05
0 1 2x 3 4 5 Did Not Reflect My Performance on the Ward	Accurately Reflected My Performance on the Ward	2.40

3. *I would be interested in using the SPE in the future for:*			*Yes*	*No*	*No response*
(a) self-evaluation	YES	NO	90%	2%	8%
(b) gaining experience	YES	NO	92%	2%	6%
(c) testing	YES	NO	52%	44%	4%

Fig. 3

not included in the students' final grade until the last year. Student evaluations during this past year have shown a definite decrease in scores for enjoyment, ease of use, and confidence that performance on simulation reflected performance on the ward.

Conclusion

The use of clinical simulations in conjunction with teaching conferences provides a framework for training students in the use of the clinical laboratory. The students' free access to the simulator during the clerkship gives them an opportunity to practice this important skill and experience the results of their decisions without jeopardizing the health of a real patient.

The major advantages of the program appear to be: *(a)* the student learns to choose those tests which provide the best diagnostic information; *(b)* he learns to interpret laboratory test results; *(c)* he learns about test availability and the amount of time required to obtain the results and how these factors can effect the "true" diagnostic value of a test; *(d)* he learns what tests cost and how this may be considered in choosing alternative tests which may provide the same information; *(e)* he learns how the total hospital-room cost affects decisions on whether to order tests sequentially or as a battery; and *(f)* he gains confidence in his ability to meld historical, physical examination, and laboratory data into a correct diagnostic conclusion.

It is still too early to determine whether this training produces physicians who can better utilize the clinical laboratory. However, students who have participated in this program clearly demonstrate a greater awareness of cost and availability of laboratory procedures and report greater confidence in their ability to use the clinical laboratory.

References

1. Friedman, R. B., Beatty, E., Korst, D., and Friedman, P., A diagnostic aid for medical education. *J. Med. Education* (June 1977).
2. Friedman, R. B., Clinical simulation in medical education—the Wisconsin experience. Chapter in *Computers in Biomedical Education*, E. Deland, Ed., May 1976.
3. DeDombal, F. T., Hartley, J. R., and Sleeman, D. H., A com-

puter assisted system for learning clinical diagnosis. *Lancet* **i,** 145 (1969).

4. DeDombal, F. T., Horracks, J. C., Staniland, J. R., and Guillou, P. J., Production of artificial "case histories" by using a small computer. *Brit. Med. J.* **ii,** 578 (1971).
5. Farquhar, B. B., Barnett, G. O., Goldfinger, S. E., and Dinnen, J. J., Computer-based examination: A technique to evaluate clinical competence. In *Self-Instruction in Medical Education*, J. P. Lysaught, Ed., University of Rochester, Rochester, N.Y., 1973, pp 91–94.
6. Feurzeig, W., Munter, P., Swets, J. A., and Breen, M., Computer-aided teaching in medical diagnosis. *J. Med. Educ.* **39,** 746 (1964).
7. Harless, W. G., The development of a computer-assisted instruction program in a medical center environment. *J. Med. Educ.* **42,** 139 (1967).
8. Harless, W. G., Drennon, G. G., Marxer, J. J., et al., CASE: A computer-aided simulation of the clinical encounter. *J. Med. Educ.* **46,** 433 (1971).
9. Hoffer, E. P., Experience with the use of computer simulation models in medical education. *Comput. Biol. Med.* **3,** 269 (1973).
10. Hubbard, J. P., Levit, E. G., Barnett, G. O., et al., Computer-based evaluation of clinical competence. *ACP Bull.*, pp 502–505 (1970).
11. Schneiderman, H. and Muller, R., The diagnosis game: A computer-based exercise in clinical problem solving. *J. Am. Med. Assoc.* **219,** 333 (1972).

PANEL DISCUSSION

Dr. Tietz: The first question is to Dr. Reece. What would be the effect of including the ordering physician's provisional diagnosis as part of the input data? Have you considered doing this?

Dr. Reece: The effect would be to improve the diagnostic output. We have considered it. In fact, we programmed Dr. Nash's data in London, England. He took seven textbooks and correlated signs and symptoms with laboratory findings, and we have programmed that and tried it out, with the thought that sooner or later we would get signs and symptoms and diagnoses. It should work very well.

Dr. Tietz: One more question for Dr. Reece. What do you mean by "standard deviation from normal"? Does this indicate standard deviation from the mean?

Dr. Reece: No. It is the standard deviation from the outer limits of normal. What we have done is to take, say 20 000 outpatients and prepare computer histograms. We arrive at a normal range, and the standard deviation is the standard deviation outside the normal limits rather than from the mean.

Dr. Tietz: Dr. Reece, do you make use of any previous laboratory data on a patient in making an interpretation of current results?

Dr. Reece: No.

Dr. Tietz: How successful are you in getting necessary information from physicians in order to calculate risk probabilities from a few laboratory tests?

Dr. Reece: Not very. We solicit information about creatinine clearances, height, and weight and also for risk factors, coronary risk, and I find that anytime I ask a clinician to provide additional information other than the test ordered, I fail.

Dr. Tietz: Dr. Griner, given the discrepancy between laboratory test charges, third-party reimbursement, and true test costs, what figures do you show your house staff to make them aware of the cost of laboratory services?

Dr. Griner: The figures that they receive are charges, to the extent that these are inpatient data and since the reimbursement formula for New York State, as far as in-hospital reimbursement is concerned, is on a cost basis (or, in fact, it now is less than a cost reimbursement). There is no profit in the laboratory. Therefore, there is a relatively close correlation between charges and costs insofar as laboratory tests go, but I would be foolish and naïve to suggest that the two are identical.

Dr. Tietz: Dr. Reece, in view of the complicated format of your report, how do you handle cumulative date reporting?

Dr. Reece: We don't. By cumulative data reporting I gather you mean tests ordered on a patient on different days during one hospital stay? If you recall, I said most of these tests were done on outpatients. We will unify a mass of information on one report ordered at one time, but we do not accumulate it for different profiles ordered at different times.

Dr. Tietz: Dr. Reece, do you know the proportion of cases in which a disease selected by computer disagreed with the admitting diagnosis, but was later shown to be correct?

Dr. Reece: No.

Dr. Tietz: Once more, Dr. Reece, have your 103 patterns been published; if so, where?

Dr. Reece: Yes, in the *American Journal of Clinical Pathology*, May 1972, but not the 102; that was the initial list. We also have an article in *Federation Proceedings* published in 1975. If anybody wishes the list of patterns, I would be happy to send it to them.

Dr. Tietz: Dr. Reece, if sequential testing is performed frequently, for example in a brittle diabetic or in a newborn with respiratory distress syndrome, do you supply an interpretive report with each of the tests requested?

Dr. Reece: Yes, we do, but the two situations you just mentioned are acute hospital situations and acid–base studies, not the sort of situations we would get into, and I should say again that our work is done mostly on outpatients.

Dr. Tietz: In your survey, did many of the indifferent answers come from one physician indifferent to all questions? In other words, were they just apathetic in general? I cannot imagine anyone being indifferent to sex- or age-related normals.

Dr. Reece: Most of the things to which they were indifferent or which they found distracting were in special parts of the program. In general, most of the physicians filled out the questionnaires thoroughly and only disliked isolated features.

Dr. Tietz: Here is a question to both Dr. Ward and Dr. Friedman. Do you think that one of your graduates needs Dr. Reece's second sheet?

Dr. Ward: No.

Dr. Reece: Dr. Ward's students are two sheets to the wind!

Dr. Friedman: I would say that the students who use the simulation probably need it as much as any other group. I don't know that the simulation would make them immune to any additional information, and I think as much information as they would have available to them with which to make a decision would increase the reliability of their decisions.

Dr. Tietz: Dr. Reece, since you got a 10% voluntary response to your questionnaire, how do you know that 90% are not indifferent or distracted?

Dr. Reece: We don't.

Dr. Tietz: Dr. Friedman, would your course of simulated case teaching be better if offered to fourth-year students?

Dr. Friedman: It is difficult to say. There are advantages and disadvantages. The disadvantage, of course, is that in the third year they are forming patterns, so this [seems to be] an appropriate time. In the fourth year they have had some experience and may be able to bring this experience to it. As Dr. Ward mentioned, after they have had some clinical experience they better realize the importance, so I would say it is probably a tossup. I guess if I had a preference it might well be a better course for the fourth year than the third. I enjoyed Dr. Ward's talk and I think perhaps an elective based around simulation in the form that he mentioned might be a nice combination of the two.

Dr. Tietz: Dr. Friedman, what method do you use to make sure the house staff are constantly aware of test availability? We try to do this but find it very difficult to make them think this way.

Dr. Friedman: We really don't have an answer to that. We have the same problem. They are not aware. The laboratory produces a monthly letter telling about changes in tests and what is available. Patterns change very slowly with the house staff—I think probably because they also change very slowly with the attending staff—and we really don't have a good way of doing it. Simulation, as I indicated, is not being used extensively on the house staff. We made it available. Interestingly, those students who went to Wisconsin know about it. As they become house-staff members, they constantly are interacting with the computer to find out if there are new cases, and if there are, they try them, but it is not something we are using in that capacity.

Dr. Tietz: One more question, Dr. Friedman. What feedback do you have on past students? Do they revert to bad habits of ordering or not?

Dr. Friedman: It is very difficult to know. I don't want to oversell the long-term gains of this program. We don't really know them; it is too soon, and I am not sure we have a good way of gauging it. I think the way we gauge it is to remember that the students only have so much time at their disposal. They tend to be the best people from the point of view of what they think they are getting out of their time, and the fact that they are doing more simulation, I think, is an indication they believe it is valuable. Also, I think the faculty believes it is important enough that they have increased its utilization in the program, but as far as long-term influences, whether we really made a dynamic change in their patterns, I am not really confident that I can ever prove that.

One thing we are looking at is stability of patterns. That is, we are looking at whether they rank high, for example in cost or length of time until diagnosis, as a group. That is, we look at where these students fall in the third year, we follow them as a group through the fourth year and then look at them versus their peers in the training program, and see whether our students are doing better. The problem, of course, is that in a way we are doing just what the factual knowledge exam does: we test them with a device we train them on. Therefore those that train on it always will have a certain advantage, and what you try to do is equalize the training. Our students had an opportunity to use it earlier, maybe it helped their patterns, but I don't have a definite long-term evaluation.

Dr. Tietz: There is a similar question directed to Dr. Friedman and Dr. Ward. How long is there a persistent difference between Wisconsin students and others when they reach internship? Dr. Ward, would you like to start that?

Dr. Ward: Well, at least as far as the Minnesota experience has been, it has been evaluated at two levels. One, in subsequent rotation, as I reported in the *Journal of Medical Education* in August of 1976. One of the very many parameters under which students are evaluated in their third and final years is the student as physician, and there appears to be some significant difference in the students who have left our program and gone to subsequent rotations—notably in medicine, and especially in pediatrics, oddly enough—in their ability to handle themselves well in the wards. The other thing that we have looked at, although this is entirely subjective, is that we get spontaneous comments from clinicians in other hospitals on our students who have been through our program and gone to other places, as well as on interns and residents who have been in our program. They appear to be much more facile in the wards, but this is entirely subjective. We have no hard data on that, whereas we do have performance as physicians, which has been published already.

Dr. Griner: I would like to make very sure there is no misconception concerning an issue that Dr. Friedman brought up, and that has to do with numbers of tests versus efficiency.

When I was talking this morning about over-utilization and our trends in terms of reduced numbers of tests, they reflect the data that we previously had that strongly suggests that in the follow-up management of patients, after a diagnosis is established, there was a tremendous amount of repetition. That is a fundamentally different issue from the need for efficiency in reaching that point, namely in the diagnostic process, which is what Drs. Ward and Friedman were discussing.

Secondly, it is, I think, inappropriate and potentially misleading to draw any conclusions that efficient laboratory application has anything to do with length of patient stay, because there are so many other variables, and until and unless one can provide data to be brought to bear on that score the question is open.

For example, in Rochester, we have a decreasing length of stay across the community, but in one hospital lab tests are going up, and in another they are going down.

Dr. Friedman: Yes. I quite agree. All I am suggesting is that one must keep this in mind as a potential possibility, and I think that there are a lot of factors involved. As to the question of how one evaluates one's students, for example in a future training program, when we are comparing them with house-staff officers who are trained in other programs, the considerations are similar. There are so many variables that may be much more important than the single course that we gave, that may be causing increases or decreases in laboratory utilization.

Dr. Reece: Dr. Hobbie and I have been trying to teach practicing physicians, many of whom are in a non-teaching atmosphere, and are quite busy, and I think that takes a different approach, one of gentle persuasion and gentle reminders. We do not have a captive audience and therefore we use different techniques.

Dr. Ward: One thing I would say in closing is that a gratifying experience we have had is the report from a small hospital in southern Minnesota which has recently been taken over by two of the graduates of our program. The administrator was recently heard to complain bitterly that his laboratory system was going broke because the two main physicians who contribute to the tests are just not ordering tests.

Optimizing the Usefulness of Simple Laboratory Tests

Thomas P. Whitehead, Ph.D., M. R. C. Path.

University of Birmingham, U.K.

The exponential increase in the number of tests performed by clinical chemistry laboratories has been a feature of the discipline for half a century.

This increase in the number of clinical chemistry investigations has not been paralleled, at least in the U.K., by a proportional increase in the number of patients investigated. Side by side with such increased investigation of individual patients there has been a slow but definite increase in the investigation of the so-called "well population." Biochemical profile techniques have formed an integral and increasing part of both the investigation of patients and of the "well population."

The increase in the number of tests performed in routine clinical chemistry laboratories has not been accompanied by equal endeavour on the part of laboratory workers to assess the efficacy of such methods of investigation. Frequently it has been easier to perform a test than to enquire why the test was necessary.

In recent years the Wolfson Research Laboratories have attempted to assess the efficacy of biochemical profile methods of investigation.

This paper describes two studies.

The first is concerned with patients who were complaining of signs or symptoms, hence their attendance at a family-practitioner clinic.

In the second study, a group of men having routine health checks at their place of employment have been investigated by biochemical profile. For the purposes of this paper, this second study is solely concerned with the changes in serum biochemical values associated with the ingestion of ethanol.

General Practice Study

The Family Practice that cooperated with the Wolfson Research Laboratories in this study contains about 7000 patients from an industrial and residential area of Birmingham, U.K.

In earlier studies blood profiles were performed irrespective of the diagnosis *(1, 2)*, but this investigation was confined to patients diagnosed as having musculoskeletal or psychoneurotic disorders, conditions chosen because they are probably the two commonest clinical conditions dealt with in family practice in the U.K.

When a patient attending the clinic was placed in either of these categories, as defined by the morbidity classification of the Royal College of General Practitioners, and on clinical grounds the doctor regarded hospital referral as necessary, then the patient was immediately referred. If not, the patient was invited to attend the practice center later, where blood was collected for the following investigations:

- hemoglobin
- packed cell volume
- mean corpuscular hemoglobin
- leukocyte count
- potassium
- creatinine
- calcium
- globulin
- aspartate aminotransferase
- glucose
- sodium
- erythrocyte count
- mean cell volume
- erythrocyte sedimentation rate
- rheumatoid factor
- urea
- urate
- albumin
- total bilirubin, alkaline phosphatase
- cholesterol
- protein-bound iodine

The hematological tests were performed with a Coulter Counter Model S in the Department of Haematology, Queen Elizabeth Medical Centre.

The electrolyte analyses were done with an Instrumentation Laboratory (Model 143) flame photometer; the protein-bound iodine was determined with an AutoAnalyzer technique (Technicon Instruments Corp.).

The other determinations were performed with another continuous-flow analysis instrument, the SMA 12/60 (Technicon Instruments Corp.).

Before the laboratory results were available to the family practitioner, he recorded the tests he would normally have requested on each patient.

When the profile results were received, the doctor also kept a record of actions such as the discharge of the patient without further action, the performance of any further tests, whether a prescription was issued, or the referral of the patient to hospital for a consultant opinion.

At this time the doctor completed a questionnaire in which he classified each abnormal laboratory result into either expected, unexpected, leading to a new or additional diagnosis, or unexpected and unexplained. The criteria for classification of a result as normal or abnormal is given in Carmalt et al. *(1)*.

Finally, at least 12 months after the original profile investigation, the general practitioner reviewed each case for which one or more laboratory results were abnormal and which had led to a new diagnosis. Patients who had been referred to hospital as a result of the laboratory investigations were also reviewed.

There were 590 patients in the psychoneurotic group and 370 patients with musculoskeletal disorders.

Results of the Study

Fewer than 1% of the tests performed would have been routinely requested by the family practitioner. This reflects the low rate of laboratory investigation in primary care in the U.K.—and the general practitioners involved in this study would be typical of those in such practice in the U.K.

After the laboratory profile was done and the results were considered by the general practitioner, about 20% of both groups required additional or repeat tests. Just fewer than half the patients (46%) had at least one abnormal result.

In both groups there was some referral to hospital as a result of the profile, 2.7% of those in the psychoneurotic group and 1.9% of the musculoskeletal-disorders group.

About one in 20 patients received additional drug therapy as a result of the profile tests, 55% of such prescriptions were for oral iron.

Table 1 lists the diagnoses made by the general practitioner as a result of the laboratory tests, disorders that were originally undetected. Such "biochemical diagnoses" as hypercholesterolemia and hyperbilirubinemia were excluded.

A patient was only characterized as having a new diagnosis if that diagnosis was regarded as clinically important by the general practitioner.

In both groups, 8% of patients had a new or additional diagnosis as a result of the laboratory tests. This was usually based upon the fact that the general practitioner had to alter in some way the patients' treatment, advice, or had to send them to hospital.

Of the 49 patients in this category in the psychoneurotic group (first three columns of numbers in Table 1), 32 patients had low blood hemoglobin, six < 100 g/liter. All responded to oral iron, except for one 65-year-old woman, who died a year later of carcinoma of the lung.

Of the 32 patients in the musculoskeletal group (last three columns of numbers in Table 1) in whom a new diagnosis was made, 15 had iron-deficient anemia; in two patients this was severe, but all 15 patients responded to therapy with oral iron.

Table 1 lists the diagnosis made, the tests that led to the new diagnosis, and whether the patient was referred to hospital.

Table 1. **Diagnoses and Hospital Referrals After Profile**

Diagnoses	*Tests used*	*Total no. patients*	*No. diagnoses presenting symptom*	*No. referred to hospital*	*Total no. patients*	*No. diagnoses presenting complaint*	*No. referred to hospital*
Diabetes mellitus	Glucose	2	1	2	6	2	5
Rheumatoid arthritis	ESR.RA Latex Globulin	2	2	0	5	4	3
Polymyalgia rheumatica	ESR Globulin	2	2	1	1	1	0
Disseminated lupus erythematosis	ESR	0			1	1	0
Ulcerative colitis	Hb	1	1	1	0		
Myxedema	PBI	2	0	0	1	0	
Thyrotoxicosis	PBI	2	1	1	1	1	1
Gout	Urate	2	2	1	1	0	0
Renal failure	Urea	0			1	1	1
Carcinoma	ESR, Hb, Globulin	3	2	3	0		
Gall stone with cholecystitis	ESR	1	0	1	0		
Iron deficiency anaemia Hb < 100 g/l	Hb	6	2	1	2	0	0
Hb > 100 g/l	Hb	26	0	0	13	0	0
Total		49	13	11	32	10	10

Also listed is the number of patients whose presenting symptom, in the family practitioner's opinion, was caused by the disorder subsequently diagnosed from the laboratory investigations. For example, two of the three patients with carcinoma who presented with psychoneurotic disease were judged to have that condition as a result of their carcinoma.

In both groups there were 17 patients who had new diagnoses other than anemia. In a one- to four-year follow-up of these 34 patients, three have died of cancer, one of renal failure, and two of unrelated conditions. The rest are well, receiving treatment, and all have improved clinically as a result of their therapy.

Twelve patients were referred to hospital as a result of laboratory-profile data and were not further investigated by the general practitioner. In four patients a persistently high erythrocyte sedimentation rate (ESR) and increased globulin was further investigated, but found to be of no clinical significance after 12 to 48 months of follow-up. No diagnosis was made in two patients referred for hypercalcemia or in a hypertensive patient with hypokalemia. A severely anemic man was shown to have lung cancer, a 56-year-old woman with dysphagia was shown to have carcinoma of the breast. A female patient had hyponatremia, which was shown to be caused by hypopituitarism.

Finally, 83 of the 960 patients had serum cholesterol values exceeding 7.8 mmol/liter (3.0 g/liter).

It is not possible to assess the cost-benefit of these investigations. It *is* important, however, that such detailed investigations of the efficacy of laboratory tests should be made, and laboratory workers should be closely involved in such surveys. "Efficacy" means here the "power to have an effect." It is obviously of importance on some occasions for the doctor to use "normal" results to reassure patients but it is salutary in these days of expensive laboratory investigation to realise that virtually all the results obtained in this survey could have been accomplished by using six tests:

- haemoglobin
- glucose
- erythrocyte sedimentation rate
- urea
- urate
- a thyroid-function test

Investigation of Biochemical Markers of Alcohol Intake

The use of biochemical profile techniques in the screening of the well population is an increasing feature of clinical chemical investigations in developed countries. In such countries the use of alcohol is commonplace, frequently in high amounts, and annual consumption by the population as a whole steadily increases. How useful are biochemical and hematological investigations in assisting the physician in determining whether an individual consumes alcohol in more than moderate amounts?

We have studied laboratory findings in relation to alcohol intake in 2034 men employed in various organisations. Nearly all were over 40 years old. The laboratory investigations took place as part of a health check-up at their place of employment. A subgroup of 146 men (7.5% of the total) have been under close clinical, laboratory, and social surveillance for several years. They were the employees of one company and under the care of one physician. They were investigated more extensively than the larger group.

The methods of collection of blood and the laboratory methods used are described in a paper by Whitehead et al. *(3)*.

Using a value of 30 U/liter as the upper limit of normal serum γ-glutamyltransferase (γ-GT, EC 2.3.2.2.), we found that almost 15% of the men studied had above-normal values. This immediately alerted us to the sensitivity of the technique; it was obvious that it was increased when only moderate amounts of alcohol were consumed, but on occasion was increased by other factors, which could be nonhepatic or hepatic.

Of the other tests in the profile the serum aspartate aminotransferase (AsT, EC 2.6.1.1) correlated well with the γ-GT (Figure 1). If 28 U/liter is the upper limit of normal, then 8.0% of the men studied had above-normal values. About half of these individuals (4.2%) had both enzyme activities increased. It can be seen from Figure 1 that serum γ-GT and serum AsT increase concomitantly.

Computer analysis of a group of individuals with values for both serum enzymes increased, as compared with a group with normal values for both enzymes, showed that the former group had significantly higher uric acid values. Figure 2 shows the distribution of uric acid values in the two groups of patients.

The drinking habits of the 146 individuals who were studied

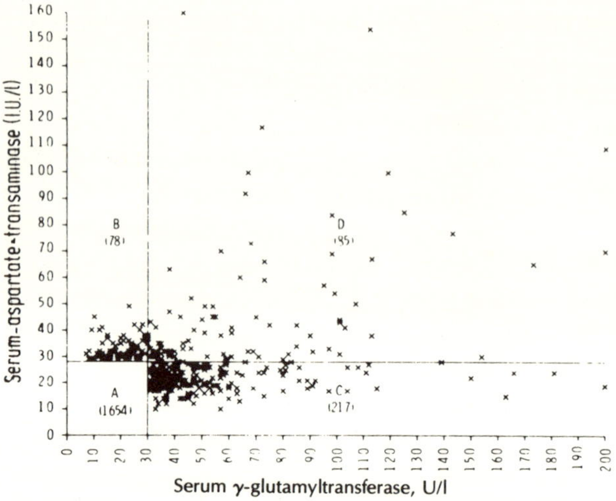

Fig. 1. Serum-γ-GT and serum-AsT in whole group of 2034 subjects.

A, both enzymes normal; B, serum-AsT raised; C, serum-γ-GT raised; D, both enzymes raised.

intensely were now compared with the results for serum γ-GT and AsT (Figure 3). The mean values of each determination are compared with "drinks per day" and the significance of the mean values between zero or one and four or more drinks per day is also shown in Table 2.

The finding that activities of both enzymes are above normal together with a urate value exceeding 400 mmol/liter (67 mg/liter) is strong evidence of a high alcohol intake.

All the men in the subgroup who showed biochemical changes typical of the larger group were studied for evidence of an effect of alcohol abuse on their professional, domestic, or social life, but none was found. Most of those with high enzyme urate values were holding very responsible positions. There therefore appears to be a group of men whose alcohol intake is such as to produce these biochemical changes but who are mentally and physically well and discharge their responsibilities efficiently. The dividing line between these men and those in whom alcohol abuse is affecting their physical, family, social, and (or) professional status can only be defined by clinical consultation. The laboratory findings may be used to support a suspicion or warn the physician of a possibility he may not have considered.

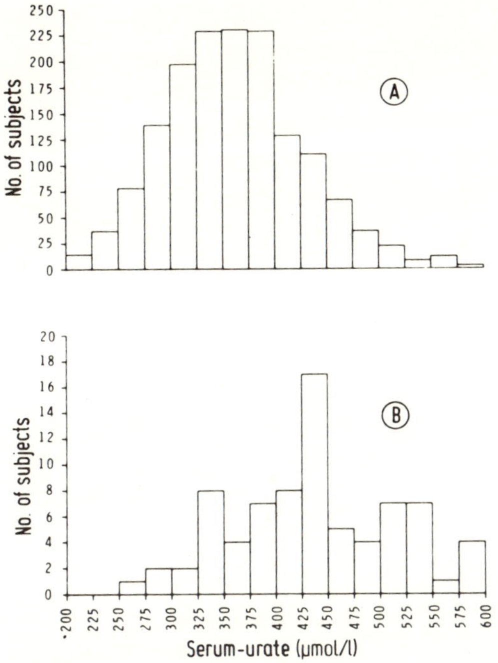

Fig. 2. (a) Serum-urate in 1654 subjects from the whole group of 2034 with normal serum-γ-GT and normal serum AsT (b) Serum-urate in the 85 subjects from the whole group of 2034 with both enzymes raised.

Conclusion

These two examples of assessing the clinical usefulness of biochemical profile techniques illustrate the necessity for close collaboration between laboratory and physician. If we are not to continue the escalating number of clinical chemical investigations, often at the expense of the individual or community concerned, then we have a duty to establish the efficacy (that is, the power to produce an effect) of the tests used, particularly at the time of their introduction. Equally, if we are to apply old tests to new clinical situations, then those responsible both in the laboratory and in the clinic should combine to establish efficacy.

Clinical chemistry has the very powerful tools of automation of chemical analysis combined with large computer data bases. They are now capable of the large-scale production of precise

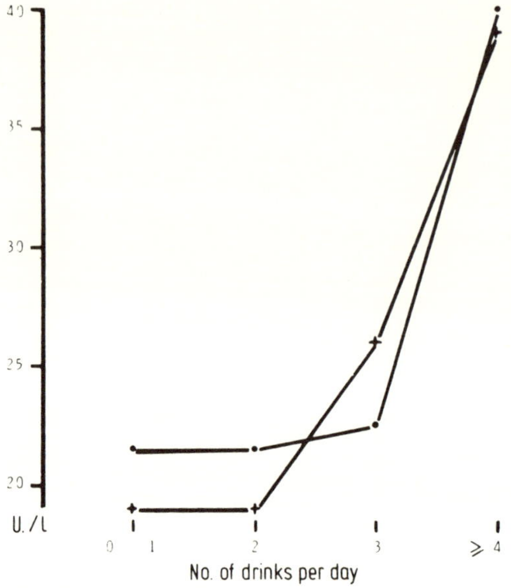

Fig. 3. Serum-γ-GT (+) and serum-AsT (·) related to daily alcohol intake in the subgroup of 146 subjects.

Table 2. **Relevant Biochemical and Hematological Findings Related to Alcohol Intake in Subgroup of 146 Subjects**

	Drinks per day				p
Mean	*0–1 (n=79)*	*2 (n=42)*	*3 (n=12)*	*4 or more (n=13)*	*(0–1 drinks per day compared with 4)*
Serum γ-G.T. (U/l)	19.2	20.5	26.4	39.9	<0.05
Serum AsT. (U/l)	21.6	21.7	22.7	40.7	<0.01
Serum urate (mmol/l)	385.2	396.6	409.2	447.6	<0.05

and frequently accurate results. Such power should also be accompanied by a responsibility to the community by careful assessment of the usefulness in patient-care terms of the tests which are embodied into clinical chemistry practice.

This work was financed by the Department of Health and Social Security, U.K.

References

1. Carmalt, M. H. B., Freeman, P. Stephens, A. J. H., and Whitehead, T. P., The value of routine multiple blood tests in patients attending the general practitioner. *Br. Med. J.* **i,** 620 (1970).
2. Percy-Robb, I. W., and Whitby, L. G., In *Automation in Analytical Chemistry, Technicon Symposia 1966,* **2,** E. Kawerau, et al., Eds., Mediad, White Plains, N. Y., 1967, p 95.
3. Whitehead, T. P., Clarke, C. A., and Whitfield, A. G. W., Biochemical and haematological marker of alcohol intake. *Lancet* **i,** 978 (1978).

The Chemyst's Tale

A Clyngcal Chemyst ther was, yclad alle in whyte,
Who diligentte dide worke both daye and nyght.
Contended he with routine and with stat,
And parfait dide he knowe of this and thatte.
Eche subtyle ferment measured he exact
And set the figures downe as solemn fact.
Fluids bubbl'd through the coyles right myrielie,
Boyled in his cauldrons to exact degree.
Balances hadde he that koude weygh precyse,
Counters, and eek other strange devyce.
For him computers wylling did magicks—
With alle wost he of Elektronicks.
Learnéd, he had passed Alchemie's gate—
None better koude makke ethyl fornicate.
Alembic and cucurbit he koude wield
(But whan he gotte machines, these he conceeled.)

Lyke in whyte array, swete Maydens dide
hym serve,
Possess'd of wondrous traits, y will observe:
Eek buxom was, well rounded in ye hamme,
Ful long of heer, and shapelie in ye gambe,
Cheeks rody, round the eyes al lyvely hewed;
Eche morn, eche her beautie Avon-newed,
Skins as whit as is the lylye flour.
But naught of daliaunce or paramour;
He onlie thot of them as beings bryght,
Adept with leech, and dial, and blynkyng lyght.

One such came up to hym one busie daye,
And seyde semely, Look you, Master, praye;
Ther canne be naught e'er seen the match of this,
So calcick is the serum and the pisse.

Quod Chemyst, By my trouthe, this manne is sick;
Anon speke this to one learn'd in Physick.
But ere we doo so, praye the tests repeate
And enter trewly alle upon the shete.
Take measure of the ferment phosphatase
Yf yts activitie doth go apace,
And assay PTH by RIA,
And telle me anon what these tests doo saye.
[For this the paycioun yette ageyn was bledde;
Ten silver pieces did yt cost; nyne went to
Overhedde.]

The Chirurgeon, none was so wyse as he,
Vers'd well in Physick and in Chirurgie;
Attir'd in skullcap, mask, and surplys green,
A bolder figure never man hath seen.
Nature nor art could better or amend
In any wise, as people did contend.
Eke day belyke he digged amongst entrailles
And cast the parts he cutte out into pailles.
He knew also of drugges as well as any can—
For sothe, he learnt it from the Detail Manne.
He cutte for stone with grate dexteritie;
For encore, sometime prostatectomie.

This paycioun he hadde cutte the daye before
And calx stones tuk out from amidst the gore,
Then Historie indited ful exact;
Insurance shetes all filled with grisly fact.
So ful were these of hieroglyph and scrawl,
Alas, was noon koude makke them out at al!
The paycioun from his hed he then dismyst
Until ther came this shete from the Chemyst.
The Chirurgeon, ensconc'd in carmin'd coat,
Resolved thereon to cutte this paycioun's throat.
[He doubted not he woud bee payde—
Hail Blue Cross, Medycare, and Medycaide!]

The mediastinum now was ful explor'd
[Excursion round the throat with gun and sword].
Two adenomas bigge as currants fetched to viewe,
Enfolded snug in thyroyd glande tissue.
The paycioun then no longer did repine,
His strength return'd, and he tuk mete and wine.
At last a cler and blyssful slep he got—
More tyme in bedde, and less astride his chambre pot.
Soon woude he able bee to work his fille;
In tyme, enow he'd earn to paye the bill.

When Chemyst met Chirurgeon on the morn
He ne hadde for his labour but a scorn:
The Chemyst dide, withal with gentil courtesie,
Make him good cheer, enquiring earnestly,
For news of how he fared, to learn by word,
Whatever of the paycioun koude bee heard.
Said Chemyst, May I veynture to suppose
Our data changed how one would dyagnose?
Quod Chirurgeon, Back to your machines.
What knowes one who has never paycioun seen?
What wenest one of paycionns must be rude
Who Urine casts and messes with the Blude.
Aye, some the like I felt betymes was wrong;
Nay, by my fay, I knew hyt alle along!

Thus endes the tale thatte y woude say to thee,
And may God save al this fair companie! Amen.

—J.S. King